GARLAND STUDIES ON

INDUSTRIAL PRODUCTIVITY

edited by

STUART BRUCHEY
UNIVERSITY OF MAINE

A GARLAND SERIES

COMPETITION AND TECHNICAL CHANGE IN THE U.S. TELEPHONE INDUSTRY

CURRENT ISSUES

NAKIL SUNG

GARLAND PUBLISHING, INC.
NEW YORK & LONDON / 1997

Library of Congress Cataloging-in-Publication Data

Sung, Nakil, 1959–
Competition and technical change in the U.S. telephone industry : current issues / Nakil Sung.
p. cm. — (Garland studies on industrial productivity)
Includes bibliographical references and index.
ISBN 0-8153-3007-3 (alk. paper)
1. Telephone companies—United States. 2. Telephone—Technological innovations—United States. 3. Competition—United States. I. Title. II. Series.
HE8815.S86 1997
384.6'06'573—dc21 97-27814

Printed on acid-free, 250-year-life paper
Manufactured in the United States of America

to my mother

Contents

List of Tables

Preface

While local telephone companies still maintain their monopolistic position, rapid technological advance in telecommunications is destroying the established market structure in the local telephone industry. The recent U.S. Telecommunications Act of 1996 aimed at eliminating any legal barrier which has suppressed technically feasible local competition. It is certain that competition in the local telephone markets is under way and will be prosperous in the future. On the other hand, economic evidence on local competition is surprisingly weak. Even recent econometric studies often indicate that telephone companies exhibit strong economies of scale or their cost function is subadditive. At the most, empirical results on the technological structure of the telephone industry are inconclusive and frequently, inconsistent with each other.

This study attempts to provide new pro-competitive evidence on the technological or cost structure of the U.S. local telephone industry. In particular, the study presents strong evidence against cost subadditivity of local telephone companies and shows that local telephone companies have been isolated from the disciplinary effects of competition in comparison with their competitive counterparts. AT&T Long Lines provides a benchmark to evaluate the relative performance of local telephone companies.

The study not only has policy implications for entry and competition in local telephone markets, but also provides a new approach to the measurement of embodied technical change. The ideas of embodiment pervade all chapters. The rate of embodied technical change is endogenously determined within the context of a cost function in this study, while the analysis of embodiment is commonly based on a production function in previous studies.

A large number of colleagues and friends assisted in one way or another in preparing this book. I am especially grateful to my former teacher, Michael Gort, for his continuing interest and encouragement. I would also like to thank Yimin Zhang and Robert W. Crandall for generously sharing with me Chessler's competition index and Bell system price indexes, respectively. This work would have been impossible without the love and encouragement of my family. I wish to express my love and appreciation to my wife, Soonye, and two daughters, Yeji and Yewon. I also thank my parents for their support.

Competition and Technical Change in the U.S. Telephone Industry

I

Overview

One of the most important public policy questions in the recent controversy on "information super-highways" is whether entry and competition in the local and intra-local access and transportation area (LATA) telephone markets are in the public interest. While competition in the U.S. long distance telephone market was introduced in the 1960s and is flourishing, local exchange carriers (LECs) still maintain their monopoly position. But rapid technological advance in telecommunications makes new communications technology available, creates new business opportunities, forms new service markets, and finally, may destroy the established market structure in the local telephone industry, as it did in the long distance market several decades ago.[1] The LECs have been responding to potential competition from cable television companies, intermodal competition with wireless communication providers, and the threat of bypass by large business users.

From the viewpoint of economics, the transition to competitive local and intra-LATA telephone markets may be justified only if LECs are not natural monopolies any more, and it can be reinforced if it is shown that LECs' productivity growth is poorer than that of their competitive counterpart. Although a number of empirical studies have focused on the technological structure of the telephone industry, their results have been inconclusive, and frequently, inconsistent with each other.[2] Some econometric studies indicate that the production structure of the telephone industry exhibits economies of scale (or cost subadditivity). This study attempts to provide new empirical evidence on these issues.

One of the major problems with the current literature has been the lack of appropriate data. Previous studies usually rely on time series data of a telephone company, mainly AT&T and Bell Canada, to estimate single- or multi-product cost functions. The small number of observations on highly correlated variables, together with the large number of parameters being estimated, may give rise to parameter estimates that are vulnerable to specification and estimation errors. The previous studies often obtain different or even opposite empirical results by adopting different forms of the cost function, different methods of estimation, and different ways of controlling for technological change, even though they use the same data. Creating good data with enough degrees of freedom is crucial to obtaining precise parameter estimates within the context of a cost function.

My empirical analysis is carried out with pooled cross-sectional time series data for 8 LECs and the AT&T Long Lines Department. The AT&T Long Lines Department as well as LECs, including local Bell operating companies (BOCs), file financial and operating data with the Federal Communications Commission (FCC), and have done so even when they were components of AT&T. The major source of data is the *Statistics of Communications Common Carriers* (*SCCC*) published by the FCC and the *Form M* report filed by the reporting companies. In particular, the *Form M* report provides detailed data for telephone plants or compensation and working hours. Annual data for the firms were collected for the period 1951-1991.

This study attempts to incorporate arguments for embodied technical change in both cost and productivity analyses.[3] While the notion of embodied technical change has attracted the attention of many economists in the last few decades due to its intuitive appeal, the theoretical and empirical work on the embodiment hypothesis has often provided mixed results. Since time series data produce fairly accurate measures of capital and labor quality for each firm, this study can explicitly take these variables into account in the cost function and the total factor productivity (TFP) growth rate regressions. In particular, within the context of cost functions embodied technical change is defined as shifts in the cost function due to improved capital and labor quality.

When capital and labor quality are considered, the remaining disembodied technical change is associated with only the level of organizational capital, mainly in the form of firm-specific information. A competitive product market plays a decisive role in monitoring and rewarding a firm's decisions on the use and accumulation of organizational capital. By taking into account the degree of competitiveness in both cost and productivity analyses in addition to input qualities,[4] I am able to trace the effect of competition on cost decline or TFP growth through its impact on the accumulation and use of organizational capital. Similarly, the relationship between competition and cost subadditivity can be examined.

The study is organized as follows. In Chapter 2, I attempt to set up a model for measuring embodied technical change on the basis of a cost function. While embodiment analysis is usually based on production functions, my model is based on a cost function because the latter approach is preferable in empirical studies on regulated industries and, also, because I am able to use firm data. The cost function contains a measure of capital quality, labor quality, and competitiveness. I estimate several alternative cost functions using cross-sectional time series data for 8 LECs to evaluate the importance of embodied technical change in the U.S. local telephone industry.

Chapter 3 examines the natural monopoly problem to provide the economic evidence on the feasibility of competition in local telephone service. Using the cost functions estimated in Chapter 2, I perform the subadditivity test developed by Evans and Heckman (1983a & 1984). While Evans and Heckman apply a local test of subadditivity, my test is global because I do not constrain the output region. In conducting the test, my focus is on the relationship between cost subadditivity (or returns to scale) and the size of a firm.

A major limitation of the subadditivity concept is its implicit assumption that technology is the same with a single firm as with numerous firms. It completely ignores the effect of a competitive product market on the firm's management decisions. Assuming away the differences in technology may severely overstate the measure of cost subadditivity. This study attempts to measure the effect of competition on cost subadditivity by considering differences in the degree of competitiveness faced by a monopolistic firm and competitive one.

In Chapter 4, I focus on the relative performance of a competitive long distance telephone company, or inter-exchange carrier (IXC), and a monopolistic local telephone company, or LEC. In the productivity literature there have been very few studies on the relative performance of a firm in competitive and non-competitive environments. This is surprising in the light of a large amount of research on the relative economic performance of private and public firms. The main reason for this gap may be that there are few cases in which a competitive and a monopolistic market coexist in the same industry. The U.S. telephone industry, therefore, offers a unique opportunity for such a study.

I start with the hypothesis that a LEC facing potential or intermodal competition shows slower productivity growth than an IXC operating in an effectively competitive environment. I conduct two types of analysis: (1) TFP growth rate measurement and (2) cost function estimation. Under some restrictive assumptions the TFP growth rate is equivalent to shifts in production or cost functions. But, if one of the assumptions is violated, the two approaches may produce different results. My hypothesis is tested using data on AT&T Long Lines and 8 LECs over the period 1951-1991.

NOTES

[1] Willig (1990) indicates that "it would be a mistake to regard the revolution in the telecommunication marketplace as a product of deregulatory policy. Instead, the fundamental force behind the revolution is and has been technological change." Therefore, "once it is recognized that the status quo cannot be maintained in the face of underlying technological change, it becomes clearer that regulatory revolution is necessary for the public interest and that appropriate deregulation should keep pace with the technological forces that are driving the markets to be competitive."

[2] For comprehensive surveys on the production structure of telephone industry, see Fuss (1983 & 1992) and Waverman (1989).

[3] A brief summary of the embodiment literature appears in Gregory and James (1973) and Gort and Wall (1993).

[4] This is based on the assumption that competition has disciplinary effects not only on competitive AT&T Long Lines but also on monopolistic LECs. In the 1960s and early 1970s AT&T Long Lines struggled with regulatory agencies to prevent entry and has competed with several actual rival companies since then. On the other hand, LECs have faced potential competition since 1970s and are competing with wireless communications companies. When we

follow contestability theory, potential competition should be accorded equal weight to that given to currently visible competition in analyses of the strength of market forces. In this sense, it is legitimate to insert the degree of competitiveness even in cost analyses of LECs. But, effective competition with actual rivalry may exhibit greater disciplinary effects in monitoring a firm's non-optimizing behavior than does potential or intermodal competition. That is the main topic of chapter 4.

II

Measuring Embodied Technical Change

2.1. INTRODUCTION

The notion of embodied technical change has been the focus of many economists in the last few decades since R.M. Solow's (1960) seminal paper. In particular, the issue is whether technical change is caused primarily by improvements in the quality of inputs or mainly due to changes in the stock of knowledge and hence independent of the growth rate of inputs employed. Also, what fraction of technological development that can potentially raise output or productivity growth must be embodied in inputs? Unfortunately, however, the fraction of output growth or productivity growth attributed to embodied technical change as compared to disembodied technical change has ranged from negligible to dominant. The theoretical and empirical work on the embodiment hypothesis has often provided mixed results. This chapter attempts to set up a model measuring embodied technical change and provide new evidence on the embodiment hypothesis.

Early evidence in support of the embodiment hypothesis was not wholly satisfactory despite its intuitive appeal. The analysis of embodiment started with a vintage production function, derived an aggregate production function after assuming the appropriate degree of separability, and estimated the rate of technical change identified in the vintage production function (the rate of embodied technical change). But the estimation of certain crucial parameters in the production function was possible only with prior knowledge of other equally important parameters.[1] This identification problem critically limited the reliability and scope of the early literature.[2]

Another reason why the early literature was not successful in identifying the vintage effect might be the lack of appropriate data. It was mainly based on macro time series data which suffer from two inherent defects: a high degree of collinearity and lack of variation in capital stock series.[3] Also, Denison (1964) indicated that the average age of plant and equipment is so insensitive to large swings in the rate of gross investment that the change in the age distribution of capital stock has little impact on output growth. Hence, either the embodiment hypothesis was not generally supported or the embodiment model led to similar predictions as the non-embodiment model with respect to output or productivity growth.

In a few instances vintage models were applied to disaggregated cross-sectional data with mixed results. For example, Gregory and James (1973) collected data from a cross-section of Australian manufacturing industries and, also, used new factory data to solve the problem of aggregating different vintages of machines into a factory vintage measure. They could not find strong support for the embodiment hypothesis because their data encompassed a short span of time (1953-1958). The settling-in and learning may take a long time before a new factory is fully integrated into the production process and is able to reach maximum efficiency. Thus the data must be collected over a longer period to fully capture the effect of embodiment.[4]

Recently, two different types of studies have revived the notion of embodied technical change. The first approach to embodiment proceeds from recent development in the price literature on quality change. Hulten (1992) combines the estimates of quality change obtained from the price literature with a version of the conventional model of the sources of growth which allows for both embodied and disembodied technical change. He measures unobserved embodiment parameters using the ratio of BLS quality-unadjusted price indexes to Gordon's (1990) quality-adjusted price indexes. He finds that about 20% of the Bureau of Labor Statistics (BLS) TFP change can be directly associated with the improvement of capital quality.

The second approach attempts to estimate the vintage effect within the framework of a production function using data which may overcome the data problems of previous studies. Assuming that the aggregate capital stock exponentially augmented by average vintage is a measure of capital stock in a technically efficient unit, Gort, Bahk, and Wall (1993) show that the average annual effect of vintage on output growth ranges from 2% to 4%. Bahk and Gort (1993) find that a 2% annual output growth attributed to time-dependent disembodied technical change disappears completely once the variables capturing embodied technical change are introduced in the model. Both studies use the same data set: pooled cross-sectional time series manufacturing plant data with the large number of observations over a long span of time (1973-1989).

Even though Hulten's attempt to merge the price literature on quality change with the conventional embodiment literature is appealing, his empirical success critically depends on the accuracy of both quality-adjusted and unadjusted price indexes. Gort and Wall (1993) indicate that when the BLS indexes are biased upward by technical change and the Gordon's indexes understate technical change, Hulten's estimates of capital quality may be severely downward biased.[5] On the other hand, while the second approach does not require prior information on price indexes, it is based on the assumption that capital quality is an exponential function of average vintage. Obviously the average vintage is just one method of summarizing the efficiency distribution of investment goods, and, also, the exponential function form is arbitrary. I follow the second approach to measure the rate of embodied technical change in LECs since there are no accurate data on quality-adjusted and unadjusted price indexes for telephone industry.

My capital and labor quality indexes are based on, but improve those developed by Gort, Bahk, and Wall (1993). Since the interaction between old and new plants in a network is important in the telephone industry, I multiply average vintage by a measure of the degree of development in a telephone network to derive an index of capital quality. Gort, Bahk, and Wall (1993) develop the index for average labor quality which is simply average wage rate. This measure has intuitive appeal because its definition is based on any attribute of labor that increases productivity of labor. Within the context of a cost function, the index of labor quality is defined as the ratio of the weighted sum to the simple sum of working hours of every occupational group in a firm, with the weights being the ratio of each group's hourly earnings to average private sector wage.

In previous studies the analysis of embodiment is commonly based on a production function owing to the level of aggregation. Output is endogenously determined at the aggregate level of national economy, industry, or plant. My study is based on a cost function because the cost function approach is preferable in empirical studies on regulated industries and, also, because I am able to use firm data.

Previous studies usually add a measure of capital quality as an augmentation term to capital stock. Then the capital stock is assumed to be measured in technically efficient units. A measure of capital quality is inserted as a separate term in my cost model in order to avoid using an arbitrary functional form. Including the index of capital quality as a separate term is more general than expressing it as an augmentation term.

The structure of this chapter is as follows. In section 2, I set forth measures of capital and labor quality and define embodied technical change within the context of a cost function. Section 3 presents several econometric specifications of a cost function. Section 4 reports estimation results, compares differences in the rates of technical change according to different specifications, and presents interesting findings on embodied technical change. Summary follows.

2.2 MODEL OF MEASURING EMBODIED TECHNICAL CHANGE

Capital Quality Index

Advances in technology affect the level and growth of output to the extent that they are carried into practice either by net capital formation or by the replacement of old plants with new ones, resulting in a shift in the efficiency distribution of plants. Successive vintages of investment embody differences in technology. Assuming that the loss of productive efficiency due to physical decay proceeds at a constant rate, the capital stock or the investment of vintage v at time t measured in technical efficiency units ($K_t(v)$) can be expressed as follows:

$$K_t(v) = \Phi(t,v)(1-\delta)^{t-v} I(v) \qquad v = T-t, T-t+1, T-t+2\cdots$$

where I(v) is the amount of vintage-specific gross investment measured in natural units.[6] Vintage (v) is the base year (T) minus the year of birth, with T being the last year of the sample period (1991). $\Phi(t,v)$ is a measure of technical efficiency. The technical efficiency of capital stock of differing vintages varies over time due to obsolescence. Since technological progress embodied in capital stock simply refers to changes in the quality of capital goods, $\Phi(t,v)$ can be regarded as a measure of the quality of capital goods.

Hulten (1992) measures $\Phi(\cdot)$ using the ratio of BLS investment price indexes to Gordon's (1990) quality-adjusted price indexes. The reliability of his measure depends crucially on the accuracy of the two price indexes with respect to quality adjustment. On the other hand, the conventional embodiment literature assumes that $\Phi(\cdot)$ is a function of the date of birth (v). If $\Phi(\cdot)=(1+\lambda)^{-v}$, the technical efficiency of capital stock increases by $(1+\lambda)$ per vintage.

In my expression $\Phi(\cdot)$ is a function not only of v but also of t because of the interaction between old and new plants. New plants may provide positive externalities to old plants in the form of improved technical efficiency.[7] In particular, these externalities may be very important in the telephone industry where every plant in service is connected through a network. The development in a telephone network affects the technical efficiency of investment goods of every vintage. Since technical change in the switching system has been dominant over that in other plants in the telephone industry, this effect may be appropriately measured by the ratio of telephones connected to the electronic switching system (ESS) to total telephones at the time t (s(t)).

Thus, I assume that $\Phi(\cdot)$ is a function of $(1+s)/a$ in which a is the age of capital stock. I insert age in place of vintage because dividing $(1+s)$ by vintage may severely overstate the technical efficiency of recent investment goods.[8] The efficiency of investment goods of the same age increases with s, ignoring their physical decay. For the period before the ESS was introduced, the index is just $1/a$.

The total amount of capital stock at time t measured in technical efficiency units is:

$$K_t = \sum_v K_t(v) = K_t^{'} \sum_v \Phi(t,v) \frac{(1-\delta)^{t-v} I(v)}{K_t^{'}}$$

where $K_t' = \Sigma(1-\delta)^{t-v} I(v)$. Then $K_t = K_t'\psi(t)$ where

$$\Psi(t) = \sum_v h(\frac{1+s(t)}{t-v}) \frac{(1-\delta)^{t-v} I(v)}{K_t^{'}} \approx H(\frac{1+s(t)}{\bar{a}})$$

Here $\bar{a}$ is the average age of an aggregate capital stock at time t, with the weights being the ratio of the surviving investments of each vintage to total investments. Thus, my capital quality index (k) has a simple form, i.e. $k = (1+s)/\bar{a}$.

Labor Quality Index

Gort, Bahk, and Wall (1993) develop the index for average labor quality (average amount of human capital) associated with labor input within the context of a production function. Their index is simply the average wage rate for each plant because differences in average wages must reflect differences in labor quality when all plants have equal access to the labor market. This measure has intuitive appeal because its definition is based on any attribute of labor that increases productivity. Thus ad hoc measures of labor quality based on education, occupation, or demographic attributes are rendered unnecessary.

As this measure cannot be used within the context of a cost function, I modify it while maintaining its basic idea. I calculate two types of labor input; the simple sum and the weighted sum of working hours of every occupational group in a firm. For the latter, the ratio of each occupational group's hourly earnings to average wage in private sector is used as a weight. Thus the weighted sum of working hours can be regarded as aggregate labor measured in technical efficiency units because relative wage rates reflect differences in the marginal products (and hence labor quality) of occupational groups under the assumption of a competitive labor market. The labor quality index (ω_t) is defined as the ratio of the weighted sum to the simple sum of working hours (L_t/L_t'). Thus the amount of aggregate labor measured in technical efficiency units (L_t) is equal to the amount of pure labor (L_t') times the labor quality index.

The labor quality index of the j-th firm (ω_t) is

$$\omega_j = \frac{\sum \frac{w_{ij}}{w_p} L_{ij}}{\sum L_{ij}} = \left(\frac{w_j}{w_p}\right)\left(\frac{\sum \frac{w_{ij}}{w_j} L_{ij}}{\sum L_{ij}}\right)$$

where w_{ij} and L_{ij} refer to the wage rate and the working hours for the i-th occupational group of the j-th firm respectively, w_j to the average wage for all groups in the j-th firm and w_p to the average wage rate in the economy. Thus this measure of labor quality is expressed as the product of average relative wage rate (the first term in the last expression) and a Jorgenson type of labor quality index (the second term in the last expression).[9] By the definition of labor quality index, quality-adjusted wage (P_L) is equal to quality-unadjusted wage (P_L') divided by the labor quality index (ω_t).[10]

Cost Function with Input Quality Indexes

Assuming the appropriate degree of separability, an aggregate production function with capital and labor quality indexes is

$$\begin{aligned} \ln Q_t &= f(\ln L_t, \ln K_t, \ln M_t; \ln\Omega_t) \\ &= f(\ln L_t'\omega_t, \ln K_t' H(k_t), \ln M_t, \ln\Omega_t) \\ &= f(\ln L_t' + \ln\omega_t, \ln K_t', \ln M_t; \ln k_t, \ln\Omega_t) \end{aligned}$$

where Q is the quantity of aggregate output, M is the quantity of intermediate materials measured in efficiency units, and Ω measures a firm's stock of knowledge or level of organizational capital.[11] The capital quality index is inserted as a separate term in the last relationship because the functional form of H(•) is unknown. The labor quality index is still expressed as an adjustment term because it is derived from the definitions of quality-adjusted and pure labor.

A cost function approach is preferable in empirical studies on regulated industries because the output level of a regulated firm is usually exogenously determined by regulation. Also, when data are collected at the firm level, a cost function reflects economic agents' optimization behavior more appropriately than a production function does. A long-run cost function dual to the above aggregate production function is

$$\ln C_t = g(\ln\frac{P_{Lt}'}{\omega_t}, \ln P_{Mt}, \ln P_{Kt}; \ln Q_t; \ln k_t; \ln\Omega_t)$$

where P_M refers to real material price and P_K to user-costs of capital stock. This cost model is different from traditional cost functions in that two input quality indexes are inserted in the cost function. Thus the effects of input qualities on cost decline can be estimated within the model.

Definitions of Embodied Technical Change

Differentiating the above cost function with respect to time, I obtain

$$\dot{C}_t = g_L \dot{P}'_{Lt} + g_M \dot{P}_{Mt} + g_K \dot{P}_{Kt} + g_Q \dot{Q}_t - g_L \dot{\omega}_t + g_K \dot{k}_t + g_\Omega \dot{\Omega}_t$$

where g_X s the derivative of g(·) with respect to lnX and the dots represent the growth rate of each variable. $g_Q=\partial \ln C_t/\partial \ln Q_t$ is a scale measure, while $g_X=\partial \ln C_t/\partial \ln P_{Xt}$ for X=L,M,K is a factor price elasticity of cost or a factor share. The last three terms measure the rate of technical change which is equivalent to annual shifts in a cost function. When the stock of knowledge is approximated by time, g_Ω becomes the rate of time-dependent disembodied technical change. $g_L \dot{\omega}_t$ (or $g_k \dot{k}_t$) measures annual percentage shifts in the cost function due to improved labor quality (or improved capital quality). Since these shifts in the production or cost function are caused by the embodiments of input qualities in physical inputs, they can be defined as embodied technical change.

Note that my definition of embodied technical change is not associated with the rate of depreciation (the rate of obsolescence plus the rate of physical decay). Differentiating lnQ with respect to lnk in the production function yields

$$\frac{\partial \ln Q_t}{\partial \ln k_t} = \frac{\partial \ln Q_t}{\partial \ln K_t} \bullet \frac{\partial \ln K_t}{\partial \ln k_t}$$

Here the second term on the right-hand side, $\partial \ln K_t/\partial \ln k_t$, represents the effect of capital quality on the amount of capital stock in technical efficiency units. When k_t is the average vintage of an aggregate capital stock, it measures change in the efficiency of capital goods from a one-year change in average vintage.[12]

In the embodiment literature, especially on growth accounting, more emphasis has been placed on the second term ($\partial \ln K_t/\partial \ln k_t$) instead of on the first term ($\partial \ln Q_t/\partial \ln k_t$). The latter is equal to a capital share under the assumption of constant returns to scale and perfect competition in output and input markets. Assuming that the capital share in growth accounting is known, several studies have been devoted to the estimation of the embodied effect of vintage. While this procedure may be legitimate at the macro level time-series data, equal emphasis should be laid on both terms at the firm level data because technological bias is common and important for firms.

Before proceeding to the next section measures of technological bias deserve a brief comment. The conventional way of measuring technical bias is

$$IB_i = \frac{\partial^2 \ln C}{\partial t\, \partial \ln P_i} = \frac{\partial S_i}{\partial t}$$

in which S_i is the i-th factor share in total costs.[13] But since I take into account embodied as well as disembodied technical change, I need an alternative definition of technical bias. The i-th factor technical bias of capital quality embodiment is defined as

$$IB_{Ki} = \frac{\partial^2 \ln C}{\partial \ln k\, \partial \ln P_i} = \frac{\partial S_i}{\partial \ln k}$$

If IB_{Ki} is positive (or negative), the improvement in capital quality is i-th factor using (or saving). The i-th factor technical bias of labor quality embodiment or disembodied technical change can be defined in a similar way.

2.3. EMPIRICAL SPECIFICATIONS

Empirically, I estimate the following five two-product long-run cost functions to determine the technological structure of LECs:

$$(Model\ 1)\ \ln C = g(\ln P_L, \ln P_M, \ln P_K; \ln Q_1; \ln Q_2; t)$$

$$(Model\ 2)\ \ln C = g(\ln \frac{P_L}{\omega}, \ln P_M, \ln P_K; \ln Q_1; \ln Q_2; \ln k, t)$$

$$(Model\ 3)\ \ln C = g(\ln \frac{P_L}{\omega}, \ln P_M, \ln P_K; \ln Q_1; \ln Q_2; \ln k)$$

$$(Model\ 4)\ \ln C = g(\ln \frac{P_L}{\omega}, \ln P_M, \ln P_K; \ln Q_1; \ln Q_2; \ln k, \ln CI)$$

$$(Model\ 5)\ \ln C = g(\ln \frac{P_L}{\omega}, \ln P_M, \ln P_K; \ln Q_1; \ln Q_2; \ln k, \ln \bar{k}, \ln CI)$$

where Q_1 refers to local output, Q_2 to toll output, CI to a measure of the competitiveness in the U.S. telephone industry, and $\bar{k}$ to the average of lagged variables of k. I omitted the subscript t and the superscript $'$ in $\ln P_L'$.

Model 1 is the conventional specification of a cost function that uses a time variable as a proxy for the state of technology. Model 2 incorporates two quality indexes as well as the time variable. While the derivatives of g(·) with respect to quality indexes measure embodied technical change, the time derivative of g(·) captures the remaining disembodied technical change.

Model 3 assumes that every technical change is embodied or that there is no disembodied technical change. This assumption seems too extreme because there may be accumulation of firm-specific information over time. A firm learns to match better its tasks with employee attributes, or to sort its employees by competence, and accumulates other form of firm-specific learning-by-doing knowledge over time.

Model 4 inserts a measure of competition in place of unmeasurable organizational capital. A competitive product market plays an important role in monitoring and rewarding a firm's decisions, especially those regarding the accumulation and use of organizational capital. Even though LECs are single sellers, intense competition in the long distance market may have spillover effects on LECs' management. The competition index is expected to capture the effect of competition on cost decline through its impact on the accumulation and use of organizational capital. I extend Chessler's competition index up to 1991 by using the market share of OCCs in toll revenues for the post-1984 period.

In model 5, I include the average of lagged capital quality indexes. Improved capital quality have two-fold effects. The decrease in the average age of an aggregate capital stock may arise from the increase in the aggregate capital stock. Thus this direct effect of capital quality improvements may be to increase total costs. But in the long term, the improvements in capital quality decreases total costs because they enhance the productivity of capital stock, thereby allowing substitution of other inputs with the more efficient capital input. The final effect depends on the relative sizes of the negative and the positive effects. It may take time for the positive effect to materialize fully because of settling-in and learning by doing. Thus the average of one-period and two-period lagged k is added to capture the possible time lag in the realization of capital quality improvements. I use the average value to increase the degrees of freedom in the estimation of a cost function, assuming that the vintage effect is materialized completely within two years.[14]

It is assumed that g(·) is twice-differentiable and can be approximated by a second-order Taylor series expansion, i.e. translog flexible cost function. The translog two-product cost function for model 4 is given by:

$$
\begin{aligned}
\ln C = {} & \alpha_0 + \sum_{i=1,2}\alpha_i \ln Q_i + \sum_{j=L,M,K}\alpha_j \ln P_j - \alpha_L \ln\omega + \alpha_k \ln k + \alpha_C \ln CI + \\
& \frac{1}{2}\sum_{i=1,2}\sum_{l=1,2}\beta_{il}\ln Q_i \ln Q_l + \frac{1}{2}\sum_{j=L,M,K}\sum_{m=L,M,K}\beta_{jm}\ln P_j \ln P_m + \frac{1}{2}\beta_{LL}(\ln\omega)^2 + \\
& \frac{1}{2}\beta_{kk}(\ln k)^2 + \frac{1}{2}\beta_{CC}(\ln CI)^2 + \sum_{i=1,2}\sum_{j=L,M,K}\beta_{ij}\ln Q_i \ln P_j - \sum_{i=1,2}\beta_{iL}\ln Q_i \ln\omega + \\
& \sum_{i=1,2}\beta_{ik}\ln Q_i \ln k + \sum_{i=1,2}\beta_{iC}\ln Q_i \ln CI - \sum_{j=L,M,K}\beta_{jL}\ln P_j \ln\omega + \sum_{j=L,M,K}\beta_{jk}\ln P_j \ln k + \\
& \sum_{j=L,M,K}\beta_{jC}\ln P_j \ln CI - \beta_{LC}\ln\omega \ln CI + \beta_{kC}\ln k \ln CI - \beta_{Lk}\ln\omega \ln k + \varepsilon
\end{aligned}
$$

where ε is a disturbance term. All variables except CI are divided by their sample means. For the translog case, this normalization of variables by sample mean permits the first-order terms to be interpreted as the elasticities at the sample mean. Conditions for linear homogeneity and symmetry are imposed, while monotonicity and concavity conditions are tested after estimation.

Assuming cost minimization given the regular constraints, Shepard's lemma can be applied to obtain the following cost share equations.

$$
S_j = \alpha_j + \sum_{i=1,2}\beta_{ij}\ln Q_i + \sum_{m=L,M,K}\beta_{jm}\ln P_m - \beta_{jL}\ln\omega + \beta_{jk}\ln k + \beta_{jC}\ln CI + \varepsilon_j
$$

where ε_i is the disturbance term. Then I jointly estimate the cost function and the two cost share equations subject to the relevant cross-equational constraints by iterating Zellner's seemingly unrelated regression method. One of the cost share equations is deleted to obtain a nonsingular covariance matrix.[15] Serially correlated errors are ignored in the estimation.[16]

2.4. EMBODIED TECHNICAL CHANGE IN THE LECS

General Empirical Findings

The empirical analysis is carried out with pooled cross-sectional time series data for 8 LECs. The major source of data is the *Statistics of Communications Common Carriers* (SCCC) published by the FCC and the *Form M* report filed by reporting telephone companies. Annual data for 8 LECs were collected for the period 1951-1991. Refer to appendix 2 for a detailed explanation of the data construction. Some important parameter estimates for five translog cost functions are provided in table 2.1, while a full set of parameter estimates for each cost model is given in the appendix 1. All the estimated cost functions satisfy the regularity conditions, i.e. monotonicity and quasi-concavity.

In model 1, of the 28 parameter estimated, 16 are significant at the 1% level, 4 are significant at the 5% level, and 1 is significant at the 10% level. In model 2, the number of statistically significant parameter estimates at the 1%, 5%, and 10% level are 29, 2, and 4 respectively out of 45. All the coefficients in model 3 except two parameters are significant at least at the 10% level. The number of significant parameter estimates at the 1%, 5%, and 10% level is 23, 7,and 2 out of 45 for model 4 and 32, 7,and 5 out of 55 for model 5.

In particular, the first-order terms of explanatory variables are highly significant for all the models except model 2. In model 2, half of the first-order terms are insignificant which makes it unreliable. A possible reason for this insignificance is that the inclusion of a time variable in addition to quality indexes may be a wrong specification for the technological structure of LECs. The time variable may capture not only the changes in the stock of firm-specific knowledge but also the changes in the quality embodied in labor and capital. In that case, the time variable overshadows embodied technical change so that the coefficient of quality indexes may become insignificant. Otherwise, most of the technical change may be embodied so that the time variable by itself has little explanatory power. In general, dropping the time variable makes the first-order terms highly significant.

The absolute size of every second-order term for all the models except model 5 is less than one, which results in stable output cost elasticities. Applying a translog cost function to a data set with small

< Table 2.1> Parameter Estimates of Cost Functions

variable	Model 1	Model 2	Model 3	Model 4	Model 5
intercept	1.090*	0.606*	0.046*	0.400*	0.222*
	(0.123)	(0.183)	(0.016)	(0.045)	(0.051)
$\ln Q_1$	0.479*	0.947*	0.712*	0.696*	0.934*
	(0.156)	(0.170)	(0.039)	(0.063)	(0.070)
$\ln Q_2$	0.670*	0.159	0.274*	0.341*	0.098
	(0.154)	(0.166)	(0.035)	(0.062)	(0.071)
$\ln P_L$	0.742*	0.745*	0.429*	0.424*	0.401*
	(0.010)	(0.012)	(0.004)	(0.010)	(0.010)
$\ln P_M$	0.065*	0.018	0.206*	0.177*	0.181*
	(0.013)	(0.015)	(0.003)	(0.008)	(0.009)
$\ln P_K$	0.193*	0.237*	0.365*	0.398*	0.418*
	(0.013)	(0.013)	(0.003)	(0.007)	(0.006)
t	-0.031*	-0.010			
	(0.008)	(0.013)			
lnCI				-0.212*	-0.141*
				(0.030)	(0.032)
$\ln\omega$		-0.745*	-0.429*	-0.424*	-0.401*
		(0.012)	(0.004)	(0.010)	(0.010)
lnk		0.384	-1.311*	-0.406*	0.255
		(0.370)	(0.080)	(0.144)	(0.314)
$\ln \bar{k}$					-0.901*
					(0.318)
$\ln Q_1 * \ln Q_2$	-0.148	-0.063	-0.368*	-0.049	-0.044
	(0.098)	(0.093)	(0.070)	(0.066)	(0.074)
$t * \ln P_L$	-0.012*	-0.012*			
	(0.000)	(0.000)			
$\ln P_L * \ln\omega$		0.111*	0.084*	-0.042***	-0.078*
		(0.012)	(0.013)	(0.023)	(0.022)
$\ln P_L * \ln k$		-0.026***	-0.287*	-0.243*	0.121**
		(0.013)	(0.020)	(0.020)	(0.047)
$\ln P_L * \ln \bar{k}$					-0.344*
					(0.045)

Note: The estimation method is Zellner's seemingly unrelated regression technique. Standard errors are in parentheses. *,**, and *** imply that the coefficient is statistically significant at the 1%, 5%, and 10% level. The number of observations is 316.

number of observations often generates unstable output cost elasticities so that no meaningful conclusions can be reached.

The principal differences between the models' estimates which are found in table 2.1 are summarized as follows.

1. In model 1 the coefficient of local output (Q_1) is smaller than that of toll output (Q_2) at the sample mean, while the opposite is true in the other models. The wrong specification of model 1 may generate biased parameter estimates because a large cost elasticity with respect to local output is consistent with the common knowledge in the telephone industry that production of an additional unit of local service costs more than that of toll service. Despite different output-cost elasticities, all the models produce closely similar estimates of overall returns to scale, which range from 1.045 in model 1 to 1.023 in model 2 (see table 2.2).[17]

2. All elasticities of cost with respect to input prices, i.e. factor shares, are positive at the sample mean. The estimates of labor share in models 1 and 2 are larger than the actual labor share, while the estimated shares come close to the actual share in the remaining models. These observations lead me to conclude that inserting a time variable in a translog cost function may be a wrong econometric specification for the technological structure of LECs.

3. The negative coefficients of technology variables imply a downward shift in the cost function. Model 1 shows that the cost function experiences a 3.1% annual downward shift at the sample mean. But the coefficient of the time variable (t) becomes smaller and statistically insignificant when I incorporate both labor and capital quality indexes (ω and k) in model 2. A decrease in the absolute size of the coefficient of t can be expected because the addition of ω and k reduces the role of t from a proxy for the state of technology to a variable measuring disembodied technical change. The insignificance of t implies either that t cannot measure disembodied technical change appropriately or that most of the technical change may be embodied.

4. Model 2 shows that a 1% increase in ω leads to a 0.75% downward shift in the cost function at the sample mean. The interaction term between $\ln\omega$ and t has a positive sign (see appendix 1), which indicates that the contribution of labor quality to productivity growth has declined over time.

5. In model 2 the coefficient of lnk is positive at the sample mean. At first glance this result seems contradictory because improved capital quality should have a positive impact on cost decline. Two explanations are possible. First, as discussed in the previous section, improvements in capital quality have both a negative and a positive effect on product growth, with the negative effect overshadowing the positive effect. Second, its sign may be misleading because the coefficients of both t and lnk are statistically insignificant. I believe that the second explanation is more plausible since dropping t in model 3 makes the coefficient of lnk significant, negative, and large. Thus incorporating a time variable and quality indexes simultaneously in a cost function may understate embodied technical change. In model 3 the coefficient of lnk is -1.311, which implies that a 1% increase in the capital quality results in a 1.3% downward shift in the cost function.

6. When I insert the competition index (CI) instead of a time variable in model 4, its coefficient is consistent with a priori expectations. A 1% increase in the competition index leads to a 0.2% downward shift in the cost function at the sample mean. But the size of the coefficient of lnk shrinks to -0.406. Either the competition index may capture part of the improvements in capital quality (or k) may partly incorporate the level of organizational capital in model 3.

7. Model 5 is different from model 4 in that the former contains the average of one-year and two-year lagged k ($\bar{k}$). In model 5 the coefficient of lnk has a positive sign and becomes insignificant, while the coefficient of $\ln\bar{k}$ has a negative sign and is highly significant. Thus it may be yesterday's improvement of capital quality and not today's that leads to downward shifts in the cost function. Thus it may take time for improved capital quality to be materialized.[18]

8. The coefficient of $\ln Q_1 * \ln Q_2$ is negative for every model which implies local cost complementarity. LECs may have at least locally efficient joint production in providing local and toll services. Economies of scope are more likely if joint production yields local cost complementarities.[19] If synergies arise from the joint production of local services and intrastate (or intra-LATA) toll services, the probability may be high that the same synergy effects occur between local or intrastate services and interstate services. If so, the prohibition of LECs from entering inter-state toll service would decrease efficiency.

9. In model 1 the interaction term between input prices and the time variable indicates a tendency for the labor share to decrease and for the capital share to increase over time. In models 2, 3, and 4 the negative sign on $\ln P_L*\ln k$ indicates that capital quality improvements in LECs are labor-saving. But the sign on $\ln P_L*\ln\omega$ is positive in models 2 and 3 and negative in model 4. Thus it is not clear whether labor quality improvements are labor-saving or not.

The Rate of Embodied Technical Change

A flexible translog functional form is advantageous in that measures of economies of scale or technical change depend upon the values of all parameters. Using this desirable property I calculate the rate of technical change for all observations. Annual shifts in the cost function caused by a technology variable is defined as the elasticity of cost with respect to that variable (g_X) times the annual percentage change in the variable (X_t). Only the average annual shift is reported in table 2.2. Total shifts, i.e. productivity growth, are just the sum of annual shifts caused by each variable. The key findings in table 2.2 are the following.

1. The estimate of productivity growth for each model ranges widely from 4.2% to 2.1%. Even though I regard model 2 as invalid owing to many statistically insignificant first-order terms, I still have four alternative models. Chapter 4 reports that the average annual TFP growth rate for the same LECs is 2.6% using gross capital stocks and 3.1% using net stocks.[20] In comparison with these estimates models 1 and 3 seem to either overstate or understate the true productivity growth. Thus I believe that models 4 and 5 represent the technological structure of LECs more appropriately.

2. The elasticity of cost with respect to labor quality in models 3,4, and 5 varies very little. Hence all the models produce nearly the same estimates of productivity growth due to improved labor quality (about 0.6%). Average annual percentage change of labor quality is 1.4% for the same period.[21]

3. The inclusion or omission of the competition index makes a big difference to the estimates of elasticity of cost with respect to capital quality. The productivity growth due to the improved capital quality is estimated at 1.5% in model 3 without the competition index and at 0.8% in model 4 with the competition index. Note that the

< Table 2.2 > Rate of Disembodied and Embodied Technical Change

source of technical change		Model 1	Model 2	Model 3	Model 4	Model 5
time	E shifts	-0.042 -4.2%	-0.032 -3.2%	n.a.	n.a.	n.a.
competition	E shifts	n.a.	n.a.	n.a.	-0.123 -1.5%	-0.113 -1.1%
labor quality	E shifts	n.a.	-0.461 -0.6%	-0.462 -0.6%	-0.462 -0.6%	-0.450 -0.6%
capital quality	E shifts	n.a.	-0.129 0.1%	-0.903 -1.5%	-0.452 -0.8%	0.519 2.2%
lagged effects	E shifts	n.a.	n.a.	n.a.	n.a.	-1.010 -3.0%
total shifts		-4.2%	-3.8%	-2.1%	-2.9%	-2.4%
returns to scale		1.045	1.023	1.030	1.052	1.035

Note: E stands for the elasticity of cost with respect to a variable. Shifts in a cost function due to change in a variable is calculated by the elasticity of cost with respect to that variable times the annual percentage change of the variable. Only the average of annual shifts for the period 1951-1991 is reported. Total shifts are the sum of the average shifts caused by each variable.

difference in model 4 and 5 lies not in the total size of the effect but in the size distribution between current and past periods.

4. The estimate of embodied technical change, which is the sum of productivity growth due to improved labor and capital qualities, is about 1.4% in models 4 and 5. About half of the productivity growth is due to the improved input qualities, while the remaining half is ascribed to disembodied technical change. Even though my estimate of embodied technical change is larger than that of previous studies, the true size of embodied technical change may be still underestimated because the competition index may capture part of the effects of improved input qualities. Dropping the competition index in model 3 increases the productivity growth due to improvements in capital quality up to 1.5%. With a better measure of organizational capital the importance of embodied technical change would be larger.

< Table 2.3 > Rate of Technical Change (by Period)

Period	source of technical change						total shifts
	competition		labor quality		capital quality		
	E	shifts	E	shifts	E	shifts	
1952-1976	-0.168	-1.2%	-0.525	-0.6%	-0.451	-1.0%	-2.8%
1977-1991	-0.050	-2.1%	-0.354	-0.5%	-0.445	-0.4%	-3.1%
Total	-0.123	-1.5%	-0.462	-0.6%	-0.452	-0.8%	-2.9%

Note: E stands for the elasticity of cost with respect to a variable. Shifts in a cost function due to change in a variable is calculated by the elasticity of cost with respect to that variable times the annual percentage change of the variable. The averages of annual shifts for two periods are reported. Total shifts are the sum of the average shifts caused by each variable. The calculation is based on model 4.

Table 2.3 reports changes in the rate of technical change before and after the introduction of competition. While there is no significant change in average total annual shifts between the two periods, there are changes in the composition of total shifts. The productivity growth ascribed to capital quality drops from 1.0% to 0.4%, even though the elasticity of cost with respect to the capital quality index is nearly the same for both periods. On the other hand, the productivity growth due to competition soars to 2.1%, while the elasticity declines from -0.168 to -0.050. The reason for changes in the composition is a sharp increase in the competition index and a drop in the capital quality index after 1977. The decrease in the importance of embodied technical change may be overestimated because the competition index increases so rapidly after 1984 that the effect of competition may be overestimated. Based on model 3 the average annual shift due to capital quality drops from 1.6% only to 1.4%.

There are another interesting findings in table 2.4 which provides the rate of embodied technical change by firm. Table 2.4 is based on model 3 rather than on models 4 or 5 because the former shows the relative contribution of labor and capital quality to productivity growth more markedly than the latter. The relative contribution of capital quality to productivity growth is 47% for the smallest firm and

< Table 2.4 > Rate of Embodied Technical Change (by Firm)

firm	maximum revenues (thousand)	source of technical change				total shifts
		labor quality		capital quality		
		E	shifts	E	shifts	
F1	$150,225	-0.444	-0.9%	-0.516	-0.8%	-1.7%
F2	$1,073,648	-0.490	-0.6%	-0.759	-1.1%	-1.7%
F3	$1,717,228	-0.453	-0.6%	-0.649	-1.4%	-2.0%
F4	$2,015,899	-0.496	-0.6%	-0.811	-1.3%	-1.9%
F5	$2,598,487	-0.485	-0.5%	-0.882	-1.5%	-2.0%
F6	$2,843,355	-0.425	-0.5%	-0.818	-0.8%	-1.3%
F7	$7,904,450	-0.443	-0.7%	-1.265	-2.3%	-3.0%
F8	$8,051,446	-0.453	-0.4%	-1.453	-2.6%	-3.0%
Total		-0.462	-0.6%	-0.903	-1.5%	-2.1%

Note: E implies the elasticity of cost with respect to a variable. Shifts in a cost function due to change in a variable is calculated by the elasticity of cost with respect to that variable times the annual percentage change of the variable. The averages of annual shifts for the period 1952-1991 are reported. Total shifts are the sum of the average shifts caused by each variable. The calculation is based on model 3. F1-F8 refer to local telephone companies.

87% for the largest firm. As the size of a firm increases, the firm's productivity growth becomes more dependent on capital quality. In other words, with an increase in the size of a firm the capital quality may play a more important role in enhancing productivity growth. Since the elasticity of cost with respect to capital quality increases with the size of a firm, a large firm may have a better ability to transform improved capital quality into reductions in the total cost than a small firm. As the U.S. telephone industry has experienced rapid innovation and is expected to continue doing so in the future, small firms may have disadvantages compared to large firms in enjoying major innovations.

2.5. SUMMARY

In this chapter I attempt to measure the rate of embodied technical change by relying on the cost function which contains measures of labor and capital qualities as a separate or an adjustment term. Within the context of a cost function embodied technical change is defined as shifts in a cost function due to improvements in labor and capital qualities.

I apply my cost models to pooled cross-sectional time series data for 8 LECs over the period 1951-1991. To evaluate the importance of embodied technical change, I compare five econometric specifications of a cost function. One of the important empirical findings is that inserting a time variable in a translog cost function may result in underestimation of the true rate of embodied technical change. Thus the conventional specification of a cost function may not appropriately reflect the technological structure of LECs.

When a competition index is inserted in place of the time variable to control for disembodied technical change, the productivity growth due to improvements in labor and capital quality indexes is 0.6% and 0.8%, respectively. Thus, two quality indexes explain nearly half the rate of technical change. Still, the true size of embodied technical change may be understated because the competition index is not a perfect measure of disembodied technical change. My results strongly indicate that capital quality improvement may have lagged effects on cost decline. It may be yesterday's improvement of capital quality and not today's that leads to downward shifts in the cost functions of LECs. Also, since large firms are better able to transform improvements in capital quality into reductions in total costs, a merger of small telephone companies may improve efficiency.

NOTES

[1] Solow (1960) estimated the rate of embodied technical change by ignoring disembodied technical change and assuming trial values for input shares and depreciation rate. On the other hand, Solow (1962), Berglas (1965), and Intriligator (1965) estimated disembodied technical change and input shares, given the rate of embodied technical change. In either case the estimates of parameters crucially depend on the accuracy of the assumed values of other parameters.

[2] After examining the identification conditions, Hall (1968) showed that if data on the prices of used machines and the interest rate

are available, then the index of embodied technical change and the rate of depreciation can be calculated from these data. Also, several authors proposed the method of circumventing this under-identification problem. For example, Wickens (1970) obtained the unconditional estimates of parameters in vintage production functions by applying nonlinear estimation methods and Boddy and Gort (1974) based estimates of embodiment on measures of obsolescence.

[3] Gregory and James (1973) indicate this problem of time series data.

[4] To examine the bias due to the time lag of settling-in and learning, Gregory and James (1973) compares the productivity gains between new factories and other factories for the time period 1956/1957-1957/1958. Because they find no significant difference between the two periods they conclude that learning and settling-in problems do not systematically reduce the productivity of new factories relative to the others. But given the results of Bahk and Gort (1993) that show it takes roughly six years for new capital to reach the production frontier, Gregory and James' data encompass too short span of time.

[5] Also, as Hulten (1992) indicates, his measure may be over- or under-stated, if differentials between the BLS and the quality-adjusted price indexes reflect other factors such as a producer's decision on optimal quality mix.

[6] This way of expressing capital in technical efficiency units is fundamentally different from the standard procedure. The common procedure in the literature on capital theory is to employ the depreciation rate (the rate of physical decay plus the rate of obsolescence) as a discount factor. This approach introduces the rate of obsolescence as an augmentation factor that reflects advances per unit in physical capabilities. Augmentation is treated separately from the physical decay of capital. See Hulten (1990 & 1992). Gort and Wall (1993) show that using the standard procedure may understate the capital stock series in overall size and its rate of growth.

[7] These externalities are similar to disembodied technical change in that both affect plants of every vintage simultaneously. But the externalities are a component of embodied technical change because they relate to increases in the efficiency of inputs (the productivity of physical capital).

[8] Age (a) is the year of measurement (t) minus the year of birth. Thus, vintage (v) and age (a) have the following relationships; v=T-t+a. When the year of measurement is used as the base year, both the vintage and the age produce exactly the same measure of efficiency. But, when time series data for the technical efficiency of investment goods are constructed, the age measure has a limitation because it depends on the year of measurement. On the other hand, the vintage measure may overstate the technical efficiency of recent investment goods because it assumes the same annual increase in the efficiency over time. Thus, I use (1+s)/a in place of (1+s)/v as a measure of capital quality.

[9] Jorgenson, Gollop, and Fraumeni (1987) define the index of the quality of hours worked as an index that transforms hours worked into the translog index of labor input. The quality index reflects changes in the composition of hours worked, classified by sex, age, education, employment class, and occupation. Because of the lack of appropriate data, my labor quality index considers only changes in the composition of hours worked classified by occupation.

[10] Denote a firm's labor expenses by B. Then $B= P_L*L= P_L'*L'$. Thus $\omega_t=L/L'= P_L'/ P_L$.

[11] When the labor and capital quality indexes are explicitly taken into account, the remaining disembodied technical change is associated with only the level of organizational capital, largely in the form of firm-specific information.

[12] This is sometimes called the embodied effect of vintage. Gort, Bahk, and Wall (1993) argue that the principal source of variation in the relative efficiency of capital of different vintages is obsolescence because maintenance outlays roughly offset the effects of physical decay at least on current production. Then the implied depreciation

rate, correctly measured, becomes roughly the dual of capital augmenting technical change. They find that the embodied effect of vintage is about 12%.

[13] For the conventional method of measuring technical bias, see Stevenson (1980).

[14] According to Bahk and Gort (1993), it takes roughly six years before new capital is fully integrated into the production process so that it reaches the efficiency frontier. Since my sample firms have such a long history that the portion of new plants in total plants in service is not large, the materialization of vintage effect would take a shorter time span. Also TFP growth rate regressions in chapter 4 show that the one- and two-year lagged growth rates of capital quality index have a positive effect on TFP growth. Thus I believe that this assumption is not unjustified.

[15] If there are n factor share equations, only n-1 of them are linearly independent, since the sum of the cost shares over all equations always equals 1 for each observation. The most common procedure for handling this singularity problem is to drop an arbitrary equation. Then this iterative Zellner's seemingly unrelated regression estimates are asymptotically equivalent to maximum likelihood estimates. Also, the results are the same regardless of the choice of the share equation to be deleted.

[16] This does not imply that serial correlation is statistically insignificant in our regressions. When I consider serially correlated errors using the three-step procedure (as used in chapter 4) which combines Cochrane-Orcutt method with Zellner's seemingly unrelated regression technique, the estimated cost functions are often inconsistent with microeconomic theory. For example, the marginal cost of toll output is negative for some observations and, also, the estimated cost function does not satisfy the quasi-concavity condition. Thus, I ignore serially correlated errors in all the models. A full set of parameter estimates using the three-step procedure is given in the appendix 1.

[17] This does not imply that LECs exhibit increasing returns to scale. For a comprehensive explanation, see the section 3 of chap 3.

[18] This is much clearer, when I added $\ln \bar{k}$ to model 3 with no competition index. Adding $\ln \bar{k}$ does not make the coefficient of lnk insignificant, but also it does decrease its size from -1.311 to -0.015. The coefficient of $\ln \bar{k}$ is -1.349. Thus the sum of the coefficient of lnk and $\ln \bar{k}$ is slightly larger than the coefficient of lnk in the equation without $\ln \bar{k}$. Thus adding the lagged variables of k does not affect the total size of the effect of capital quality on productivity growth, but it does influence its size distribution between current and past periods.

[19] Calculation of important multiproduct cost measures such as the degrees of economies of scope or product-specific economies of scale requires evaluating the cost function at points at which one or more output levels are zero. Several studies circumvent this problem by employing a Box-Cox transformation of output variables, i.e. a generalized translog cost function. But, as Fuss (1992) indicates, "the researcher must be able to estimate the technology of a firm which produces only" one service in order to specify economies of scope or product-specific economies of scale. Thus the precise estimate of economies of scope seems impossible in the U.S. telephone industry in view of the nonavailability of data. Every LEC produces both local and intrastate (or intraLATA) toll services.

[20] TFP growth rate is equivalent to shifts in a production or cost function under the assumption of constant returns to scale and perfect competition in output and input markets. If producers sell at prices that equal marginal costs and if there are no economies of scale and no regulatory biases, then TFP growth rate correctly represents productivity growth as measured by shifts in a production or cost function. In the U.S. telephone industry the assumption of a competitive product market or marginal cost pricing is violated. Even though the TFP growth rate is not necessarily the same as shifts in the production or cost function, the former may give some guidelines as to which measure of shifts is appropriate.

[21] The section 3 of chap 4 reports average annual percentage change of labor and capital qualities for 5 periods.

III

Firm Size and Cost Subadditivity

3.1. INTRODUCTION

One of the most important public policy questions in the recent controversy on 'information super-highways' is whether entry and competition in the local and intra-LATA telephone markets are in the public interest. The local telephone industry has become more competitive over time, even though LECs are still single sellers in their market. LECs have been responding to potential competition from cable television companies, intermodal competition from wireless communication providers, and the threat of bypass by large business users. Even though a number of empirical studies have focused on the question of whether a telephone industry is a natural monopoly, the findings of these studies are inconclusive and frequently, inconsistent with each other. This chapter conducts cost subadditivity tests for LECs to provide new economic evidence on this issue.

At the time of the breakup of the Bell system in 1984, it was common belief that substantial and effective competition is unlikely or impossible for LECs, especially in local access. The economic rationale for the divestiture of BOCs from the AT&T was that LECs exhibited the characteristics of a natural monopoly. For local service, at least until recently, it was generally considered wasteful to allow two or more rival suppliers since that would entail duplication of facilities. Thus, despite intense competition in the long distance telephone market, LECs have still maintained their monopoly position. But the availability of new technology, as well as the advancement in old technology, has consistently created new business opportunities and made possible new alternative services in local telephone markets. Competition in local telephone markets is imminent or already exists on several different fronts.[1]

First, wireless telephone service providers, including cellular phone companies, have been competing with LECs for telephone calls since 1980s. A number of telephone calls which would otherwise be handled by LECs are routed by wireless communications. Cellular operators are now developing digital wireless communications technology that is expected to increase their capacities 3 to 15 times.[2] The digital technology enables them to overcome the problem of limited capacities which has been a big bottleneck for the growth in demand for cellular services. Also, by the end of this century the advent of personal communications services (PCSs) is expected to significantly expand the opportunities for wireless local access.

Second, a decrease in the price of telecommunication equipment due to rapid technological advance in ESS and fiber optics makes it possible for even small businesses to install their own communications facilities. Many large business users have already switched to private networks for transmitting intra-company calls. Business users may bypass the local network of LECs by connecting directly from a long distance network of IXCs to their internal switchboard such as private branch exchange (PBX). The uniform pricing of access service to residential and business customers, which is a legacy of outdated regulation, provides incentives even for uneconomic bypass.[3]

Finally, as telecommunications and broadcasting technologies converge, the suppliers of cable television services which serve half of all homes become the most likely candidates for direct rivalry with LECs. Cable television companies already have cable wires to give them direct and widespread access to subscribers' premises and these facilities can be modified to carry telephone messages.[4] Meanwhile, LECs are seeking the installation of fiber-optic cable for delivery of full-motion switched video, data, and voice. The so-called broadband integrated services digital network (b-ISDN) creates new business opportunities for LECs. The competition between cable television companies and LECs is already under way.

Contrary to the developments in the telephone industry, a number of econometric studies indicate that the telephone industry exhibits economies of scale or cost subadditivity. Most of these empirical studies rely on time series data for telephone companies, mainly AT&T and Bell Canada, to estimate single product or multi-product cost functions.[5] Even though they use the same data for AT&T or Bell Canada, they obtain differing and conflicting empirical results by adopting different functional forms or different estimation methods and controlling for technological change in different ways.[6]

Shin and Ying (1992) is the first main study which focuses only on LECs. They start with the observation that a major problem with previous studies is the choice of data and level of aggregation. Obviously the most serious problem with time series data is the small number of observations on highly correlated variables, as compared to the large number of parameters being estimated. They construct pooled cross-sectional time series data for LECs,[7] estimate a three-product (access, local, and toll) translog cost function, and conduct the Evans-Heckman subadditivity test. With enough degrees of freedom to generate precise estimates, their results show that the cost function for LECs is not globally subadditive. Even though the marginal costs of hypothetical firms are negative for some output levels, dropping those observations strengthens their finding.

Following Shin and Ying (1992), I construct pooled cross-sectional time series data for LECs with enough number of observations to conduct the Evans-Heckman subadditivity test. My study is based on more reliable data than that of Shin and Ying, especially with respect to capital stock and output variables.[8] The capital stock is the weighted sum of real net stocks of five types of telephone plants, with the weights being the capital costs shares of each type of plant. Each net stock is constructed by the commonly used perpetual inventory method. Different price indexes and different rates of depreciation for each type of telephone plant produce a better measure of aggregate capital stock. The output measures are deflated revenues. Given the availability of data, deflated revenues are a better measure than physical units of output, i.e. number of calls, in the telephone industry.[9] The deflators are drawn from the Bell system price indexes for the period 1951-1971 and from the Producer Price Indexes for the period 1972-1991.

The empirical analysis is carried out with the same data as in chapter 2. The basic cost model for the test is model 4 in chapter 2, while model 1 is used in order to compare my results with those of previous studies. Movements along a cost curve are appropriately separated from shifts in the cost curve because model 4 incorporates both embodied and disembodied technical change.

The concept of cost subadditivity implicitly assumes that technology is the same with a single firm as with numerous firms. It ignores completely the effect of a competitive product market on a firm's management decisions. The existence of numerous firms presupposes competition which may change the form of a cost function. Competition affects the level and composition of output, the level and composition of inputs, the rate and direction of innovation, the use and accumulation of organizational capital and so on. Assuming away differences in technology may severely overstate the measure of cost subadditivity.

But, given the lack of appropriate data, it is understandable why previous studies on cost subadditivity assume the same technology for a monopolistic firm as for a competitive firm. Since my cost model contains a measure of competitiveness in the U.S. telephone industry, I can examine the effect of competition on cost subadditivity. Assuming different degrees of competitiveness for a monopolistic and a competitive firm may be one approximation to estimating and comparing different cost functions.

The structure of this chapter is the following. In section 2, I present the methodology of the cost subadditivity test. Section 3 provides empirical results where the main focus is on returns to scale and subadditivity. Summary follows.

3.2. METHODOLOGY OF THE TEST

During the last decade there has been a wave of deregulation in public utilities that were once thought to be characterized by substantial monopoly attributes. The economic rationale for deregulation is that these public utilities are not natural monopolies any more. An industry or a firm is said to be a natural monopoly if the cost function is strictly subadditive over the entire relevant range of outputs. Thus the transition towards competitive local and intra-LATA telephone markets can be justified only if the cost function of LECs is not subadditive any more.

The cost function is subadditive when any combination of outputs can be produced more cheaply by a single firm than by any group of two or more firms. There are two approaches to the investigation of cost subadditivity. The first approach is to examine the sufficient conditions for cost subadditivity on the basis of the parameter values of empirically estimated cost functions.[10] It is well known that economies of scale are neither necessary nor sufficient for monopoly to be the most efficient form of productive organization. Even economies of scale and economies of scope are not a sufficient condition for subadditivity. A condition for subadditivity is the existence of both decreasing average incremental costs for each product and economies of scope. Also, the combination of declining ray average costs and trans-ray supportability or cost complementarity conditions gives another set of sufficient conditions for cost subadditivity.

The second approach to the cost subadditivity test is developed by Evans and Heckman (1983b & 1984). I adopt this approach because it is more direct and empirically less demanding.[11] The Evans-Heckman test is straightforward. A cost function is subadditive at an output vector if a single firm's production of the quantity in question is cheaper than that of any two-firm alternative. Specifically, the cost function is subadditive, if

$$g(\alpha Q_1, \beta Q_2) + g((1-\alpha)Q_1, (1-\beta)Q_2) > g(Q_1, Q_2)$$

where g(•) represents a long-run cost function, Q_1 refers to local output, and Q_2 to toll output. The scalars, α and β, are real values between 0 and 1. As monopoly outputs are divided between two hypothetical firms, all other variables in the cost function are left unchanged. They conduct this subadditivity test using the translog cost function estimated for the Bell System. Their results indicate that for all observations in the admissible region there exists a two-firm alternative that would lower costs.

As commonly pointed out, the Evans and Heckman test is local because they restrict the subadditivity test to the admissible region.[12] To avoid extrapolating the cost function to unobserved output configurations, the output of two hypothetical firms is required to be no less than the smallest output vector of the Bell system and, also, must lie within the sample range of local-toll output ratios. Whereas, I apply the test of subadditivity globally because my data set is much larger and of a much wider range than Evans and Heckman's. My sample firms range from 2 large LECs, to 5 midsize LECs, and 1 small LEC. As there is no LEC which produces only one type of output, I do not impose zero values for one or two output variables. In the simulation, the scalars, α and β, range over (0.1, 0.2, 0.3,...0.9) and do not include zero.

One major limitation of the concept of cost subadditivity is its implicit assumption that technology is the same with a single firm as with numerous firms.[13] The existence of numerous firms presupposes competition which may change the form of a cost function. Competition has effects on the level and composition of output, the level and composition of inputs, the rate and direction of innovation and so on. The great benefit from competition is its disciplinary effect on the accumulation and use of organizational capital, mainly in the form of firm-specific information.

Monitoring costs of a modern corporation is frequently associated with supervising organizational capital; for example, matching tasks with the attributes of individual employees, sorting employees by competence, and other aspects of firm-specific learning-by-doing. It is costly or difficult for shareholders to monitor and reward the optimal use and accumulation of organizational capital in a monopolistic market because a monopoly has no comparable counterparts. Competition keeps these monitoring costs down by forcing a firm to use and accumulate organizational capital efficiently.

It is difficult or almost impossible to conduct an empirical test of cost subadditivity without assuming the same technology for a competitive firm as for a monopolistic firm because reliable data on both types of firm are seldom available for the cases of interest. In that case, the absence of the competition effect in previous studies may be understandable. But assuming away differences in technology may severely overstate the measure of cost subadditivity. Since my cost model contains a measure of the competitiveness in the U.S. telephone industry, I can investigate the effect of competition on cost subadditivity. Assuming different degrees of competitiveness for a monopolistic and a competitive firm may be one approximation to estimating and comparing different cost functions.

Specifically, I examine the change in the probability of subadditivity by increasing the competition index for two hypothetical firms and leaving unchanged that for a monopolistic firm. Now the cost function is subadditive, if

$$g(\alpha Q_1, \beta Q_2; (1+\delta)CI) + g((1-\alpha)Q_1, (1-\beta)Q_2; (1+\delta)CI) > g(Q_1, Q_2; CI)$$

where CI represents the competition index and δ ranges over (0.1, 0.2, 0.3, 0.4,...). The savings from having a monopoly (S) is calculated as follows:

$$S = \frac{g(\alpha Q_1, \beta Q_2; (1+\delta)CI) + g((1-\alpha)Q_1, (1-\beta)Q_2; (1+\delta)CI) - g(Q_1, Q_2 : CI)}{g(Q_1, Q_2; CI)}$$

where a positive (negative) value indicates subadditivity (supadditivity).

3.3. EMPIRICAL RESULTS

Estimates of Returns To Scale

The empirical analysis in this chapter is carried out with the same data as in chapter 2. The basic cost model for the test is model 4 in chapter 2 which is regarded as the best specification, while model 1 is used as complementary so that my results can be compared with those of previous studies. Movements along a cost curve are appropriately separated from shifts in the cost curve because model 4 incorporates both embodied and disembodied technical change.[14] Refer to chapter 2 for general empirical findings for both models.

The sum of output cost elasticities is 1.149 for model 1 and 1.037 for model 4 at the sample mean. Thus the overall returns to scale for each model is 0.870 and 0.964 respectively at the sample mean. One attractive feature of flexible functional forms such as a translog cost function is that it allows a measure of scale economies to vary, depending upon input prices and the level of outputs. Thus the estimate of scale economies is different for every observation. Using this desirable property, I calculate average overall returns to scale for every sample firm in table 3.1. The average overall returns to scale for 8 LECs is 1.045 in model 1 and 1.052 in model 4 which are larger than those at the sample mean. These estimates are much smaller than those in previous studies which often provide evidence on strong economies of scale in the U.S. telephone industry. Considering that imposing the translog specificiation on data may lead to an upward bias in the estimate of the degree of scale economies,[15] I conclude that LECs exhibit constant returns to scale technology.

< Table 3.1 > Output Cost Elasticity and Returns to Scale

(1) Model 1

firm	maximum revenues (thousand)	output-cost elasticity		overall returns to scale
		local output	toll output	
F1	$150,225	0.591	0.320	1.098
F2	$1,073,648	0.542	0.399	1.064
F3	$1,717,228	0.532	0.412	1.060
F4	$2,015,899	0.517	0.430	1.057
F5	$2,598,487	0.503	0.451	1.049
F6	$2,843,355	0.477	0.472	1.055
F7	$7,904,450	0.404	0.598	1.000
F8	$8,051,446	0.383	0.628	0.990
Total		0.492	0.466	1.045

(2) Model 4

firm	maximum revenues (thousand)	output-cost elasticity		overall returns to scale
		local output	toll output	
F1	$150,225	0.842	0.131	1.029
F2	$1,073,648	0.704	0.248	1.052
F3	$1,717,228	0.706	0.234	1.066
F4	$2,015,899	0.648	0.290	1.068
F5	$2,598,487	0.642	0.300	1.063
F6	$2,843,355	0.667	0.276	1.061
F7	$7,904,450	0.567	0.394	1.044
F8	$8,051,446	0.547	0.425	1.031
Total		0.662	0.290	1.052

Note: F1-F8 refer to local telephone companies.

The coefficients of the squared terms of the output variable are positive at the sample mean in model 1. Thus, as firm size (as measured by the sum of local and toll output) increases, there may be a tendency for the degree of overall returns to scale to decrease. On the other hand, in model 4 the coefficient of $0.5*(\ln Q_1)^2$ is negative and the coefficient of $0.5*(\ln Q_2)^2$ is positive. But, since the absolute size of the latter is larger than that of the former, the negative relationship between firm size and the degree of returns to scale seems to hold in model 4. This result confirms Nerlove's (1963) classical finding that the degree of returns to scale tends to fall as the size of an electric utility company increases.

Table 3.1 shows the same negative relationship between a firm's size and the degree of returns to scale. In model 1 the smallest (largest) firm has the largest (smallest) degree of scale economies. As the size of a firm increases, any advantage of increasing returns to scale appears to diminish. The average returns to scale of the two largest firms is nearly constant. The relationship is not as clear in model 4 as in model 1. But when the smallest firm is dropped, the same tendency seems to hold in model 4.

Cases of Cost Subadditivity in the LECs

In this subsection I report the results of the subadditivity tests. The number of vector combinations for one observation is 81 since the scalars, α and β, range over (0.1, 0.2, 0.3,...,0.9). The costs of 81 two-firm alternatives are compared with those of a monopolistic firm for each observation. The gain is calculated for 25,596 possible cases using a simulation technique.[16] The results are reported in table 3.2, of which the key findings are as follows:

1. Of the 25,596 possible cases, 18,104 cases (70.7% of total cases) lead to cost subadditivity in model 1. The average gain of total cases is 2.16%, which implies that a monopoly is, on average, more efficient. But this conclusion is reversed when embodied technical change is considered. In model 4 14,788 cases (57.8%) lead to cost subadditivity. If the outputs of an existing monopoly were divided among two firms, a competitive alternative would lower total costs for 42.2% of total cases. Since the average gain is -0.22%, the competitive alternative is, on average, more efficient. Differences in the probabilities between the two models indicate that the question of

< Table 3.2 > Cost Subadditivity Cases

(1) Model 1

firm	A	B	C (%)	average monopoly gain (%)			
				max	min	mean	SE
F1	2835	2717	95.8	43.1	-2.9	6.9	5.4
F2	3321	2843	85.6	13.8	-4.3	3.4	3.2
F3	3321	2751	82.8	11.6	-5.7	3.1	3.2
F4	3321	2638	79.4	9.6	-6.4	2.6	3.3
F5	3321	2645	79.6	11.7	-6.9	2.3	3.0
F6	2835	2307	81.4	61.3	-9.1	3.6	6.2
F7	3321	1248	37.6	5.5	-14.5	-1.7	3.8
F8	3321	955	28.8	4.4	-13.8	-2.1	3.5
Total	25596	18104	70.7	61.3	-14.5	2.2	3.9

(2) Model 4

firm	A	B	C (%)	average monopoly gain (%)			
				max	min	mean	SE
F1	2835	1938	68.4	34.3	-10.1	2.2	5.3
F2	3321	2019	60.8	8.3	-22.1	0.3	4.4
F3	3321	2301	69.3	10.4	-16.7	1.4	4.4
F4	3321	2057	61.9	10.1	-19.1	0.3	5.2
F5	3321	2067	62.2	7.7	-23.5	-0.1	5.1
F6	2835	1869	65.9	25.4	-27.3	0.8	7.0
F7	3321	1348	40.6	8.2	-27.1	-2.6	6.7
F8	3321	1189	35.8	6.9	-31.1	-3.6	7.0
Total	25596	14788	57.8	34.3	-31.1	-0.2	5.6

Note: F1-F8 are the same as those in table 3.1. A=number of total possible cases, B=number of subadditivity cases, and C=the ratio of A to B. SE stands for standard error. The monopoly gain is calculated by (costs of two firm alternative - costs of a monopoly)*100/costs of a monopoly.

how to control for technical change may have an effect on cost subadditivity tests. No due account of technical change, whether embodied or disembodied, may result in an upward bias in the measure of cost subadditivity.

2. My results are not as strong as those of Shin and Ying (1992) who report 20.0%-38.9% cost subaddivitity cases, even though their estimate of overall returns to scale is larger than mine. Thus the small degree of returns to scale does not guarantee the high probability of cost subadditivity cases. This is legitimate because scale economies are neither necessary nor sufficient for cost subadditivity.

3. The table shows that the size of a firm may have the same negative relationship with the probability of subadditivity as with the degree of returns to scale.[17] As the size of a firm increases, the percentage of cost subadditivity cases decreases. In model 1 the percentage of subadditivity cases is 95.8% for the smallest firm and 28.8% for the largest firm. Model 4 shows the same tendency as model 1. In model 4 the percentage of subadditivity cases is 68.4% for the smallest firm and 35.8% for the largest firm. The average gain tends to decrease with the size of a firm in both models. Clearly the two largest LECs have nearly no savings from being monopolies.

This finding may explain the seemingly contradictory results of current literature. For example, Gentzoglanis (1993) cannot reject the hypothesis that Alberta Government Telephones is a natural monopoly, even though he uses the translog specification of a cost function and conducts the same Evans-Heckman test. His conclusion may not be contradictory because my results indicate that such small telephone companies as Alberta Government Telephones may benefit from having a monopoly.

One problem of a translog cost function is that it may violate certain regularity conditions over large regions of output space. Röller (1990a) argues that if the estimated cost structure is constrained to be proper over a region that includes the axis (i.e. stand alone costs), the Evans-Heckman test does not reject the hypothesis of natural monopoly of the Bell System. To check the robustness of my results, I calculate the marginal costs of each hypothetical output vector. Whenever any marginal cost is negative, I drop that output vector from the subadditivity test.

Imposing the condition of positive marginal costs reduces the number of possible cases by 44.8% in model 1 and by 46.2% in model

< Table 3.3 > Cost Subadditivity Cases After Dropping Output Vectors with Negative MC

(1) Model 1

firm	A	B	C (%)	average monopoly gain (%)			
				max	min	mean	SE
F1	2053	1935	94.3	12.8	-2.9	5.1	2.7
F2	2837	2411	85.0	8.6	-4.3	3.1	2.9
F3	2889	2369	82.0	9.0	-5.7	2.8	3.0
F4	1825	1326	72.7	7.5	-6.4	1.4	2.8
F5	1736	1360	78.3	6.1	-6.9	1.8	2.5
F6	1279	1063	83.1	8.3	-8.6	2.7	3.0
F7	997	583	58.5	5.4	-14.5	-0.4	3.6
F8	758	489	64.5	4.1	-5.2	0.5	1.6
Total	14374	11536	80.3	12.8	-14.5	2.5	2.8

(2) Model 4

firm	A	B	C (%)	average monopoly gain (%)			
				max	min	mean	SE
F1	1363	728	53.4	7.2	-10.0	0.5	3.4
F2	2829	1689	59.7	8.3	-22.1	0.2	4.5
F3	2743	1837	67.0	10.4	-16.7	1.2	4.5
F4	1929	1097	56.9	8.6	-17.7	-0.6	4.2
F5	1877	1215	64.7	7.7	-15.8	0.4	4.0
F6	1097	661	60.3	7.2	-25.7	-0.5	5.3
F7	1199	709	59.1	8.2	-22.4	0.4	5.2
F8	752	435	57.8	6.9	-8.5	1.0	3.3
Total	13789	8371	60.7	10.4	-25.7	0.3	4.3

Note: F1-F8 are the same as those in table 3.1. A=number of total possible cases, B=number of subadditivity cases, and C=the ratio of A to B. SE stands for standard error. The monopoly gain is calculated by (costs of two firm alternative - costs of a monopoly)*100/costs of a monopoly.

< Table 3.4 > Competition Effects on Cost Subadditivity Cases

difference in competition index	A	B	C (%)	average monopoly gain (%)		
				max	min	average
0%	12960	7595	58.6	10.4	-31.1	-0.3
10%	12960	6783	52.3	10.6	-31.9	-0.9
20%	12960	6092	47.0	10.8	-32.7	-1.3
30%	12960	5769	44.5	11.0	-33.6	-1.7
40%	12960	5513	42.5	11.2	-34.4	-2.1
50%	12960	5275	40.7	12.1	-35.2	-2.4
100%	12960	4845	37.4	16.2	-38.1	-3.4

Note: A=number of total possible cases, B=number of subadditivity cases, and C=the ratio of A to B. The monopoly gain is calculated by increasing the competition index for two hypothetical firms by 10%, 20%, 30%, 40%, 50%, and 100%, leaving unchanged the competition index for a monopoly.

4, respectively. Table 3.3 shows that this condition has few effects on the conclusions. In model 1 the 80.3 % of 14,374 possible cases, which are slightly higher than before, lead to cost subadditivity. The average gain (2.53%) is slightly larger than before, too. In model 4 the 60.7% of total cases have cost subadditivity. Also the percentage of subadditivity cases still seems to decrease with a firm's size, even though this tendency is not as clear as before in model 4.

The Table 3.4 reports the results from applying the same cost subadditivity test over the period 1972-1991 while increasing the competition index for two hypothetical firms by 10%, 20%, 30%, 40%, 50%, and 100%. I chose 1972 as the starting year because there was no change in the competition index before 1972. The number of possible cases is 12,960. The principal findings in table 3.4 are the following:

1. When the same competition index is applied to both alternatives, the probability of subadditivity cases and the average gain from having a monopoly is 58.6% and -0.30% for this period respectively. These measures are a little higher than those for the whole period.

2. As expected, both the percentage of subadditivity cases and the average gain from having a monopoly tend to decrease as I plug higher indexes of competition in the cost function of two competitive firms. The probability of subadditivity cases continues to drop from 58.6% for no differential, to 47.0% for a 20% differential, and to 40.7% for a 50% differential in the competition index between a competitive and a monopolistic firm.

3. Even when the competition index for competitive firms increases by 100% the probability of subadditivity drops only to 37.4%.[18] There seem to be diminishing marginal returns to the effect of intensified competition on the probability of subadditivity. For example, the probability of subadditivity decreases by 17.9% due to the first 50% increase in the competition index and only by 3.3% due to the second 50% increase in the competition index. Thus, it appears that the cost function may still be subadditive for some observations, mainly at small output levels, even though the competition index increases much more.

3.4. SUMMARY

Competition in local telephone market seems imminent and may already exist. Wired LECs have competed with wireless communication providers for capturing local calls, have been subject to the threat of bypass by business users, and may compete with large cable television companies in the near future. Since the empirical findings of previous studies on the natural monopoly problem of the U.S. telephone industry are inconclusive, this chapter conducts cost subadditivity tests for LECs to provide new economic evidence on this policy question.

My empirical results show that the LECs exhibit constant returns to scale technology and that the cost functions of the LECS are not always subadditive. Different methods of controlling for technical change produce both different estimates of scale economies and different probabilities of cost subadditivity. The percentage of subadditivity cases as well as the estimate of scale economies depends on the size of a firm. As the size of a firm increases, both the degree of returns to scale and the probability of cost subadditivity tend to decrease. In particular, the two largest firms have nearly no cost savings from having a monopoly. Even though more intense competition reduces the number of subadditivity cases, competition may not eliminate all the possibilities of subadditivity.

These findings have important policy implications. Since small LECs have slight increasing returns to scale, a merger of small LECs may increase the efficiency of a combined firm. While small LECs may have enjoyed cost savings from having a monopoly until now, there may be no economic rationale for large LECs such as regional holding companies (RHCs) to maintain a monopoly position. It appears that economies of scale and scope may pervade at least portions of some local telephone services. The operation of local loop and the provision of access to it may continue to constitute a natural monopoly. The market mechanism will determine the boundaries between a natural monopoly area and a competitive area.

NOTES

1 For a detailed explanation on the feasibility of competition in the U.S. local telephone industry, see Baumol and Sidak (1994).

2 Catherine Arnst, "The New Wireless Looks a Bit Pricey," Business Week / December 5, 1994.

3 Using the data at the central office level for New England Telephone, Palmer (1992) shows that business service revenues subsidize residential service provision at 65% of the suburban central offices and at 54% of the central offices tested.

4 For example, Time Warner is upgrading its systems with fiber-optic cable and digital switches to create stand-alone local phone businesses. See Lander, M., Lewyn, M. and Robello, K., "Now, Time Warner Is a Phone Company," Business Week / November 21, 1994.

5 For illustration, the estimate of overall returns to scale for AT&T (or Bell system) is 1.8 or higher in Nadiri and Schankerman (1981), 1.40-2.65 in Christensen, Cummings, and Schoech (1983), 1.28-2.90 in Röller(1990a), and 1.12-1.38 in Diewert and Wales

(1991). The estimate for Bell Canada is 1.37-1.59 in Denny, Fuss, and Waverman (1981), 1.62 in Kiss, Karabadjian, and Lefebvre (1983), and 1.56 in Bernstein (1989). For a comprehensive survey on the scale economies or cost subadditivity of AT&T and Bell Canada, see Fuss (1983 & 1992) and Waverman (1989).

[6] Evans and Heckman (1983a & 1984), the first major study on cost subadditivity of the Bell system, reject the hypothesis that AT&T is a natural monopoly. Using the same data as Evans and Heckman, Charnes, Cooper, and Sueyoshi (1988), Röller (1990a), and Diewert and Wales (1991) obtain opposite results.

[7] The data for LECs are available because both BOCs and independent operating companies were distinct financial units even when they were commonly owned by AT&T or independent holding companies. They have provided both local and intra-state (or intra-LATA) long distance services within their territories. A possible problem with the data may lie in handling the complicated separation and settlement process (and later access charge). The allocation of non-traffic sensitive (NTS) costs of local access among LECs and IXCs was determined through the political decision-making process between federal and state regulatory agencies. The difficulty in solving this problem may have effects on the reliability of any firm data in the U.S. telephone industry.

[8] Shin and Ying (1992) make no attempt to measure inputs in effective efficiency units. They construct a real capital stock series by dividing gross communications plant by the 20-year average communication equipment implicit price deflator. The accounting value of gross stock may severely overstate the true economic value of capital stock. They argue that their measures of telephone outputs (access lines, number of local calls, and number of toll calls) are better than deflated revenues. If the appropriate data is available, they may be right, particularly because the separation of local access from local service is conceptually appealing. They use the number of total telephones as a proxy for the number of access lines which have been reported since 1984 in the *SCCC*. The number of total telephones may be a poor proxy because the number of telephones per access line may

increase over time and varies across firms. Also the number of calls may be a poor measure of output in the telephone industry because telephone calls are not homogenous.

[9] According to Denny, Fuss, and Waverman (1981), using the number of calls understates output growth predominantly, while the rate of increase in price is overstated. Thus they conclude that the best available output measure in telephone industry is constant (or deflated) dollar revenues.

[10] Refer to Panzar (1992) for a formal explanation of the sufficient conditions for cost subadditivity.

[11] Evans and Heckman (1983b) indicate that the Baumol, Panzar, and Willig test of natural monopoly is difficult to implement because of two problems. First, estimates of economies of scope and product-specific scale economies (or decreasing average incremental costs) require information about the stand-alone costs of separate production. Appropriate data on stand-alone costs are seldom available. Second, when data are available, the test is frequently inconclusive.

[12] Sueyoshi and Anselmo (1986) illustrate Evans and Heckman's admissible region.

[13] In this respect Evans and Heckman (1993a) indicate that "researchers who use scale economy evidence (or evidence of subadditive costs) to make policy recommendations concerning whether an industry should be subject to regulations . . . often ignore the facts that (a) regulations are costly to administer and may induce the regulated firm to act inefficiently, . . . and (c) entry barriers erected by regulators may stifle the development of more efficient technologies." But their test is still based on the assumption that competitive firms have the same technology as a regulated monopoly.

[14] A lot of studies have emphasized the importance of distinguishing movements along a production or cost curve from shifts in the production or cost function. A model that postulates

constant returns to scale would overstate the productivity growth resulting from technical change, when confronted with data generated by a firm or an industry subject to increasing returns. On the contrary, a model which understates technical change may overestimate a movements along a cost curve (economies of scale).

[15] Guilkey and Lovell (1980) reach this conclusion through the Monte Carlo study.

[16] Even though half of total cases are duplicated, dropping duplicated output vector combinations does not change the probability of subadditivity.

[17] This confirms Shin and Ying (1992)'s suspicion, even though they do not report the percentage of subadditivity cases by the size of a firm. According to them, "the degree to which a monopoly is relatively more efficient seems to lessen as the firm's size, as measured by access lines, increases. However the evidence is far from clear-cut." My evidence clearly shows the negative relationship between the percentage of subadditivity cases and a firm's size.

[18] Since my competition index soars from 1.72 in 1971 to 485.48 in 1991 a 100% increase may not be extreme.

IV

Competition and Productivity Growth

4.1. INTRODUCTION

This chapter attempts to compare productivity growth for a competitive firm with that for a monopolistic firm. It is sometimes argued that a monopolistic firm cannot operate as efficiently as its competitive counterpart. The rationale underlying this belief is that a firm's technical efficiency is a prerequisite for its survival in a competitive market. Competition provides discipline and reduces managerial incentives to slack off. On the other hand, studies in the Schumpeterian tradition maintain that a large firm in a concentrated market is superior to a small firm operating in a competitive market in making innovation or technological progress.[1] A monopolistic firm has greater demand for innovation because its market power increases the ability to profit from the innovation and, also, provides larger supply of innovation because of superior access to capital, ability to pool risks, and economies of scale in the maintenance of research and development. Since there has been no clear theoretical consensus as to which market organization is conducive to technological progress the efficiency comparison becomes an empirical issue.

There have been surprisingly few studies on the relative performance of a firm in competitive and non-competitive environments, as compared with the large volume of literature on that of private and public firms.[2] One reason for this gap may be that there are few cases in which a competitive and a monopolistic firm coexist in the same industry.[3] The deregulation of the U.S. telephone industry provides an excellent experiment for testing hypotheses about the effect of competition. New entry in the long distance telephone market was allowed in the early 1960s, while entry and competition in the local telephone market is still under consideration. IXCs operates in an intensely competitive market, while LECs maintain their monopoly position. Hence I focus on differences in productivity growth between AT&T Long Lines (current AT&T Communications) and several LECs.[4]

In this comparison attention should be paid to two points. First, even the local telephone market has become competitive over time, even though LECs are still single sellers in their market. While AT&T Long Lines faces intense competition from Other Common Carriers (OCCs) such as MCI and Sprint, LECs have been responding to potential competition from big cable television companies, intermodal competition from wireless communication providers, and the threat of bypass by large business users. Also, cost-cutting efforts in long distance telephone companies may have spilled over into monopolistic LECs. Second, both AT&T Long Lines and LECs have been heavily regulated by the FCC or state regulatory agencies,[5] whether the rate-of-return or the price cap is assigned.[6]

Despite these similarities, competitiveness in a long distance market is so different from that in a local market that I expect differences in economic performance. Certainly, effective competition with actual rivalry exhibits greater disciplinary effects in monitoring a firm's non-optimizing behavior than does potential or intermodal competition. In particular, the disciplinary effect of potential competition or intermodal competition may be limited, when every participant know that it takes time for competition to be materialized, or where there exists only weak substitution between goods at different modes. Thus, my hypothesis is that the AT&T Long Lines operating in an effectively competitive market environment is more efficient than the LECs under potential or intermodal competition.

For a successful comparison of the relative performance between AT&T Long Lines and LECs I focus on differences in the annual rate of productivity growth ($\partial \ln TFP/\partial t$ or $\partial \ln C/\partial t$) to control for differences in the demand and supply conditions between the AT&T Long Lines and LECs. The demand for long distance services is still increasing, while the demand for local services is already saturated. Also, the major technological advances in the telephone industry occurred more frequently in the long distance market. For example, microwave technology and direct distance dialling (DDD) were mainly introduced to provide long distance services and ESS or fiber-optic cables have been installed faster in long distance markets than in local markets.

Economic evidence on differences in the productivity growth of a competitive IXC and a monopolistic LEC may have important policy implications. In the recent controversy on 'information super-highways' one of the most important policy questions is whether to permit entry and competition in the local and intra-LATA telephone markets. The transition toward competitive local and intra-LATA telephone markets can be reinforced if it is shown that LECs' productivity growth is poorer than that of their competitive counterpart.

There exist several industry-based productivity studies which measure the productivity growth in the U.S. telephone industry,[7] including the BLS estimate. One major drawback of these studies is that they do not distinguish long distance markets from local markets. Ying and Shin (1993) is a careful firm-level productivity study which aims at estimating the spillover effect of the breakup on cost-saving by LECs.[8] But their data cover too short a period (1976-1987) to construct reliable capital stock series.[9]

I test my hypothesis by conducting two types of analysis; (a) TFP growth rate measurement and (b) the estimation of a cost function. Under some restrictive assumptions TFP (or productivity residual) growth rate is exactly equivalent to shifts in a production or cost function. But if any assumption is violated the two approach may produce different results.

This chapter is organized as follows. Section 2 describes briefly the differences in competitiveness in the long distance and local telephone markets and examines the disciplinary effects of competition on a firm's management decisions. Section 3 provides evidence on the effect of competition on TFP growth, output growth, and input quality growth. I perform the TFP growth rate regressions in section 4, while section 5 examines the effects of competition on cost decline. Summary follows.

4.2. COMPETITION AND PRODUCTIVITY GROWTH IN THE TELEPHONE INDUSTRY

At the empirical level, the prediction on the efficiency differentials based on market structure becomes unclear, if firms under consideration have different type of ownership, or if they are regulated, or, are exposed to different degrees of competition. Thus it is appropriate to start with an explanation of differences in competitiveness in the long distance and the local telephone market.

A Brief History of the U.S. Telephone Industry

The U.S. telephone industry consisted of several vertically integrated common carriers before the divestiture of AT&T (alias the Bell system). AT&T was the largest firm which accounted for about 80% of industry revenues through a group of 22 BOCs and AT&T Long Lines. The BOCs offered both local and intrastate toll services, while AT&T Long Lines provided inter-state toll services. Even though the monopoly position of AT&T was protected due to the common belief that it was a natural monopoly, rapid technological changes in the telephone industry began to disintegrate this private monopoly-public regulation harmony.

In the historic Above-890 decision (1959) the FCC permitted private firms to operate microwave facilities for their own use. The availability of microwave technology induced new entrants to provide not only self-services but also commercial services. Finally, in 1969 MCI was granted authority to build the microwave relay system for long distance private line services to subscribers and thus, to compete with AT&T in a limited way. But it was not until the Execunet decision (1977) that competition in the message toll services (MTS), which are basic and major long distance telephone services,[10] was established. This decision opened competition in the telephone industry in a real sense.

Even though competition had begun, the survival of new entrants was highly uncertain because of unequal competition stemming from the vertical integration of the Bell system and the horizontal monopoly power of BOCs. After the 7 year antitrust suit initiated by the Justice Department, AT&T reached a settlement with the Department in 1982, resulting in the so-called modified final judgement (MFJ). Following the MFJ, AT&T divested its 22 BOCs, which were regrouped into 7 regional holding companies (RHOCs), and the market boundary between IXCs and LECs was redrawn at the local access and transportation areas (LATAs).[11]

While the monopoly position of LECs is nearly unchallenged, as compared to intense competition in the long distance market, technological change opens the possibility of competition even in the local market. First, wireless telephone service providers, including cellular phone companies, has been competing with LECs for telephone calls. A lot of telephone calls which would otherwise be handled by LECs are routed by wireless communication companies. Second, a drop in the price of telecommunication equipment due to rapid technological advance in ESS and fiber optic cable makes it possible for even small businesses to install their own communications facilities. Thus, even a small increase in price may induce business users to bypass LECs' networks. Finally, as telecommunications and broadcasting technologies converge, cable television operators become the most likely candidates for new entrants in the local telephone market. Cable television companies serve more than half of all homes, while telephone companies serve nearly all homes. Competition between cable television companies and LECs is technically possible and is already under way.

In sum, through the 1970s and 1980s LECs have faced potential competition which will be realized in the near future. They are already competing with wireless communications companies. On the other hand, in the 1960s and early 1970s, AT&T Long Lines struggled with regulatory agencies to prevent entry and has competed with several actual rival companies since then.

The Effects of Competition on Productivity Growth

There may be no clear theoretical consensus as to whether a private firm in an effectively competitive market (AT&T Long Lines) operates more efficiently than a private firm under potential or intermodal competition (LECs). But even though the efficiency comparison is basically an empirical issue, there are several reasons why AT&T Long Lines may outperform the LECs. The principal-agent problems are less severe for a competitive firm than for a monopolistic firm because the shareholders of the former can easily obtain information on the performance of their firm. Thus competition keeps costs down to the lowest attainable level. On the other hand, potential competition or intermodal competition may have only limited disciplinary effects, when it is well known that the materialization of competition takes time, or where there exists only weak substitution between wired and wireless communication services.

The existence of a second competitor, even when the incumbent is subject to regulation, can improve the incumbent's technical efficiency in several ways.

First, competition encourages a firm to choose the profit-maximizing composition of output (or rate structure) which stimulates demand and enlarges the total size of a market. Competition induces rivals to search for segments of the market not previously served and to elicit more demand from existing customers by offering a variety of services (or rates). Regulatory agencies often distorted rate structure and hence output composition.[12] Under competitive pressures a competitive firm is obliged to choose its output composition to maximize profits. As a matter of fact, competition in the long distance market leads to a wide variety of services and discount plans which benefit every consumer. Even though an increase in demand does not induce shifts in a cost function, it leads to an increase in the TFP growth rates of firms with idle capacities.

Second, whether incentives to innovate are greater in competitive markets or in monopolistic markets is still an empirical issue. But a regulated monopoly may retard the use of inventions in actual production activities, even though it may engage in more inventive activity than an unregulated carrier in order to protect its monopoly position.[13] Regulation reduces the motivation for a firm to innovate by capturing its efficiency gain, while innovation is a prerequisite for survival in intense competition with actual rivals.[14] Thus active rivalry gives strong incentives to make innovations in the form of new services, new methods of production and distribution, new methods of management and financing, new marketing methods, and so on.

Third, productive capital stock or qualified labor accelerates productivity growth of a firm. Embodied technical change may result from innovation in production or management or may be just the outcome of rapid capital turnover (i.e. modernization of plant and equipment) or a more skilled workforce. A competitive profit-maximizing firm chooses an optimal input quality/quantity mix for its survival, while a monopolistic firm has room for managerial slack in choosing the optimal level. Considering a rapid drop in the price of telecommunication equipments and a moderate increase in wage rates, high capital/moderate labor quality is likely to be an optimal input quality mix for a competitive firm. But this conclusion may be restrictive. Classical rate-of-return regulation may systematically induce a regulated firm to invest more heavily in plant and equipment than they would if they were unregulated. Overinvestment may be materialized in the form of higher capital quality.[15]

The last and greatest benefit from actual rivalry is its impact on the use and accumulation of firm-specific organizational capital. Organizational capital may take the form of matching tasks with the attributes of individual employees, the sorting of employees by competence, and other aspects of firm-specific learning-by-doing. Monitoring costs of a modern corporation are frequently associated with supervising this organizational capital. It is costly for shareholders to monitor and reward managers' decisions on organizational capital in a monopolistic market because a monopoly has no comparable counterparts. A competitive market provides more information about organizational capital and incentives for its efficient use and accumulation.[16] Hence competition keeps monitoring costs down by forcing managers to reduce non-optimizing use of organizational capital.[17] The efficient use and accumulation of organizational capital leads to shifts in a cost function, even when there is no embodied technical change.

Productivity growth, which is often called technical change or technological progress, implies shifts in a production function or cost function. Sometimes productivity growth is defined as the rate of change of an index of outputs minus the rate of change of an index of inputs (Divisia index or Törnqvist index of TFP growth rate). My analysis in the following two sections is based on the latter definition of productivity growth. I measure the effect of competition on shifts in a cost function in section 5.

4.3. TFP GROWTH, OUTPUT GROWTH, AND INPUT QUALITY GROWTH

TFP Growth Rate

In this section I compare the TFP growth rates, output growth rates, and input quality growth rates of AT&T Long Lines and LECs before and after the introduction of competition to identify the effect of competition on these variables. The empirical analysis is carried out with pooled cross-sectional time series data for 8 LECs and AT&T

Long Lines. Refer to appendix 2 for a detailed explanation of the data construction.

The conventional Törnqvist index of TFP growth rate is:

$$\Delta \ln TFP = \Delta \ln Q - \Delta \ln X =$$
$$\frac{1}{2}\sum_{i=1}^{m}(r_{it} + r_{i,t-1})\ln(\frac{Q_{it}}{Q_{i,t-1}}) - \frac{1}{2}\sum_{j=1}^{n}(s_{jt} + s_{j,t-1})\ln(\frac{X_{jt}}{X_{j,t-1}})$$

where Q and Q_i refer to aggregate output and i-th output, X and X_j to aggregate input and j-th input, r_{it} to revenue share of i-th output in total revenues during period t and s_{jt} to cost share of j-th input in total costs during period t. In calculating the TFP growth rate three outputs (local, toll, and miscellaneous outputs) are aggregated into an index of aggregate output using revenue shares as weights and three inputs (labor, capital, and material inputs) into an index of aggregate input using cost shares as weights.

I use two types of capital stock, i.e. net stock and gross stock. Net stock was constructed by the commonly used perpetual inventory method. The construction method for gross stock is basically the same as in Gort, Bahk and Wall (1993). I use Winfrey S-3 distribution as a retirement pattern of investment goods, while Gort, Bahk, and Wall assume simultaneous exit of investment goods of the same vintage.

Table 4.1 shows the average annual rate of TFP growth for each firm at three stages. The first stage (1952-1968) represents the period of no competition in the telephone industry. The second stage (1968-1976) is a kind of transition period during which new entrants provided private line services but were not allowed to enter the MTS market. Competition in the long distance market was fully permitted at the last stage. The average annual rate for the last stage doesn't include the 1984 rate because that year represents a big structural change caused by the breakup of the Bell system and should be regarded as an exceptional case. The key findings in table 4.1 are summarized as follows.

1. In general, the TFP growth rates based on net stock are greater than those based on gross stock. Even though using different types of capital stock produces different measures of TFP growth rate, it generates very similar patterns in the change of TFP growth rate.

2. Whatever type of capital stock is adopted, AT&T Long Lines experiences a higher TFP growth rate than any LEC. The average rate for AT&T Long Lines is 3.6% in term of gross stock and 3.7% in

< Table 4.1 > Average Annual TFP Growth Rate (by Firm)

(unit: %)

firm		1952-1968	1969-1976	1977-1991	(1985-1991)	whole period
AT&T Long Lines	A	9.3	6.3	7.7	7.8	8.1
	B	2.8	3.3	4.7	5.7	3.6
	C	2.8	3.5	4.9	6.8	3.7
BOC1	A	8.6	7.5	5.0	3.0	7.1
	B	2.3	2.0	2.7	1.6	2.3
	C	3.0	2.8	2.9	1.5	2.9
BOC2	A	6.2	6.2	4.1	2.7	5.4
	B	2.0	1.6	3.2	2.4	2.4
	C	2.6	2.4	3.4	1.9	2.8
BOC3	A	6.2	6.0	3.5	1.6	5.2
	B	2.1	2.3	3.0	1.9	2.4
	C	2.6	2.8	3.4	1.7	2.9
BOC4	A	6.7	6.1	3.9	0.4	5.6
	B	2.2	2.8	2.3	1.5	2.3
	C	2.9	3.0	2.9	2.4	2.9
BOC5	A	7.4	7.4	4.4	0.2	6.3
	B	2.9	1.4	2.7	2.0	2.5
	C	3.8	1.7	3.3	2.4	3.2
BOC6	A	6.4	5.7	4.6	2.3	5.6
	B	2.4	1.5	3.0	1.5	2.4
	C	3.1	2.1	3.0	0.4	2.8
IDP1	A	13.1	6.0	6.9	3.3	8.7
	B	2.1	4.2	3.9	4.5	3.4
	C	2.8	5.1	4.0	5.4	3.9
IDP2	A	6.6	8.0	4.0	1.7	5.9
	B	2.9	3.4	2.6	2.2	2.9
	C	3.8	3.2	2.9	1.9	3.2
Average of 8 LECs	A	7.7	6.6	4.6	1.9	6.2
	B	2.4	2.4	2.9	2.2	2.6
	C	3.1	2.9	3.2	2.2	3.1

Note: A, B, and C are average annual percentage change in the aggregate output, the TFP growth rate based on gross stock, and the TFP growth rate based on net stock, respectively. BOC denotes a local Bell operating company and IDP an independent company. The 1984 rate is excluded in calculating average annual rates.

term of net stock over the whole period (1952-1991), while that for LECs is only 2.6% and 3.1% respectively.[18]

3. The TFP growth rate of AT&T Long Lines increases from stage to stage, while LECs show nearly no change in the TFP growth rate at every stage. In particular, the higher average rate for AT&T Long Lines results from a big jump in the TFP growth rate after the breakup. On the other hand, every LEC except IDP 1 experiences a surprisingly big drop in the TFP growth rate after 1984 despite the expected spillover effect of competition.

Output Growth Rate

The big difference in the TFP growth rates of AT&T Long Lines and LECs in the last stage is partly due to the big difference in demand growth rate in the two groups. Increased demand may lead to increases in the TFP growth of firms because of economies of scale or density, or when the firms have under-utilized capacities. The average annual rate of demand growth for AT&T Long Lines declines a little over time but is still stable at 7.7% in the last stage.[19] On the other hand, the average annual rate of demand growth for LECs continues to decline from 7.7% in the first stage to 4.6% in the last stage, especially to 1.9% after the breakup. Table 4.2 shows that the average annual percentage changes in constant-dollar local and toll revenues of LECs continue to drop, confirming the decline in the demand growth rate of LECs.

Identifying the effect of competition on demand growth does not seem easy because the growth in demand for telephone services has slowed down over time. Thus, simply comparing demand growth rates before and after the introduction of competition may be misleading.[20] Also, the competition effect might be interwoven with a structural change in the toll market caused by the breakup. After 1984, AT&T Communications began to provide inter-LATA toll services as well as interstate toll services. Thus, one may think that the higher demand growth rate for AT&T Long Lines may be the outcome of the market reorganization due to the breakup.

But because I focus on the growth rate, this institutional change would affect only the 1984 rate which is excluded in the calculation of average rates in tables 4.1 and 4.2. Also, it is a common belief in the telephone industry that the interstate toll market has grown at least as fast as, or faster than the inter-LATA toll market. Thus the

< Table 4.2 > Average Annual Percentage Change in Revenues

(unit: %)

Period	current-dollar revenues			constant-dollar revenues		
	average of LECs		AT&T	average of LECs		AT&T
	local	toll	toll	local	toll	toll
1952-1958	9.6	8.9	9.5	6.8	7.6	8.2
1959-1968	6.5	11.4	10.9	6.5	12.4	12.0
1969-1976	9.3	12.7	9.9	5.1	9.5	6.9
1977-1983	10.4	13.0	11.4	5.4	9.6	8.0
1985-1991	4.6	1.1	5.4	0.8	4.6	9.0
1951-1991	7.9	9.7	9.6	5.0	9.1	9.0

Note: The current-dollar revenues are deflated by appropriate price indexes to derive the constant-dollar revenues. The base year of the price indexes is 1972. The access revenues of LECs are allocated to local and toll revenues for the period 1985-1991. The access expenses of AT&T Long Lines are subtracted from both toll revenues and operating expenses. The 1984 rate is excluded in calculating average annual rates.

institutional change, if any, should have negative or zero effects on the growth rate of demand for AT&T Long Lines. The fact that the average growth rate of demand for AT&T Long Lines declines to 6.3% in the second stage and increases to 7.7% with intensified competition leads to the conclusion that competition may stimulate demand and enlarge the size of interstate or inter-LATA toll markets or that the growth rate of demand for AT&T Long Lines could begin to slow down without competition as it did for local services. Competition stimulates the demand for long distance services partly because of a cut in the price of toll services and partly because of the introduction of a variety of new price plans and marketing efforts.[21]

One of the main reasons why the demand growth for LECs slowed down after 1984 is that their market attained maturity. For example, even the average annual percentage change of current-dollar local and toll revenues in table 4.2 declines after 1984 to 4.6% and 1.1% respectively. This may explain LECs' eagerness to enter into both inter-LATA telephone markets and cable television industry and, also, their decision to give up the monopoly position.

Input Quality Growth Rate

Other reasons for the improved performance of AT&T Long Lines may be higher input qualities as well as the efficient use and accumulation of organization capital. As expected, competition has nearly no effects on labor quality. Even though the level of labor quality for AT&T Long Lines is always higher than that for LECs, average annual change patterns are nearly the same for every firm. I report change patterns in two capital quality measures; average vintage and the capital quality index used in previous chapters. In table 4.3 both capital quality measures for AT&T Long Lines increase faster for the period 1977-1991 than those for LECs. The average annual percentage change in average vintage is -10.8% for AT&T Long Lines and -5.3% for LECs over the period 1984-1991. The average annual percentage changes in the second index for AT&T Long Lines and for LECs are 5.0% and 0.6% over the period 1984-1991.[22]

If the relative price of capital input drops due to rapid technical change in telecommunication equipments and if competition stimulates demand growth and innovation, the high capital quality-moderate labor quality mix may be optimal for AT&T Long Lines. Obviously improvements in capital quality is one of the main reasons for the differences in TFP growth rate in AT&T Long Lines and LECs.

I reconfirm the above findings by several financial performance statistics for LECs in table 4.4. Operating revenues per employee and gross income per employee grow steadily in the LECs despite a moderate drop in both indexes after the breakup. Also, operating expenses per line decline by 2.7% after 1984. This indicates that the LECs attempted to cut operating expenses facing a decrease in the growth of demand for local services. But these efforts have limitations in that operating revenues per telephone line decline. The LECs invested a lot in telephone plants through the 1970s due to a rapid increase in the demand for local services. These plants are included in aggregate capital stocks of the 1980s. Thus, while LECs attempt to adjust themselves to new environments, they may have no choice but to layoff employees and curtail investment.

< Table 4.3 > Average Annual Level and Percentage Change in Input Qualities

(1) Average Annual Level

period	labor quality		average vintage		capital quality	
	AT&T	average of LECs	AT&T	average of LECs	AT&T	average of LECs
1951-1958	129.2	116.0	47.3	46.5	0.093	0.100
1959-1968	141.2	121.7	36.2	36.6	0.116	0.110
1969-1976	167.9	131.9	26.9	26.8	0.121	0.130
1977-1983	185.8	151.5	20.4	19.3	0.159	0.168
1984-1991	202.6	173.2	10.2	13.3	0.285	0.194
1951-1991	163.7	138.5	28.8	28.5	0.153	0.140

(2) Average Annual Percentage Change

(unit:%)

period	labor quality		average vintage		capital quality	
	AT&T	average of LECs	AT&T	average of LECs	AT&T	average of LECs
1952-1958	-0.1	1.0	-2.6	-3.0	2.3	2.0
1959-1968	1.1	0.1	-3.4	-2.9	3.0	1.2
1969-1976	3.3	2.8	-2.9	-3.7	-1.4	2.2
1977-1983	0.5	1.3	-5.3	-4.8	8.3	3.7
1984-1991	1.1	1.7	-10.8	-5.3	5.0	0.6
1952-1991	1.2	1.4	-5.0	-3.9	3.3	1.8

Note: The labor quality index is the ratio of the weighted sum to the simple sum of working hours of occupational groups, with the weights being the relative wage rate of the occupational groups. Capital quality index is (1+s)/average age, where s is the ratio of telephone connected to the ESS to total telephones. Both average vintage and capital quality index are constructed using gross capital stock.

< Table 4.4 > Average Annual Percentage Change in Performance Indexes for LECs

(unit: %)

period	▲A	▲B	▲C	▲D
1952-1958	9.0	13.0	4.0	2.0
1959-1968	5.9	6.6	2.8	2.6
1969-1976	9.9	8.7	6.1	7.0
1977-1983	11.8	10.7	9.5	10.2
1984-1991	7.8	6.3	-3.5	-2.7
1952-1991	8.6	8.6	3.5	3.7

Note: The indexes are calculated as follows:

A=operating revenues/number of employees

B=net operating revenues/number of employees

C=operating revenues/number of telephones

D=operating expenses/number of telephones

The number of telephones is extrapolated using the growth rate of access lines for the period 1985-1991. ▲ implies the growth rate.

4.4. THE TFP GROWTH RATE REGRESSION

Basic Regressions

Until now, I have examined the indirect effects of competition on TFP growth rate in that competition affects productivity residuals through its impact on both output growth and input quality growth. In this section I conduct TFP growth rate regressions to measure the contribution of output growth and input quality growth to TFP growth nd, more importantly, to identify the direct effects of competition on TFP growth rate. A measure of competition is inserted in the regressions to examine the direct effects of competition. The productivity residual which is left over after controlling for both demand and technical change is associated with organizational capital. Thus the direct effects of competition on TFP growth is materialized through its impact on the accumulation and use of organizational capital.

The basic regression equation is:

$$\dot{TFP} = \beta_0 + \beta_Q \dot{Q} + \beta_t t + \beta_\omega \dot{\omega} + \beta_k \dot{k} + \beta_{cv} CV + \beta_{ATT} ATT + \beta_{CV.ATT} ATT \bullet CV$$

where Q refers to aggregate output, t to time trend, ω to labor quality index, k to capital quality, and CV to a measure of the degree of competitiveness in the U.S. telephone industry. The dots on a variable imply the change rate of the variable. ATT is equal to 1 for AT&T Long Lines and to 0 for LECs.[23]

Q measures demand conditions that may affect TFP growth rates because of economies of scale and density. t measures disembodied technical change, whether industrywide or firm-specific, while ω and k are firm-specific labor and capital quality indexes. The coefficient of CV measures the direct effects of competition on firms' productivity growth after controlling for output scale or density and technological advancement. Hence, the coefficient of the CV•ATT term represents differences in the effects of competition on the TFP growth rates of AT&T Long Lines and LECs due to different levels of organizational capital. Also, I assume that both AT&T Long Lines and LECs exhibit the same relationships between TFP growth rate and its components.

The TFP growth rate regressions are carried out with pooled cross-sectional time series data for AT&T Long Lines and 8 LECs. Within the context of my analysis, gross capital stock is more appropriate than net capital stock since the latter may sweep away all vintage (or age) effects.[24] I use two measures of competition. The dummy variable for 1977 is used to represent an epoch of competition because competition in the long distance telephone markets began in 1977. The competition index developed by Chessler is my second measure of competition. I extend Chessler's index up to 1991 using the market share of OCCs in toll revenues for the post-1984 period. I transform it into logarithmic form.

Regression results are reported in table 4.5. In every regression equation the coefficients of the intercept term and the ATT dummy measure constant productivity growth of LECs and AT&T Long Lines obtained by controlling for other factors. The competition dummy variable is used in models 1 and 3 and the competition index in

< Table 4.5 > TFP Growth Rate Regression for All Firms

variable	Model 1	Model 2	Model 3	Model 3
intercept	- 0.025* (0.005)	- 0.027* (0.005)	- 0.009* (0.003)	- 0.015* (0.004)
▲lnQ	0.577* (0.035)	0.579* (0.037)	0.551* (0.039)	0.580* (0.041)
t	0.001* (0.000)	0.001* (0.000)		
▲lnω	0.076*** (0.052)	0.063 (0.051)	0.108* (0.052)	0.085*** (0.052)
▲lnk	- 0.288* (0.044)	- 0.267* (0.043)	- 0.333* (0.052)	- 0.297* (0.052)
▲lnk1			0.058 (0.059)	0.066 (0.059)
▲lnk2			0.071 (0.059)	0.080*** (0.059)
▲lnk3			- 0.094** (0.054)	- 0.093** (0.054)
1977	0.007*** (0.005)		0.018* (0.003)	
ATT	- 0.002 (0.005)	- 0.002 (0.006)	- 0.001 (0.006)	0.000* (0.006)
ATT*1977	0.014*** (0.009)		0.013*** (0.009)	
lnCI		0.001 (0.002)		0.005 (0.001)
ATT*lnCI		0.002 (0.002)		0.002 (0.002)
adjusted R-square	0.453	0.448	0.425	0.423
number of observations	339	339	312	312

Note: The estimation method is ordinary least square method. Q refers to the Törnqvist index of aggregate output, t to time, ω to labor quality index, and k to capital quality index. k1,k2,k3 are one-year, two-year, and three-year lagged k. CI is the competition index. ATT=1 for AT&T Long Lines and =0 for other firms. 1977=1 for the period 1977-1991 and =0 for the period 1952-1976. ▲ represents the rate of change of a variable. Standard errors are in parentheses. *,**,and *** imply that the coefficient is statistically signifcant at the 5%,10%,and 20% level.

models 2 and 4. In models 1 and 2, I take into account both disembodied and embodied technical change. In models 3 and 4, I drop the time variable (t), the effect of which is mostly captured by quality indexes, and add lagged growth rates of capital quality index since it may take time for improvements in capital quality to be materialized. To determine the order of lagged variables, I inserted 8 lagged variables and eliminated the statistically insignificant variables one by one until the last lagged variable has a statistically significant coefficient. I found that the three-year lagged variable has a statistically significant coefficient. I report only the results using the second capital quality index because adopting average vintage as a proxy for capital quality produced nearly the same results. The key findings in table 4.5 follow.

1. An increase in demand has a positive effect on TFP growth in every equation. Its coefficients, which range from 0.55 to 0.58, are large, have the expected signs, and are statistically significant at the 1% level.

2. When due account of embodied technical change is taken, the coefficient of a time variable becomes nearly zero. This implies that quality indexes may capture most of the technical change or that technical change may be mostly embodied. This is consistent with the findings in section 4 of chapter 2 and will be confirmed in a cost function analysis.

3. The coefficients of the labor quality index are statistically significant, except for model 2, and have the expected sign. Improvements in human capital lead to an increase in productivity residuals.

4. All the coefficients of the capital quality index have negative signs and are statistically significant. At first glance this seems contradictory because improvements in capital quality should have positive effects on TFP growth rates. But an increase in the capital quality index has the same two-fold effects on TFP growth rates as it does on cost decline. First, it has negative effects on TFP growth rates because a decrease in average age, which implies an increase in the index, leads to an increase in capital stock in the same period. On the other hand, it enhances the productivity of capital stock, allowing substitution of other inputs with capital inputs, and hence increases TFP growth rates. The final effect depends on the relative sizes of the positive and the negative effects.

5. When three lagged variables of capital quality index are placed, the one-year and two-year lagged variables have a positive sign. This implies that there may be at least one or two year lags before improvements in capital quality are fully realized. The negative sign of the three-year lagged variable indicates that the materialization may not take a long time. Since my sample firms have such a long history that the portion of new plants in total plants in service is not large, the materialization of improved capital quality would take a short time span.

6. The competition dummy (1977) has a positive and statistically significant coefficient, which implies even LECs experience a slight increase in productivity growth after 1977 with output scale and technological advancement being controlled. More important, the ATT-1977 interaction term is statistically significant in both models and indicates that AT&T Long Lines shows a higher TFP growth rate than LECs by 1.4% (model 1) or 1.3% (model 3) after the introduction of competition in the long distance market. The statistically insignificant coefficients of the ATT dummy suggest that there might be no differences in productivity growth of AT&T Long Lines and LECs without competition.

7. When the competition index (lnCI) is used to measure the degree of competition, the coefficients of both the lnCI and the ATT*lnCI term are small and statistically insignificant. Since the competition index is constant up to 1971, shows a slight increase from 1972 to early 1980s, and increases very rapidly after 1984, these results may not be surprising.[25] But, as expected, both the lnCI and the ATT*lnCI term have positive coefficients.

Complementary Regressions

To test the robustness of my results, I carry out the same TFP growth rate regressions for AT&T Long Lines and for 8 LECs, separately, and attempt to identify differences in the coefficient of the competition dummy variable (1977) for AT&T Long Lines and LECs. Table 4.6 reports regression results. In model 1 the intercept, demand growth rate, and competition dummy variable are inserted as explanatory variables. Since all variables representing technical change are dropped in model 1, the coefficient of the dummy variable captures a structural change in the TFP growth rate before and after 1977 with demand fluctuations being controlled.[26] Model 2 in the

< Table 4.6 > TFP Growth Rate Regression for Subsamples

variable	Model 1		Model 2	
	AT&T Long Lines	LECs	AT&T Long Lines	LECs
intercept	- 0.020**	- 0.009*	- 0.014	- 0.010*
	(0.011)	(0.003)	(0.015)	(0.003)
▲lnQ	0.583*	0.453*	0.645*	0.550*
	(0.108)	(0.037)	(0.148)	(0.040)
▲lnω			0.026	0.121*
			(0.157)	(0.056)
▲lnk			- 0.259***	- 0.354*
			(0.153)	(0.056)
▲lnk1			0.187	0.036
			(0.184)	(0.062)
▲lnk2			- 0.149	0.105**
			(0.186)	(0.062)
▲lnk3			- 0.260**	- 0.058
			(0.149)	(0.058)
1977	0.022*	0.017*	0.041*	0.018*
	(0.010)	(0.003)	(0.013)	(0.003)
adjusted R-square	0.445	0.335	0.347	0.431
number of observations	39	300	36	276

Note: The estimation method is ordinary least square method. Q refers to the Törnqvist index of aggregate output, t to time, ω to labor quality index, and k to capital quality index. k1,k2,k3 are one-year, two-year, and three-year lagged k. 1977=1 for the period 1977-1991 and =0 for the period 1952-1976. ▲ represents the rate of change of a variable. Standard errors are in parentheses. *,**,and *** imply that the coefficient is statistically signifcant at the 5%,10%,and 20% level.

table has the same specification as model 3 in table 4.5 except that it drops the ATT and ATT*1977 term.

As table 4.6 shows, I obtain similar results with respect to demand fluctuations and disembodied or embodied technical changes and, also, reach stronger conclusions on the superior performance of AT&T Long Lines over LECs. The key findings in table 4.6 which do not have counterparts in table 4.5 follow.

1. Model 1 shows that AT&T Long Lines experiences a bigger jump in the TFP growth rate than LECs in 1977. The productivity residual of AT&T Long Lines (LECs) jumps by 2.2% (1.7%). The coefficients are statistically significant at the 10 % level for both groups.[27] Thus, competition may have a bigger impact on productivity residuals in AT&T Long Lines than in the LECs.[28]

2. In model 2, the coefficient of the labor quality index for LECs is statistically significant and has the expected sign, while that for AT&T Long Lines is negative and statistically insignificant. Two explanations are possible. If the labor market for AT&T Long Lines was not competitive, the index cannot measure labor quality appropriately. The other explanation is that the number of observations may be too small to obtain a precise estimate. The second reason is more plausible because the index produced significant estimates for LECs.

3. I obtain the same results for LECs as for all firms with respect to capital quality. Improvements in capital quality may decrease today's TFP growth rate but increase tomorrow's. On the other hand, in AT&T Long Lines, the one-year lagged variable has a positive sign as in LECs, but the sign of the two-year and three-year lagged variables is negative. Capital quality may be more quickly materialized in AT&T Long Lines than in LECs. But, more plausibly, this phenomenon may occur due to the small number of observations.

4. In model 2, the coefficient of the competition dummy variable in AT&T Long Lines (0.041) is larger than that in LECs (0.018).[29] This suggests that competition exhibits more effects on TFP growth rates of AT&T Long Lines than on those of LECs through its impact on organizational capital. The organizational capital of AT&T Long Lines becomes more efficient than LECs due to effective competition in long distance markets.

4.5. THE EFFECTS OF COMPETITION ON COST DECLINE

Empirical Specification of a Cost Function

In this section a cost function is analyzed to examine differences in the effects of competition on shifts in the cost function in AT&T Long Lines and LECs. The cost functions are estimated simultaneously with two factor cost share equations subject to the relevant cross-equation constraints. The differences in competition effects are tested by introducing competition variables as additional regressors in the cost functions. A single-product cost function is adopted because AT&T Long Lines produces only toll outputs.

Specifically, I estimate the following four cost functions which treat the measurement of competition effects in different methods:

$$(Model\ 1)\ \ln C = g(\ln P_L, \ln P_M, \ln P_K, \ln Q, t; ATT)$$

$$(Model\ 2)\ \ln C = g(\ln P_L, \ln P_M, \ln P_K, \ln Q, t; ATT, 1977)$$

$$(Model\ 3)\ \ln C = g(\ln \frac{P_L}{\omega}, \ln P_M, \ln P_K, \ln Q, \ln k, \ln CI; ATT)$$

$$(Model\ 4)\ \ln C = g(\ln \frac{P_L}{\omega}, \ln P_M, \ln P_K, \ln Q, \ln k, t; ATT, 1977)$$

where Q refers to aggregate output, P_L to wage rate, P_M to material price, P_K to user-costs of capital, t to time variable, ω to labor quality index, k to capital quality index, and CI to competition index. ATT is a firm dummy variable which equals 1 for AT&T Long Lines and 0 for LECs. 1977 is a competition dummy variable which is equal to 1 for the period 1977-1991 and 0 for the period 1951-1976. Models 1 and 2 represent the conventional specification of a cost function which views a time variable as a proxy for the state of technology, while models 3 and 4 contain input quality indexes. Refer to chapter 2 for a full explanation of these specifications of a cost function.

The time derivative of cost ($\partial \ln C/\partial t$) measures the rate of technical change in models 1 and 2 or disembodied technical change in model 4. The interaction between a time variable and a competition dummy variable (1977) accounts for a structure change in shifts in the cost function before and after 1977. Thus the interaction term between time, 1977, and the firm dummy (t*ATT*1977) captures the difference in competition effects on cost decline in AT&T Long Lines and LECs. In model 3, where the time variable is replaced with the competition index, the firm dummy and its interaction term with the competition index are introduced as additional regressors to estimate the difference in competition effects.

It is assumed that g(.) is twice-differentiable and can be well approximated by a second-order Taylor series expansion, i.e. translog flexible cost function. For example, a translog cost function for model 4 is given by:

$$
\begin{aligned}
\ln C = {} & \alpha_0 + \alpha_Q \ln Q + \sum_{i=L,M,K} \alpha_i \ln P_i - \alpha_L \ln \omega + \alpha_k \ln k + \alpha_t t + \\
& \frac{1}{2}\beta_{QQ}(\ln Q)^2 + \frac{1}{2}\sum_{i=L,M,K}\sum_{j=L,M,K}\beta_{ij}\ln P_i \ln P_j + \frac{1}{2}\beta_{LL}(\ln \omega)^2 + \\
& \frac{1}{2}\beta_{kk}(\ln k)^2 + \frac{1}{2}\beta_{tt}t^2 + \sum_{i=L,M,K}\beta_{Qi}\ln Q \ln P_i - \beta_{QL}\ln Q \ln \omega + \\
& \beta_{Qk}\ln Q \ln k + \beta_{Qt} t \ln Q - \sum_{i=L,M,K}\beta_{iL}\ln P_j \ln \omega + \sum_{i=L,M,K}\beta_{ik}\ln P_i \ln k + \\
& \sum_{i=L,M,K}\beta_{it} t \ln P_i - \beta_{Lt} t \ln \omega + \beta_{Kt} t \ln k - \beta_{Lk}\ln \omega \ln k + \gamma_{ATT} ATT + \\
& \gamma_{1977} 1977 + \gamma_{ATT.1977} ATT \cdot 1977 + \gamma_{t.1977} t \cdot 1977 + \gamma_{t.ATT.1977} t \cdot ATT \cdot 1977 + \varepsilon
\end{aligned}
$$

where ε is a disturbance term. All variables except a time variable and dummy variables are divided by their sample means. For the translog case, normalization of all variables by sample mean prior to estimation permits the first-order terms to be interpreted as the elasticities at the sample mean. Conditions for linear homogeneity and symmetry are imposed, while monotonicity and concavity conditions are tested after estimation.

Assuming cost minimization given the regular constraints, Shepard's lemma can be applied to obtain the following cost share equations.

$$
S_j = \alpha_j + \beta_{Qj}\ln Q + \sum_{i=L,M,K}\beta_{ji}\ln P_i - \beta_{jL}\ln \omega + \beta_{jk}\ln k + \beta_{jt} t + \varepsilon_j
$$

where ε_j is the disturbance term. Then I jointly estimate the cost function and the two cost share equations by iterating Zellner's seemingly unrelated regression method. One of the cost share equations is deleted to obtain a nonsingular covariance matrix.

After the cost functions are estimated, differences in the competition effects on productivity growth in AT&T Long Lines and LECs are examined. In model 2 productivity growth, i.e. an annual time-dependent shift in the cost function, is calculated as follows.

$$\frac{\partial \ln C}{\partial t} = \beta_t + \beta_{Qt} \ln Q + \sum_{i=L,M,K} \beta_{it} \ln P_i + \beta_{tt} t + \gamma_{t,1977} 1977 + \gamma_{t,ATT,1977} ATT \cdot 1977$$

The negative sign of $\beta_{t,ATT,1977}$ indicates that AT&T Long Lines experiences a bigger jump in productivity growth than do LECs after 1977. Similarly, in model 3 the effect of the competition index on cost decline is expressed as follows.

$$\frac{\partial \ln C}{\partial \ln CI} = \alpha_c + \beta_{Qc} \ln Q + \sum_{i=L,M,K} \beta_{ic} \ln P_i + \beta_{cc} \ln CI + \gamma_{ATT,c} ATT$$

If $\beta_{ATT,c}$=0, the hypothesis that AT&T Long Lines shows greater competition effects may be rejected.

Basic Estimation Results

The empirical analysis is carried out with pooled cross-sectional time series data for AT&T Long Lines and 8 LECs. Some important parameter estimates for the cost functions are provided in table 4.7, while a full set of parameter estimates for each cost model is given in appendix 1. All the estimated cost functions satisfy the regularity conditions, i.e. monotonicity and quasi-concavity. Also, the marginal cost for every model is positive for each observation. These imply that my empirical translog cost functions are consistent with microeconomic theory.

In model 1, of the 23 parameter estimated, 20 are significant at the 1% level, 1 are significant at the 5% level, and 2 are insignificant at the 10% level. In model 2 the number of statistically significant parameter estimates at the 1% and 5% level is 20 and 1 respectively out of 26. The number of significant parameter estimates at the 1%, 5%, and 10% level becomes 24, 1, and 4 out of 38 in model 3 and 33, 2, and 1 out of 41 in model 4.

In particular, the first-order terms for explanatory variables are highly significant for all models except for one parameter in model 3. The size of every second-order term for all models is small, which

< Table 4.7 > Parameter Estimates of Total Cost Functions

variable	Model 1	Model 2	Model 3	Model 4
intercept	1.506*	1.502*	0.968*	1.709*
	(0.036)	(0.039)	(0.032)	(0.092)
lnQ	1.173*	1.141*	1.027*	1.134*
	(0.018)	(0.018)	(0.015)	(0.024)
$\ln P_L$	0.694*	0.701*	0.413*	0.666*
	(0.012)	(0.012)	(0.010)	(0.014)
$\ln P_M$	0.155*	0.130*	0.213*	0.153*
	(0.014)	(0.016)	(0.010)	(0.018)
$\ln P_K$	0.151*	0.169*	0.374*	0.181*
	(0.013)	(0.014)	(0.009)	(0.015)
t	- 0.035*	- 0.040*		-0.065*
	(0.002)	(0.004)		(0.008)
lnCI			- 0.362*	
			(0.022)	
$\ln\omega$			- 0.413*	-0.666*
			(0.010)	(0.014)
lnk			- 0.100	0.945*
			(0.111)	(0.261)
ATT	- 0.468*	- 0.744*	- 0.322*	-0.619*
	(0.032)	(0.021)	(0.025)	(0.020)
1977		- 0.087		0.161
		(0.138)		(0.143)
ATT*1977		0.284		0.571**
		(0.208)		(0.256)
t*ATT	- 0.021*			
	(0.002)			
lnCI*ATT			- 0.122*	
			(0.011)	
t*1977		- 0.001		-0.011***
		(0.005)		(0.005)
t*ATT*1977		- 0.018*		-0.029*
		(0.006)		(0.008)

Note: The estimation method is Zellner's seemingly unrelated regression technique. Q refers to aggregate output, P_L to wage rate, P_M to material price, P_K to user-cost of capital, t to time, ω to labor quality index, k to capital quality index, and CI to competition index. ATT=1 for AT&T Long Lines and = 0 for LECs. 1977=1 for the period 1977-1991 and =0 for the period 1951-1976. The number of observations is 357. Standard errors are in the parentheses. *,**, and *** imply that the coefficient is statistically significant at the 1%, 5%, and 10% level.

results in stable elasticities. Applying a translog cost function to a data set with small number of observations often generates such unstable cost elasticities that no meaningful conclusions can be reached.

The principal findings from parameter estimates are summarized as follows.

1. Each model produces closely similar estimates of output cost elasticity which ranges from 1.173 in model 1 to 1.027 in model 3 at the sample mean. The telephone companies exhibit decreasing or constant returns to scale.

2. All elasticities of cost with respect to input prices, i.e. factor shares, are positive at the sample mean. In model 1 the interaction term of input prices with the time variable indicates a tendency for labor share to decrease, for material share to increase, and for capital share to increase over time.

3. The negative coefficients of technology variables imply a percentage downward shift in cost function. Model 1 (Model 2) shows that the cost function experiences a 3.5% (4.0%) annual percentage downward shift at the sample mean.

4. When the competition index is used instead of a time variable in model 3 its coefficient is consistent with our expectations. A 1% increase in the competition index leads to a 0.36% downward shift in the cost function at the sample mean. A 1% increase in ω (k) leads to a 0.36% (0.13%) downward shift in the cost function at the sample mean. As discussed in chapter 2, the size of the coefficient of k may be understated because the competition index may capture part of improvements in capital quality.

5. In model 4 the coefficient of k is positive at the sample mean, just as it was in chapter 2. Its sign may be misleading because incorporating t with quality indexes may understate embodied technical change.

6. In model 3 the negative sign of $\ln P_L * \ln k$ indicates that capital quality improvements are labor-saving. The sign of $\ln P_L * \ln \omega$ is negative, which implies that labor quality improvements are labor-saving.

Now the differences in the competition effects on productivity growth in AT&T Long Lines and LECs are examined. The key findings are as follows:

1. In model 1 the coefficient of t*ATT is highly significant and has a negative sign. That is, AT&T experiences a 2.1% higher annual percentage shift in the cost function at the sample mean than LECs over the period 1951-1991.

2. Model 2 shows that the coefficient of t*1977 is insignificant at the 10 % level and its absolute size is very small. On the other hand, the coefficient of t*ATT*1977 is significant, negative, and large. These results indicate that AT&T experiences a big jump in productivity growth after 1977, while there may be no significant change in LECs' productivity growth after 1977.

3. In model 3 the sign of lnCI*ATT is negative, which indicates that the effect of the competition index on productivity growth is bigger in AT&T than in LECs.

4. Even though model 4 may be a wrong specification, it provides the same conclusion as model 2. While the coefficients of both t*1977 and t*ATT*1977 are significant and have the correct signs, the absolute size of the latter is greater than that of the former.

Complementary Estimation Results

Until now I ignored serially correlated errors in the estimation of translog cost functions. Since serial correlation is found to be statistically significant in all my regressions, I consider serially correlated errors to test the robustness of my estimates. My estimation follows the same three-step procedure as Ehrlich, Gallais-Hamonno, Liu, and Lutter (1994). First, I use the ordinary least squares residuals to estimate the serial correlation in the residuals of the cost function (ρ_c) and the share equations (ρ_s).[30] Second, I transform the data by adopting Cochrane-Orcutt method and recalculate the ordinary least squares estimates using the transformed data. Third, I jointly estimate the cost function and the two cost share equations by Zellner's seemingly unrelated regression method on the basis of the transformed data.

Important parameter estimates for the cost functions using the three-step procedure are reported in table 4.8, while a full set of parameter estimates for each cost model is given in appendix 1. All the estimated cost functions except model 1 still satisfy the regularity conditions. Even though I use the different estimation method, I obtain very similar estimates and, also, reach the same conclusions on the superior performance of AT&T Long Lines over LECs. For

< Table 4.8 > Parameter Estimates of Total Cost Functions Using the Three-Step Procedure

variable	Model 1	Model 2	Model 3	Model 4
intercept	1.528*	1.661*	0.817*	1.634*
	(0.085)	(0.103)	(0.044)	(0.106)
lnQ	1.168*	1.168*	1.002*	1.163*
	(0.034)	(0.035)	(0.025)	(0.033)
lnP_L	0.568*	0.569*	0.383*	0.545*
	(0.020)	(0.020)	(0.014)	(0.026)
lnP_M	0.153*	0.148*	0.238*	0.117*
	(0.030)	(0.030)	(0.019)	(0.038)
lnP_K	0.279*	0.283*	0.379*	0.338*
	(0.019)	(0.019)	(0.012)	(0.023)
t	- 0.043*	- 0.057*		- 0.060*
	(0.006)	(0.009)		(0.009)
lnCI			- 0.213*	
			(0.026)	
$ln\omega$			- 0.383*	- 0.545*
			(0.014)	(0.026)
lnk			- 0.081	0.357***
			(0.117)	(0.233)
ATT	- 0.427*	- 0.783*	- 0.563*	- 0.718*
	(0.084)	(0.045)	(0.068)	(0.038)
1977		0.207		0.274
		(0.240)		(0.226)
ATT*1977		0.478**		0.176
		(0.224)		(0.244)
t*ATT	- 0.018*			
	(0.004)			
lnCI*ATT			- 0.047**	
			(0.021)	-0.013***
t*1977		- 0.010		(0.009)
		(0.009)		-0.009
t*ATT*1977		- 0.019*		(0.008)
		(0.008)		
ρ_C	0.780	0.790	0.831	0.736
ρ_S	0.856	0.856	0.914	0.889

Note: ρ_C and ρ_C are the first-order autocorrelation coefficient in the cost function and the share equations, respectively. The number of observations is 348. The definition of other variables are the same as in table 4.7.

instance, the coefficient of t*ATT in model 2, lnCI*ATT in model 3, and t*ATT*1977 in model 4 becomes smaller, but still negative and statistically significant, after considering serially correlated errors. Whether considering serial correlation or not does not affect my conclusions.

My empirical specifications of a cost function implicitly assume that AT&T Long Lines and LECs have historically the same access to improvements in technology and input markets. This assumption may be too extreme to provide precise predictions on technical change in AT&T Long Lines. But this limitation may be removed by adding the interaction terms between the dummy variable for AT&T Long Lines and the first-order terms in the cost functions.

Some important parameter estimates for the modified cost functions are provided in table 4.9, while a full set of parameter estimates for each cost model is given in appendix 1. In model 3-2, of the 44 parameter estimated, 35 are significant at the 1% level, 1 is significant at the 5% level, and 1 is significant at the 10% level. In model 4-2 the number of statistically significant parameter estimates at the 1% and 5% level is 36 and 2 out of 48, respectively. As in the previous cases, the first-order terms for explanatory variables are highly significant for two models and the size of every second-order term is small. The general findings from the modified models are nearly the same as from the previous models.

Also, the conclusions from the previous estimation results still hold, even though the absolute size of the relevant parameter estimates becomes smaller. That is, the sign of lnCI*ATT and t*ATT*1977 is negative and statistically significant. In sum, the findings in the analysis of cost functions are consistent with those in the TFP growth rate regressions. Competition has a greater effect on productivity growth for AT&T than for LECs.

4.6. SUMMARY

I compare the productivity growth for AT&T Long Lines with that for 8 LECs by measuring TFP growth rates and by estimating cost functions. The empirical analysis is carried out with pooled cross-sectional time series data for AT&T Long Lines and 8 LECs. In particular, I attempt to construct a reliable capital stock series which, I believe, are crucial in the empirical analysis. I explicitly take into

< Table 4.9 > Parameter Estimates of Total Cost Functions, Including the ATT-First Order Interaction Terms

variable	Model 3-2		Model 4-2	
	parameter estimate	standard error	parameter estimate	standard error
intercept	0.970*	0.034	1.538*	0.089
lnQ	1.049*	0.015	1.096*	0.025
$\ln P_L$	0.428*	0.010	0.727*	0.010
$\ln P_M$	0.193*	0.009	0.084*	0.017
$\ln P_K$	0.379*	0.009	0.190*	0.016
t			- 0.053*	0.008
lnCI	- 0.338*	0.022		
lnω	- 0.428*	0.010	- 0.727*	0.010
lnk	- 0.172	0.111	0.919*	0.245
ATT	- 0.359*	0.047	- 0.154	0.150
lnQ*ATT	- 0.239*	0.034	- 0.041	0.092
$\ln P_L$*ATT	- 0.068*	0.009	- 0.095*	0.005
$\ln P_M$*ATT	0.069*	0.008	0.070*	0.008
$\ln P_K$*ATT	- 0.001	0.009	0.025*	0.008
lnω*ATT	0.068*	0.009	0.095*	0.005
lnk*ATT	0.535*	0.131	0.510*	0.139
lnCI*ATT	- 0.091*	0.019		
1977			0.200	0.135
ATT*1977			0.757**	0.306
t*ATT			- 0.026*	0.009
t*1977			- 0.013**	0.005
t*ATT*1977			- 0.028*	0.010

Note: The estimation method is Zellner's seemingly unrelated regression technique. The definition of the variables is the same as in table 4.7. The number of observations is 357.

account embodied as well as embodied technical change in both productivity and cost function analysis.

Whatever the concept of capital stock used, AT&T Long Lines experiences higher TFP growth rate than any other LEC. Moreover, the average annual TFP growth rate for AT&T Long Lines increases from stage to stage, while LECs exhibit nearly no change in their average rate for the three stages. In particular, the higher overall average rate for AT&T Long Lines is achieved by a big jump in the TFP growth rate after the breakup. Also, I provide strong evidence that competition may enlarge market demand. Under competitive pressures, AT&T Long Lines adopts high capital quality and moderate labor quality.

TFP growth rate regressions suggest that AT&T Long Lines experiences a greater structural change in the TFP growth rate than LECs before and after the introduction of competition, with demand fluctuations being controlled. When I take into account embodied or disembodied technical change as well as demand fluctuations, it is shown that competition forces AT&T Long Lines to use organizational capital more efficiently than LECs. I, also, obtain several interesting findings on the relationships between input quality growth and TFP growth.

The estimation of cost functions confirms the conclusions from the productivity analysis. The results suggest that the telephone companies exhibit nearly constant returns to scale and, also, indicates a tendency for labor share to decrease and for capital share to increase over time. The cost models provide the same relationships between input qualities and technical change as the TFP growth rate regressions. The cost analysis indicates that AT&T shows a larger annual percentage shift in the cost functions than LECs over the whole period. Moreover, AT&T experiences a big jump in the shifts after 1977 while there is no significant change for LECs.

During the period in which the economic performance of AT&T Long Lines dominated that of LECs, the only major difference in their environment was the existence of actual rivalry in the long distance market. AT&T Long Lines operating in an effectively competitive market environments is more efficient than LECs under potential or intermodel competition. If entry and competition is allowed into local and intra-LATA telephone markets, consumers may enjoy the benefits from the efficient management of LECs.

NOTES

[1] For a comprehensive survey of empirical studies on the Schumpeterian hypotheses, see Cohen and Levin (1990). They indicate that the empirical results bearing on the Schumpeterian hypotheses are inconclusive.

[2] Boardman and Vining (1989) survey the literature on the relationship of ownership types to productivity.

[3] Studies on the relative performance of a competitive and a monopolistic firm often rely on international comparison which may be vulnerable to strong country effects. For example Caves, Christensen, and Swanson (1981) examines the relative performance of competitive Canadian railroads and regulated U.S. railroads.

[4] Data on IXCs except AT&T Long Lines are not available partly because they don't report to the Federal Communications Commission (FCC) and partly because their history is too short to construct time-series data. Even when AT&T Long Lines was a component of the Bell system, its operation was independent of other arms and BOCs. Also, it files financial and operating data with the FCC separately from AT&T.

[5] Crandall (1988) indicates that the great surprise in telephone industry deregulation has been the absence of deregulation because the most important intrastate and inter-state telephone services continue to be subject to formal rate regulation. But the regulation of AT&T may have positive effects on competition in the long-distance market because its objective is to protect new entrants from the giant incumbent.

[6] The FCC began to use a price cap regulation in order to set AT&T's prices for interstate services in 1989. Also, many states have recently adopted a price-cap framework in their regulation of the intrastate telephone services provided by either IXCs or LECs.

[7] In particular, Oum (1990) and Crandall (1991) examine the effect of competition or the breakup of the Bell system on the productivity growth of U.S. telephone industry.

[8] There exist a few firm-level productivity studies which rely on time series data for a vertically integrated company, mainly AT&T and Bell Canada. For example, Kiss (1983) and Denny, Fuss, and Waverman (1981) decompose the TFP growth rate of Bell Canada using econometrically estimated cost functions. Nadiri and Schankerman (1981) and Nadiri and Prucha (1989) measure the productivity growth of the Bell system by estimating several alternative cost models.

[9] For a more detailed description of the data problem with Ying and Shin (1993), see the section 1 of Chapter 3.

[10] In 1977 AT&T's revenues from MTS were $11.1 billion, or 83% of total toll revenues, while its private line service revenues were only $1,184 million, or 9.0% of total toll revenues. Thus it is not until the introduction of competition in the MTS market that competition in the long distance telephone market was virtually materialized.

[11] Following the breakup, assets, employees, and markets of BOCs were transferred to AT&T Communications (former AT&T Long Lines). AT&T Communications' gross plant doubled from $9.95 billion to $17.45 billion, its number of employees tripled from 47,518 to 116,757, and the operating revenues increased from $5.07 billion to $35.04 billion. Thus the question of how to handle this transfer was very important in our data development and process.

[12] Stigler (1971) indicates that regulators often choose rates to maximize political support. Regulators set low rate for consumers who are likely to be most vocal in protest and set high rate for more passive and powerless consumers.

[13] Shephard (1971) takes the case of microwave technology as an example of the Bell system's preemptive innovation in response to competition. The Bell system mounted a large and rapid innovation effort between 1946 and 1950 to create a nationwide microwave relay system and to preempt the microwave field. At that time several firms created or planned microwave capacity between major eastern cities and applied for an FCC franchise.

[14] Stanbury (1990) provides an illustration where MCI and Sprint led AT&T in building fiber-optic networks and AT&T had to accelerate the write-off of obsolete equipment.

[15] But as Gort and Wall (1988) indicate, if investor foresight reduces the power of regulators for good as well as for bad the Averch-Johnson effect may be small.

[16] Stanbury (1990) indicates that new entrants increases the amount of information available to a regulator. Information may take a wide variety of forms; the lowest attainable level of costs, the boundaries of any natural monopoly, the potential and actual viability of new services and so on. The main problem a regulatory agency often faces is that it is difficult to acquire enough information to pass judgement on telephone rates. In particular, the regulatory agency has little ability to distinguish true cost change from managerial slack. Obviously potential entrants cannot offer such information.

[17] There exist a few formal models on the disciplinary effect of product market competition on managerial slack. For illustration, Hart (1983) develops a model according to which average managerial slack is lower under competition, if there is a common component to firms' costs and managers are rewarded as a function of their own profits.

[18] As compared with these estimates, Nadiri and Schankerman (1981) calculate a 4.09% average annual percentage change of TFP for the Bell system over the period 1947-1976. Oum (1990)'s estimate for the U.S. telephone industry is 2.9% over the period 1951-1987 and Crandall (1991)'s estimate is 3.5% over the period 1961-1988.

[19] How to deal with access revenues and expenses may lead to different measures of demand growth. To measure output growth rate as well as TFP growth rate in a consistent way before and after the breakup, LECs' access revenues were allocated to local and toll revenues, while AT&T's access expenses were subtracted from both toll revenues and operating expenses.

[20] Taylor and Taylor (1993) conclude that regulated competition in the interstate toll market has not led to an expansion of demand,

when they find that toll demand grew no more than would be based on changes in price, population, and consumer income. But their conclusion may be misleading because they do not take into account the overall decline in the growth rate of demand for telephone services.

[21] The estimation of the demand growth rate equation which is not reported here confirms this hypothesis. The parameter estimates for AT&T Long Lines' toll demand growth equation are the following.

$$\dot{Q} = 0.137 - 0.508\dot{P} + 0.516\dot{Y} - 2.255\dot{POP} - 0.003t + 0.017D_1 + 0.015D_2$$

where Q refers to toll demand, P to real toll price, Y to real GNP per capita, POP to the size of the U.S. population, t to time trend, D_1 to the dummy for 1977, and D_2 to the dummy for 1984. The dots represent the rate of change of a variable. The negative sign of t indicates that the demand growth rate for AT&T Long Lines declines over time. Also the dummies for 1977 and 1984 have positive signs which imply upward shifts in the growth rate. When I applied this equation to the LECs' local and toll demand growth rate by replacing POP with the total number of telephones, the coefficient of the dummy for 1984 turned out to be negative in both equations which implies downward shifts in the growth rate.

[22] Differences in capital quality may be confirmed by examining the rate of change of real investment. Improvements in capital quality are carried out by net capital formation or by the replacement of old plants with the new ones. Average annual percentage change of real investment is 12.0% for AT&T Long Lines and 1.9% for LECs over the period 1977-1983. Average percentage changes for two groups are 9.2% and 0.4% for the period 1984-1991. AT&T Long Lines maintain its rate of increase in investments in competitive environments, while LECs facing a drop in demand growth curtail investments.

[23] Other firm dummy variables are found to be often statistically insignificant and hence are excluded from the regression.

[24] In this respect I follow the Kendrick's approach. The basic distinction between Kendrick's and Jorgenson's measure of productivity is that Jorgenson attempts to incorporate changes in the quality and utilization of labor and capital input directly in his measures. In the Kendrick's approach the productivity residual can be explained through statistical and econometric methods and does not need to be explained away in the direct measure of productivity growth.

[25] The statistical insignificance of the lnCI and the ATT*lnCI variable may occur because the lnCI term is nonstationary, while other variables are stationary.

[26] This regression plays a role in adjusting the Törnqvist index for the market imperfections and non-constant returns to scale technology.

[27] When the competition index is used, its coefficient is significant and slightly larger for AT&T Long Lines.

[28] But this conclusion is at most restrictive because omitted explanatory variables make the parameter estimate of the competition dummy biased.

[29] When the competition index is used in place of the competition dummy in this regression equation, its coefficient is 0.009 in AT&T Long Lines and 0.004 in LECs. Both coefficients are highly significant.

[30] I impose the restriction that all share equations have the same autoregressive parameter.

V

Conclusions

At the time of the breakup of the Bell system in 1984, it appeared that substantial and effective competition would be unlikely or impossible for LECs, particularly in local access. The economic rationale for the divestiture of BOCs from AT&T was that LECs exhibited natural monopoly characteristics. For local service, at least until recently, it has generally been deemed wasteful to allow two or more rival suppliers since that would entail duplication of facilities. Thus, despite intense competition in long-distance telephone markets, LECs still maintain their monopoly position. But, the availability of new technology and the advancement in old technology create new business opportunities and make possible new alternative services in local telephone markets. Competition in local market may arise or already exists on several different fronts.

This study attempts to provide new economic evidence on two important and current policy questions: the natural monopoly problem and productivity growth of monopolistic LECs. Strong evidence against cost subadditivity of LECs and on efficiency differentials between a competitive IXC and a monopolistic LEC may provide the economic rationale for allowing entry and competition in local and intra-LATA telephone markets. AT&T Long Lines, a competitive firm in the industry, provides a benchmark to evaluate the relative performance of LECs.

The key conclusions of the study are the following: The LECs exhibit nearly constant returns-to-scale technology and their cost functions are not always subadditive. The question of how to deal with technical change may affect the estimate of scale economies and the measure of cost subadditivity. The estimate of the degree of scale economies and the percentage of cost subadditivity cases depend on the size of a firm. As the size of a firm increases, both the overall returns to scale and the percentage of cost subadditivity cases rapidly decrease. In particular, the largest LECs have nearly no cost savings from having monopoly positions. There may be no economic rationale for large LECs such as 7 RHCs to maintain a monopoly position. Even though more intense competition continues to reduce the number of subadditivity cases, it may not eliminate all the possibilities of subadditivity. It appears that economies of scale and scope may pervade at least portions of local telephone services. Competition will determine the boundaries between a natural monopoly area and a competitive area.

Whatever the concept of capital stock used, LECs experience lower TFP growth rate than AT&T Long Lines. Moreover, the average annual TFP growth rate for AT&T Long Lines increases from stage to stage, while LECs exhibit nearly no change in their average rate for the three stages. The higher overall average rate for AT&T Long Lines is achieved by a big jump in the TFP growth rate after the breakup of the Bell system. There is strong evidence that competition may enlarge market demand. Under competitive pressure, AT&T Long Lines adopts higher capital quality and moderate labor quality.

The TFP growth rate regressions show that AT&T Long Lines experiences a bigger structural change in the TFP growth rate than LECs before and after competition. When disembodied and embodied technical change as well as demand fluctuations are controlled for, it is shown that competition forces AT&T Long Lines to use its organizational capital more efficiently. The analysis of cost functions confirms the conclusions from the productivity analysis. AT&T Long Lines shows a larger average annual shift in the cost function than LECs over the sample period. Also AT&T Long Lines experiences a big jump in the shifts after 1977, while there is no significant change for LECs. Using the competition index leads to the same conclusion as using the competition dummy variable.

This study not only has policy implications for entry and competition in local and intra-LATA telephone markets, but also provides a new approach to the embodiment hypothesis. Although the embodiment problem is the main issue in chapter 2, the ideas of embodiment pervade the subsequent chapters which deal with policy questions. Although the notion of embodied technical change has attracted the attention of many economists in the last few decades, the fraction of output growth or productivity growth attributed to embodied technical change, as compared to disembodied technical change, has ranged from negligible to dominant.

The principal conclusions on embodied technical change are summarized as follows: Adding a time variable in a translog cost function may underestimate the true rate of embodied technical change. The conventional specification of a cost function which views a time variable as a proxy for technology may not reflect the true technological structure of the U.S. telephone industry.

When the competition index is inserted to measure the effect of competition on technical change through its impact on the use and accumulation of organizational capital, the technical change due to improvements in labor and capital quality is 0.6% and 0.8% respectively. Two input quality indexes explain nearly half of the rate of technical change. Still, the importance of embodied technical change may be understated because the competition index may capture part of embodied technical change. Capital quality may have lagged effects. It may be yesterday's improvement in capital quality and not today's that leads to downward shifts in a cost function. Large firms may have better ability to transform improvements in capital quality into an actual cut in total costs.

The TFP growth rate regressions confirm the results from the cost models. In particular, improved labor and capital quality have a positive effect on TFP growth rate. Also, the lagged effects in the materialization of capital quality are confirmed.

Appendix 1

Parameter Estimates of Cost Functions

Table A.1 : Parameter Estimates of Two-Product Cost Functions

(1) Model 1

variable	parameter estimate	SE	variable	parameter estimate	SE
intercept	1.090*	0.123	lnQ_1*lnP_L	- 0.020*	0.005
lnQ_1	0.479*	0.156	lnQ_1*lnP_M	0.021*	0.006
lnQ_2	0.670*	0.154	lnQ_1*lnP_K	- 0.001	0.006
lnP_L	0.742*	0.010	$t*lnQ_1$	- 0.003	0.005
lnP_M	0.065*	0.013	lnQ_2*lnP_L	0.019*	0.005
lnP_K	0.193*	0.013	lnQ_2*lnP_M	- 0.012***	0.006
t	- 0.031*	0.008	lnQ_2*lnP_K	- 0.007	0.006
$0.5*(lnQ_1)^2$	0.090	0.108	$t*lnQ_2$	- 0.003	0.005
$0.5*(lnQ_2)^2$	0.228**	0.093	lnP_L*lnP_M	0.069*	0.009
$0.5*(lnP_L)^2$	- 0.041**	0.016	lnP_L*lnP_K	- 0.028**	0.014
$0.5*(lnP_M)^2$	- 0.054*	0.011	$t*lnP_L$	- 0.012*	0.000
$0.5*(lnP_K)^2$	0.043*	0.016	lnP_M*lnP_K	- 0.015	0.010
$0.5*t^2$	- 0.001**	0.000	$t*lnP_M$	0.005*	0.000
lnQ_1*lnQ_2	- 0.148	0.098	$t*lnP_K$	0.007*	0.000

Note: The estimation method is Zellner's seemingly unrelated regression technique. Q_1 refers to local output, Q_2 to toll output, P_L to wage rate, P_M to material price, P_K to user-cost of capital, and t to time. The estimation is carried out with pooled cross-sectional time series data for 8 LECs. The number of observations is 316. *,**, and *** imply that the coefficient is statistically significant at the 1%, 5%, and 10% level. SE stands for standard error.

(2) Model 2

variable	parameter estimate	SE	variable	parameter estimate	SE
intercept	0.606*	0.183	$\ln Q_1 * \ln k$	0.323***	0.175
$\ln Q_1$	0.947*	0.170	$\ln Q_2 * \ln P_L$	0.025*	0.005
$\ln Q_2$	0.159	0.166	$\ln Q_2 * \ln P_M$	- 0.023*	0.006
$\ln P_L$	0.745*	0.012	$\ln Q_2 * \ln P_K$	- 0.002	0.006
$\ln P_M$	0.018	0.015	$t * \ln Q_2$	0.014**	0.006
$\ln P_K$	0.237*	0.013	$\ln Q2 * \ln \omega$	- 0.025*	0.005
t	- 0.010	0.013	$\ln Q2 * \ln k$	- 0.422*	0.162
$\ln \omega$	- 0.745*	0.012	$\ln P_L * \ln P_M$	0.061*	0.008
$\ln k$	0.384	0.370	$\ln P_L * \ln P_K$	0.050*	0.012
$0.5*(\ln Q_1)^2$	0.034	0.103	$t * \ln P_L$	- 0.012*	0.000
$0.5*(\ln Q_2)^2$	0.113	0.088	$\ln P_L * \ln \omega$	0.111*	0.012
$0.5*(\ln P_L)^2$	- 0.111*	0.012	$\ln P_L * \ln k$	- 0.026***	0.013
$0.5*(\ln P_M)^2$	- 0.029*	0.010	$\ln P_M * \ln P_K$	- 0.032*	0.009
$0.5*(\ln P_K)^2$	- 0.018	0.015	$t * \ln P_M$	0.007*	0.001
$0.5*t^2$	- 0.001***	0.001	$\ln P_M * \ln \omega$	- 0.061*	0.008
$0.5*(\ln \omega)^2$	- 0.111*	0.012	$\ln P_M * \ln k$	- 0.090*	0.017
$0.5*(\ln k)^2$	2.516*	0.547	$t * \ln P_K$	0.005*	0.001
$\ln Q_1 * \ln Q_2$	- 0.063	0.093	$\ln P_K * \ln \omega$	- 0.050*	0.012
$\ln Q_1 * \ln P_L$	- 0.027*	0.005	$\ln P_K * \ln k$	0.116*	0.015
$\ln Q_1 * \ln P_M$	0.035*	0.006	$t * \ln \omega$	0.012*	0.000
$\ln Q_1 * \ln P_K$	- 0.008	0.006	$t * \ln k$	- 0.031**	0.015
$t * \ln Q_1$	- 0.019*	0.006	$\ln \omega * \ln k$	0.026***	0.013
$\ln Q_1 * \ln \omega$	0.027*	0.005			

Note: The estimation method is Zellner's seemingly unrelated regression technique. Q_1 refers to local output, Q_2 to toll output, P_L to wage rate, P_M to material price, P_K to user-cost of capital, t to time, ω to labor quality index, and k to capital quality index. The estimation is carried out with pooled cross-sectional time series data for 8 LECs. The number of observations is 316. *,**, and *** imply that the coefficient is statistically significant at the 1%, 5%, and 10% level. SE stands for standard error.

(3) Model 3

variable	parameter estimate	SE	variable	parameter estimate	SE
intercept	0.046*	0.016	lnQ_1*lnP_K	- 0.048*	0.006
lnQ_1	0.712*	0.039	$lnQ_1*ln\omega$	- 0.057*	0.008
lnQ_2	0.274*	0.035	lnQ_1*lnk	0.609*	0.150
lnP_L	0.429*	0.004	lnQ_2*lnP_L	- 0.062*	0.008
lnP_M	0.206*	0.003	lnQ_2*lnP_M	0.023*	0.007
lnP_K	0.365*	0.003	lnQ_2*lnP_K	0.040*	0.006
$ln\omega$	- 0.429*	0.004	$lnQ2*ln\omega$	0.062*	0.008
lnk	- 1.311*	0.080	$lnQ2*lnk$	- 0.843*	0.138
$0.5*(lnQ_1)^2$	0.369*	0.080	lnP_L*lnP_M	0.016***	0.008
$0.5*(lnQ_2)^2$	0.386*	0.062	lnP_L*lnP_K	0.067*	0.013
$0.5*(lnP_L)^2$	- 0.084*	0.013	$lnP_L*ln\omega$	0.084*	0.013
$0.5*(lnP_M)^2$	0.014	0.010	lnP_L*lnk	- 0.287*	0.020
$0.5*(lnP_K)^2$	- 0.036**	0.016	lnP_M*lnP_K	- 0.031*	0.008
$0.5*(ln\omega)^2$	- 0.084*	0.013	$lnP_M*ln\omega$	- 0.016*	0.008
$0.5*(lnk)^2$	1.372*	0.376	lnP_M*lnk	0.047**	0.017
lnQ_1*lnQ_2	- 0.368*	0.070	$lnP_K*ln\omega$	- 0.067*	0.013
lnQ_1*lnP_L	0.057*	0.008	lnP_K*lnk	0.240*	0.014
lnQ_1*lnP_M	- 0.009	0.007	$ln\omega*lnk$	0.287*	0.020

Note: The estimation method is Zellner's seemingly unrelated regression technique. Q_1 refers to local output, Q_2 to toll output, P_L to wage rate, P_M to material price, P_K to user-cost of capital, ω to labor quality index, and k to capital quality index. The estimation is carried out with pooled cross-sectional time series data for 8 LECs. The number of observations is 316. *,**, and *** imply that the coefficient is statistically significant at the 1%, 5%, and 10% level. SE stands for standard error.

(4) Model 4

variable	parameter estimate	SE	variable	parameter estimate	SE
intercept	0.400*	0.045	$\ln Q_1$*lnk	0.138	0.153
$\ln Q_1$	0.696*	0.063	$\ln Q_2$*$\ln P_L$	- 0.046*	0.008
$\ln Q_2$	0.341*	0.062	$\ln Q_2$*$\ln P_M$	0.014**	0.006
$\ln P_L$	0.424*	0.010	$\ln Q_2$*$\ln P_K$	0.032*	0.006
$\ln P_M$	0.177*	0.008	lnQ2*lnCI	0.016	0.017
$\ln P_K$	0.398*	0.007	lnQ2*lnω	0.046*	0.008
lnCI	- 0.212*	0.030	lnQ2*lnk	- 0.227	0.142
lnω	- 0.424*	0.010	$\ln P_L$*$\ln P_M$	- 0.028**	0.014
lnk	- 0.406*	0.144	$\ln P_L$*$\ln P_K$	- 0.014	0.018
0.5*$(\ln Q_1)^2$	- 0.025	0.074	$\ln P_L$*lnCI	0.002	0.003
0.5*$(\ln Q_2)^2$	0.123**	0.061	$\ln P_L$*lnω	- 0.042***	0.023
0.5*$(\ln P_L)^2$	0.042**	0.023	$\ln P_L$*lnk	- 0.243*	0.020
0.5*$(\ln P_M)^2$	- 0.011	0.013	$\ln P_M$*$\ln P_K$	0.039*	0.010
0.5*$(\ln P_K)^2$	- 0.025	0.018	$\ln P_M$*lnCI	0.010*	0.003
0.5*$(\ln CI)^2$	0.036*	0.009	$\ln P_M$*lnω	0.028**	0.014
0.5*$(\ln \omega)^2$	0.042***	0.023	$\ln P_M$*lnk	0.036**	0.017
0.5*$(\ln k)^2$	1.135*	0.333	$\ln P_K$*lnCI	- 0.012*	0.002
$\ln Q_1$*$\ln Q_2$	- 0.049	0.066	$\ln P_K$*lnω	0.014	0.018
$\ln Q_1$*$\ln P_L$	0.041*	0.008	$\ln P_K$*lnk	0.207*	0.014
$\ln Q_1$*$\ln P_M$	- 0.001	0.007	lnCI*lnω	- 0.002	0.003
$\ln Q_1$*$\ln P_K$	- 0.040*	0.006	lnCI*lnk	- 0.064	0.045
$\ln Q_1$*nCI	- 0.043**	0.017	lnω*lnk	0.243*	0.020
$\ln Q_1$* lnω	- 0.041*	0.008			

Note: The estimation method is Zellner's seemingly unrelated regression technique. Q_1 refers to local output, Q_2 to toll output, P_L to wage rate, P_M to material price, P_K to user-cost of capital, CI to competition index, ω to labor quality index, and k to capital quality index. The estimation is carried out with pooled cross-sectional time series data for 8 LECs. The number of observations is 316. *,**, and *** imply that the coefficient is statistically significant at the 1%, 5%, and 10% level. SE stands for standard error.

(5) Model 5

variable	parameter estimate	SE	variable	parameter estimate	SE
intercept	0.222*	0.051	lnQ_2*lnP_M	0.006	0.007
lnQ_1	0.934*	0.070	lnQ_2*lnP_K	0.036*	0.005
lnQ_2	0.098	0.071	lnQ2*lnCI	0.091*	0.021
lnP_L	0.401*	0.010	lnQ2*lnω	0.041*	0.008
lnP_M	0.181*	0.009	lnQ2*lnk	0.525***	0.291
lnP_K	0.418*	0.006	lnQ2*lnkl	- 1.210*	0.305
lnCI	- 0.141*	0.032	lnP_L*lnP_M	- 0.050*	0.014
lnω	- 0.401*	0.010	lnP_L*lnP_K	- 0.028***	0.017
lnk	0.255	0.314	lnP_L*lnCI	0.008**	0.003
lnkl	- 0.901*	0.318	$lnP_L*ln\omega$	- 0.078*	0.022
$0.5*(lnQ_1)^2$	0.014	0.079	lnP_L*lnk	0.121**	0.047
$0.5*(lnQ_2)^2$	0.088	0.070	lnP_L*lnkl	- 0.344*	0.045
$0.5*(lnP_L)^2$	0.078*	0.022	lnP_M*lnP_K	0.048*	0.009
$0.5*(lnP_M)^2$	0.002	0.014	lnP_M*lnCI	0.009*	0.003
$0.5*(lnP_K)^2$	- 0.020	0.017	$lnP_M*ln\omega$	0.050*	0.014
$0.5*(lnCI)^2$	0.022**	0.009	lnP_M*lnk	- 0.073***	0.042
$0.5*(ln\omega)^2$	0.078*	0.022	lnP_M*lnkl	0.095**	0.040
$0.5*(lnk)^2$	7.573*	2.463	lnP_K*lnCI	- 0.017*	0.002
$0.5*(lnkl)^2$	11.891*	2.563	$lnP_K*ln\omega$	0.028***	0.017
lnQ_1*lnQ_2	- 0.044	0.074	lnP_K*lnk	- 0.048	0.031
lnQ_1*lnP_L	0.036*	0.008	lnP_K*lnkl	0.250*	0.030
lnQ_1*lnP_M	0.008	0.007	lnCI*lnω	- 0.008**	0.003
lnQ_1*lnP_K	- 0.044*	0.005	lnCI*lnk	0.146	0.098
lnQ_1*nCI	- 0.110*	0.020	lnCI*lnkl	- 0.179***	0.099
$lnQ_1*ln\omega$	- 0.036*	0.008	lnω*lnk	- 0.121**	0.047
lnQ_1*lnk	- 0.612**	0.303	lnω*lnkl	- 8.725*	2.496
lnQ_1*lnkl	1.192*	0.311	lnk*lnkl	0.344*	0.045
lnQ_2*lnP_L	- 0.041*	0.008			

Note: The estimation method is Zellner's seemingly unrelated regression technique. Q_1 refers to local output, Q_2 to toll output, P_L to wage rate, P_M to material price, P_K to user-cost of capital, CI to competition index, ω to labor quality index, k to capital quality index, and kl to the average of one-year and two-year lagged k's. The estimation is carried out with pooled cross-sectional time series data for 8 LECs. The number of observations is 316. *,**, and *** imply that the coefficient is statistically significant at the 1%, 5%, and 10% level. SE stands for standard error.

Table A.2 : Parameter Estimates of Two-Product Cost Functions Considering Serially Correlated Errors

(1) Model 1

variable	parameter estimate	SE	variable	parameter estimate	SE
intercept	1.265*	0.163	$\ln Q_1 * \ln P_L$	- 0.020**	0.009
$\ln Q_1$	0.435*	0.160	$\ln Q_1 * \ln P_M$	0.022**	0.012
$\ln Q_2$	0.737*	0.165	$\ln Q_1 * \ln P_K$	- 0.002	0.008
$\ln P_L$	0.643*	0.020	$t * \ln Q_1$	- 0.002	0.005
$\ln P_M$	0.062*	0.027	$\ln Q_2 * \ln P_L$	0.017**	0.009
$\ln P_K$	0.294*	0.018	$\ln Q_2 * \ln P_M$	- 0.011	0.012
t	- 0.047*	0.011	$\ln Q_2 * \ln P_K$	- 0.005	0.008
$0.5*(\ln Q_1)^2$	- 0.029	0.088	$t * \ln Q_2$	- 0.007	0.006
$0.5*(\ln Q_2)^2$	0.146**	0.087	$\ln P_L * \ln P_M$	- 0.018***	0.012
$0.5*(\ln P_L)^2$	0.085*	0.013	$\ln P_L * \ln P_K$	- 0.066*	0.010
$0.5*(\ln P_M)^2$	- 0.024***	0.016	$t * \ln P_L$	0.042*	0.010
$0.5*(\ln P_K)^{21}$	0.024**	0.012	$\ln P_M * \ln P_K$	- 0.009*	0.001
$0.5*t^2$	0.000	0.000	$t * \ln P_M$	0.006*	0.001
$\ln Q_1 * \ln Q_2$	- 0.048	0.085	$t * \ln P_K$	0.003*	0.001
			ρ_c	0.740	
			ρ_s	0.814	

Note: The estimation method is a two-step Cochrane-Orcutt procedure and Zellner's seemingly unrelated regression technique. Q_1 refers to local output, Q_2 to toll output, P_L to wage rate, P_M to material price, P_K to user-cost of capital, and t to time. The estimation is carried out with pooled cross-sectional time series data for 8 LECs. The number of observations is 308. ρ_c and ρ_s are the first-order autocorrelation coefficient in the cost function and the share equations, respectively. *,**, and *** imply that the coefficient is statistically significant at the 1%, 5%, and 10% level. SE stands for standard error.

(2) Model 2

variable	parameter estimate	SE	variable	parameter estimate	SE
intercept	0.943*	0.187	$\ln Q_1 * \ln k$	- 0.113	0.152
$\ln Q_1$	0.658*	0.167	$\ln Q_2 * \ln P_L$	0.021**	0.009
$\ln Q_2$	0.489*	0.171	$\ln Q_2 * \ln P_M$	- 0.016***	0.012
$\ln P_L$	0.664*	0.022	$\ln Q_2 * \ln P_K$	- 0.005	0.008
$\ln P_M$	0.035	0.027	$t * \ln Q_2$	0.001	0.006
$\ln P_K$	0.302*	0.018	$\ln Q2 * \ln \omega$	- 0.021**	0.009
t	- 0.032*	0.013	$\ln Q2 * \ln k$	- 0.004	0.148
$\ln \omega$	- 0.664*	0.022	$\ln P_L * \ln P_M$	- 0.015	0.012
$\ln k$	0.280	0.323	$\ln P_L * \ln P_K$	- 0.059*	0.010
$0.5*(\ln Q_1)^2$	- 0.072	0.091	$t * \ln P_L$	- 0.009*	0.001
$0.5*(\ln Q_2)^2$	0.084	0.090	$\ln P_L * \ln \omega$	- 0.074*	0.014
$0.5*(\ln P_L)^2$	0.074*	0.014	$\ln P_L * \ln k$	- 0.002	0.018
$0.5*(\ln P_M)^2$	- 0.018	0.016	$\ln P_M * \ln P_K$	- 0.033*	0.010
$0.5*(\ln P_K)^2$	0.026**	0.013	$t * \ln P_M$	0.007*	0.001
$0.5*t^2$	0.000	0.001	$\ln P_M * \ln \omega$	0.015	0.012
$0.5*(\ln \omega)^2$	0.074*	0.014	$\ln P_M * \ln k$	- 0.071*	0.022
$0.5*(\ln k)^2$	0.562***	0.431	$t * \ln P_K$	0.003*	0.001
$\ln Q_1 * \ln Q_2$	0.003	0.088	$\ln P_K * \ln \omega$	0.059*	0.010
$\ln Q_1 * \ln P_L$	- 0.025*	0.010	$\ln P_K * \ln k$	0.072*	0.015
$\ln Q_1 * \ln P_M$	0.027*	0.012	$t * \ln \omega$	0.009*	0.001
$\ln Q_1 * \ln P_K$	- 0.003	0.008	$t * \ln k$	- 0.016***	0.012
$t * \ln Q_1$	- 0.007	0.006	$\ln \omega * \ln k$	0.002	0.018
$\ln Q_1 * \ln \omega$	0.025*	0.010	ρ_c	0.735	
			ρ_s	0.780	

Note: The estimation method is a two-step Cochrane-Orcutt procedure and Zellner's seemingly unrelated regression technique. Q_1 refers to local output, Q_2 to toll output, P_L to wage rate, P_M to material price, P_K to user-cost of capital, t to time, ω to labor quality index, and k to capital quality index. The estimation is carried out with pooled cross-sectional time series data for 8 LECs. The number of observations is 308. ρ_c and ρ_s are the first-order autocorrelation coefficient in the cost function and the share equations, respectively. *,**, and *** imply that the coefficient is statistically significant at the 1%, 5%, and 10% level. SE stands for standard error.

(3) Model 3

variable	parameter estimate	SE	variable	parameter estimate	SE
intercept	- 0.072*	0.015	$\ln Q_1 * \ln P_K$	- 0.006	0.007
$\ln Q_1$	0.895*	0.036	$\ln Q_1 * \ln \omega$	- 0.033*	0.010
$\ln Q_2$	0.109*	0.032	$\ln Q_1 * \ln k$	- 0.181***	0.141
$\ln P_L$	0.406*	0.008	$\ln Q_2 * \ln P_L$	- 0.035*	0.010
$\ln P_M$	0.204*	0.010	$\ln Q_2 * \ln P_M$	0.039*	0.012
$\ln P_K$	0.390*	0.006	$\ln Q_2 * \ln P_K$	- 0.004	0.007
$\ln \omega$	- 0.406*	0.008	$\ln Q2 * \ln \omega$	0.035*	0.010
$\ln k$	- 0.800*	0.081	$\ln Q2 * \ln k$	- 0.090	0.135
$0.5*(\ln Q_1)^2$	0.234*	0.075	$\ln P_L * \ln P_M$	- 0.045*	0.012
$0.5*(\ln Q_2)^2$	0.228*	0.062	$\ln P_L * \ln P_K$	- 0.055*	0.009
$0.5*(\ln P_L)^2$	0.101*	0.013	$\ln P_L * \ln \omega$	- 0.101*	0.013
$0.5*(\ln P_M)^2$	0.012	0.015	$\ln P_L * \ln k$	- 0.020	0.020
$0.5*(\ln P_K)^2$	0.022**	0.011	$\ln P_M * \ln P_K$	0.033*	0.009
$0.5*(\ln \omega)^2$	0.101*	0.013	$\ln P_M * \ln \omega$	0.045*	0.012
$0.5*(\ln k)^2$	1.511*	0.456	$\ln P_M * \ln k$	- 0.060*	0.024
$\ln Q_1 * \ln Q_2$	- 0.208*	0.066	$\ln P_K * \ln \omega$	0.055*	0.009
$\ln Q_1 * \ln P_L$	0.033*	0.010	$\ln P_K * \ln k$	0.081*	0.014
$\ln Q_1 * \ln P_M$	- 0.027**	0.012	$\ln \omega * \ln k$	0.020	0.020
			ρ_c	0.680	
			ρ_s	0.898	

Note: The estimation method is a two-step Cochrane-Orcutt procedure and Zellner's seemingly unrelated regression technique. Q_1 refers to local output, Q_2 to toll output, P_L to wage rate, P_M to material price, P_K to user-cost of capital, ω to labor quality index, and k to capital quality index. The estimation is carried out with pooled cross-sectional time series data for 8 LECs. The number of observations is 308. ρ_c and ρ_s are the first-order autocorrelation coefficient in the cost function and the share equations, respectively. *,**, and *** imply that the coefficient is statistically significant at the 1%, 5%, and 10% level. SE stands for standard error.

(4) Model 4

variable	parameter estimate	SE	variable	parameter estimate	SE
intercept	0.275*	0.047	$\ln Q_1*\ln k$	-0.230***	0.142
$\ln Q_1$	0.833*	0.072	$\ln Q_2*\ln P_L$	- 0.020**	0.010
$\ln Q_2$	0.199*	0.070	$\ln Q_2*\ln P_M$	0.026**	0.012
$\ln P_L$	0.424*	0.013	$\ln Q_2*\ln P_K$	- 0.006	0.007
$\ln P_M$	0.193*	0.016	lnQ2*lnCI	0.041**	0.020
$\ln P_K$	0.383*	0.009	lnQ2*lnω	0.020**	0.010
lnCI	- 0.119*	0.030	lnQ2*lnk	0.144	0.134
lnω	- 0.424*	0.013	$\ln P_L*\ln P_M$	- 0.040*	0.013
lnk	- 0.145	0.144	$\ln P_L*\ln P_K$	- 0.060*	0.009
$0.5*(\ln Q_1)^2$	- 0.037	0.078	$\ln P_L*\ln CI$	- 0.012*	0.004
$0.5*(\ln Q_2)^2$	0.056	0.072	$\ln P_L*\ln\omega$	- 0.100*	0.013
$0.5*(\ln P_L)^2$	0.100*	0.013	$\ln P_L*\ln k$	0.000	0.020
$0.5*(\ln P_M)^2$	0.005	0.017	$\ln P_M*\ln P_K$	0.036*	0.010
$0.5*(\ln P_K)^2$	0.024**	0.012	$\ln P_M*\ln CI$	0.014*	0.004
$0.5*(\ln CI)^2$	- 0.004	0.009	$\ln P_M*\ln\omega$	0.040*	0.013
$0.5*(\ln\omega)^2$	0.100*	0.013	$\ln P_M*\ln k$	- 0.070*	0.024
$0.5*(\ln k)^2$	- 0.507	0.395	$\ln P_K*\ln CI$	- 0.001	0.003
$\ln Q_1*\ln Q_2$	0.001	0.073	$\ln P_K*\ln\omega$	0.060*	0.009
$\ln Q_1*\ln P_L$	0.018**	0.010	$\ln P_K*\ln k$	0.071*	0.014
$\ln Q_1*\ln P_M$	- 0.013	0.013	lnCI*lnω	0.012*	0.004
$\ln Q_1*\ln P_K$	- 0.005	0.007	lnCI*lnk	0.047	0.044
$\ln Q_1*nCI$	- 0.061*	0.020	lnω*lnk	0.000	0.020
$\ln Q_1*\ln\omega$	- 0.018**	0.010	ρ_c	0.773	
			ρ_s	0.892	

Note: The estimation method is a two-step Cochrane-Orcutt procedure and Zellner's seemingly unrelated regression technique. Q_1 refers to local output, Q_2 to toll output, P_L to wage rate, P_M to material price, P_K to user-cost of capital, CI to competition index, ω to labor quality index, and k to capital quality index. The estimation is carried out with pooled cross-sectional time series data for 8 LECs. The number of observations is 308. ρ_c and ρ_s are the first-order autocorrelation coefficient in the cost function and the share equations, respectively. *,**, and *** imply that the coefficient is statistically significant at the 1%, 5%, and 10% level. SE stands for standard error.

(5) Model 5

variable	parameter estimate	SE	variable	parameter estimate	SE
intercept	0.167*	0.054	$\ln Q_2*\ln P_M$	0.023**	0.012
$\ln Q_1$	0.960*	0.079	$\ln Q_2*\ln P_K$	- 0.004	0.007
$\ln Q_2$	0.082	0.081	lnQ2*lnCI	0.085*	0.023
$\ln P_L$	0.436*	0.012	lnQ2*lnω	0.019**	0.010
$\ln P_M$	0.180*	0.015	lnQ2*lnk	0.398**	0.184
$\ln P_K$	0.384*	0.008	lnQ2*lnkl	- 0.565*	0.205
lnCI	- 0.057**	0.033	$\ln P_L*\ln P_M$	- 0.040*	0.013
lnω	- 0.436*	0.012	$\ln P_L*\ln P_K$	- 0.059*	0.009
lnk	0.189	0.221	$\ln P_L*\ln CI$	- 0.016*	0.004
lnkl	- 0.727*	0.240	$\ln P_L*\ln\omega$	- 0.099*	0.013
$0.5*(\ln Q_1)^2$	- 0.070	0.088	$\ln P_L*\ln k$	- 0.013	0.023
$0.5*(\ln Q_2)^2$	0.015	0.085	$\ln P_L*\ln kl$	0.045***	0.027
$0.5*(\ln P_L)^2$	0.099*	0.013	$\ln P_M*\ln P_K$	0.031*	0.010
$0.5*(\ln P_M)^2$	0.009	0.017	$\ln P_M*\ln CI$	0.019*	0.005
$0.5*(\ln P_K)^2$	0.028*	0.012	$\ln P_M*\ln\omega$	0.040*	0.013
$0.5*(\ln CI)^2$	- 0.020**	0.010	$\ln P_M*\ln k$	- 0.033	0.028
$0.5*(\ln\omega)^2$	0.099*	0.013	$\ln P_M*\ln kl$	- 0.102*	0.034
$0.5*(\ln k)^2$	2.384**	1.353	$\ln P_K*\ln CI$	- 0.003	0.003
$0.5*(\ln kl)^2$	4.415*	1.562	$\ln P_K*\ln\omega$	0.059*	0.009
$\ln Q_1*\ln Q_2$	0.043	0.085	$\ln P_K*\ln k$	0.046*	0.016
$\ln Q_1*\ln P_L$	0.017**	0.010	$\ln P_K*\ln kl$	0.057*	0.019
$\ln Q_1*\ln P_M$	- 0.010	0.013	lnCI*lnω	0.016*	0.004
$\ln Q_1*\ln P_K$	- 0.007	0.007	lnCI*lnk	0.006	0.064
$\ln Q_1*nCI$	- 0.104*	0.023	lnCI*lnkl	0.095***	0.073
$\ln Q_1*\ln\omega$	- 0.017**	0.010	lnω*lnk	0.013	0.023
$\ln Q_1*\ln k$	- 0.477*	0.188	lnω*lnkl	- 3.334*	1.382
$\ln Q_1*\ln kl$	0.528*	0.207	lnk*lnkl	- 0.045***	0.027
$\ln Q_2*\ln P_L$	- 0.019**	0.010	ρ_C	0.714	
			ρ_S	0.866	

Note: The estimation method is a two-step Cochrane-Orcutt procedure and Zellner's seemingly unrelated regression technique. Q_1 refers to local output, Q_2 to toll output, P_L to wage rate, P_M to material price, P_K to user-cost of capital, CI to competition index, ω to labor quality index, k to capital quality index, and kl to the average of one-year and two-year lagged k's. The estimation is carried out with pooled cross-sectional time series data for 8 LECs. The number of observations is 292. ρ_C and ρ_S are the first-order autocorrelation coefficient in the cost function and the share equations, respectively. *,**, ***, and SE imply the same as before.

Table A.3 : Parameter Estimates of Single-Product Cost Functions

(1) Model 1

variable	parameter estimate	SE	variable	parameter estimate	SE
intercept	1.506*	0.036	lnQ*lnPM	0.019*	0.002
lnQ	1.173*	0.018	lnQ*lnPK	- 0.015*	0.002
$\ln P_L$	0.694*	0.012	t*lnQ	- 0.006*	0.000
$\ln P_M$	0.155*	0.014	lnPL*lnPM	0.100*	0.011
$\ln P_K$	0.151*	0.013	lnPL*lnPK	0.087*	0.016
t	- 0.035*	0.002	lnPM*lnPK	- 0.084*	0.012
$0.5*(\ln Q)^2$	0.044*	0.005	t*lnPL	- 0.011*	0.000
$0.5*(\ln P_L)^2$	- 0.187*	0.019	t*lnPM	0.003*	0.001
$0.5*(\ln P_M)^2$	- 0.015	0.013	t*lnPK	0.008*	0.001
$0.5*(\ln P_K)^2$	- 0.003	0.018	ATT	- 0.468*	0.032
$0.5*t^2$	0.000**	0.000	t*ATT	- 0.021*	0.002
$\ln Q*\ln P_L$	- 0.004*	0.002			

Note: The estimation method is Zellner's seemingly unrelated regression technique. Q refers to local output, P_L to wage rate, P_M to material price, P_K to user-cost of capital, and t to time. ATT = 1 for AT&T Long Lines and = 0 for LECs. The estimation is carried out with pooled cross-sectional time series data for AT&T Long Lines Department and 8 LECs. The number of observations is 357. *,**, and *** imply that the coefficient is statistically significant at the 1%, 5%, and 10% level. SE stands for standard error.

(2) Model 2

variable	parameter estimate	SE	variable	parameter estimate	SE
intercept	1.502*	0.039	$\ln Q*\ln P_K$	- 0.015*	0.002
$\ln Q$	1.141*	0.018	$t*\ln Q$	- 0.005*	0.000
$\ln P_L$	0.701*	0.012	$\ln P_L*\ln P_M$	0.108*	0.012
$\ln P_M$	0.130*	0.016	$\ln P_L*\ln P_K$	0.082*	0.016
$\ln P_K$	0.169*	0.014	$\ln P_M*\ln P_K$	- 0.065*	0.013
t	- 0.040*	0.004	$t*\ln P_L$	- 0.011*	0.000
$0.5*(\ln Q)^2$	0.035*	0.005	$t*\ln P_M$	0.004*	0.001
$0.5*(\ln P_L)^2$	- 0.189*	0.019	$t*\ln P_K$	0.007*	0.001
$0.5*(\ln P_M)^2$	- 0.042*	0.015	ATT	- 0.744*	0.021
$0.5*(\ln P_K)^2$	- 0.016	0.019	1977	- 0.087	0.138
$0.5*t^2$	0.000	0.000	ATT*1977	0.284	0.208
$\ln Q*\ln P_L$	- 0.004**	0.002	t*1977	- 0.001	0.005
$\ln Q*\ln P_M$	0.019*	0.002	t*ATT*1977	- 0.018*	0.006

Note: The estimation method is Zellner's seemingly unrelated regression technique. Q refers to local output, P_L to wage rate, P_M to material price, P_K to user-cost of capital, and t to time. ATT = 1 for AT&T Long Lines and = 0 for LECs. 1977 = 1 for the period 1977-1991 and = 0 for the period 1951-1976. The estimation is carried out with pooled cross-sectional time series data for AT&T Long Lines Department and 8 LECs. The number of observations is 357. *,**, and *** imply that the coefficient is statistically significant at the 1%, 5%, and 10% level. SE stands for standard error.

(3) Model 3

variable	parameter estimate	SE	variable	parameter estimate	SE
intercept	0.968*	0.032	lnQ*lnω	0.013*	0.002
lnQ	1.027*	0.015	lnQ*lnk	- 0.014	0.034
lnP_L	0.413*	0.010	lnP_L*lnP_M	- 0.057*	0.014
lnP_M	0.213*	0.010	lnP_L*lnP_K	0.015	0.017
lnP_K	0.374*	0.009	lnP_L*lnCI	0.003	0.004
lnCI	- 0.362*	0.022	$lnP_L*ln\omega$	- 0.041***	0.023
lnω	- 0.413*	0.010	lnP_L*lnk	- 0.254*	0.020
lnk	- 0.100	0.111	lnP_M*lnP_K	0.067*	0.013
$0.5*(lnQ)^2$	0.007	0.005	lnP_M*lnCI	0.006***	0.004
$0.5*(lnP_L)^2$	0.041***	0.023	$lnP_M*ln\omega$	0.057*	0.014
$0.5*(lnP_M)^2$	- 0.010	0.016	lnP_M*lnk	0.096*	0.020
$0.5*(lnP_K)^2$	- 0.082	0.018	lnP_K*lnCI	- 0.009*	0.003
$0.5*(lnCI)^2$	0.076*	0.007	$lnP_K*ln\omega$	- 0.015	0.017
$0.5*(ln\omega)^2$	0.041***	0.023	lnP_K*lnk	0.158*	0.020
$0.5*(lnk)^2$	0.708**	0.310	lnCI*lnω	- 0.003	0.004
$lnQ*lnP_L$	- 0.013*	0.002	lnCI*lnk	- 0.114*	0.040
$lnQ*lnP_M$	0.024*	0.002	lnω*lnk	0.254*	0.020
$lnQ*lnP_K$	- 0.011*	0.002	ATT	- 0.322*	0.025
lnQ*nCI	- 0.022*	0.004	lnCI*ATT	- 0.122*	0.011

Note: The estimation method is Zellner's seemingly unrelated regression technique. Q refers to local output, P_L to wage rate, P_M to material price, P_K to user-cost of capital, CI to competition index, ω to labor quality index, and k to capital quality index. ATT = 1 for AT&T Long Lines and = 0 for LECs. The estimation is carried out with pooled cross-sectional time series data for AT&T Long Lines Department and 8 LECs. The number of observations is 357. *,**, and *** imply that the coefficient is statistically significant at the 1%, 5%, and 10% level. SE stands for standard error.

(4) Model 4

variable	parameter estimate	SE	variable	parameter estimate	SE
intercept	1.709*	0.092	$\ln P_L * \ln P_M$	0.062*	0.011
lnQ	1.134*	0.024	$\ln P_L * \ln P_K$	0.100*	0.017
$\ln P_L$	0.666*	0.014	$t * \ln P_L$	- 0.010*	0.001
$\ln P_M$	0.153*	0.018	$\ln P_L * \ln \omega$	0.161*	0.018
$\ln P_K$	0.181*	0.015	$\ln P_L * \ln k$	- 0.053*	0.018
t	- 0.065*	0.008	$\ln P_M * \ln P_K$	- 0.045*	0.012
$\ln \omega$	- 0.666*	0.014	$t * \ln P_M$	0.003*	0.001
lnk	0.945*	0.261	$\ln P_M * \ln \omega$	- 0.062*	0.011
$0.5*(\ln Q)^2$	0.025*	0.005	$\ln P_M * \ln k$	0.034	0.022
$0.5*(\ln P_L)^2$	- 0.161*	0.018	$t * \ln P_K$	0.007*	0.001
$0.5*(\ln P_M)^2$	- 0.017	0.014	$\ln P_K * \ln \omega$	- 0.100*	0.017
$0.5*(\ln P_K)^2$	- 0.055*	0.021	$\ln P_K * \ln k$	0.019	0.020
$0.5*t^2$	0.001*	0.000	$t * \ln \omega$	0.010*	0.001
$0.5*(\ln \omega)^2$	- 0.161*	0.018	t*lnk	- 0.034*	0.011
$0.5*(\ln k)^2$	1.283**	0.505	$\ln \omega * \ln k$	0.053*	0.018
$\ln Q * \ln P_L$	- 0.011*	0.002	ATT	- 0.619*	0.020
$\ln Q * \ln P_M$	0.023*	0.002	1977	0.161	0.143
$\ln Q * \ln P_K$	- 0.012*	0.002	ATT*1977	0.571**	0.256
lnQ*t	- 0.005*	0.001	t*1977	- 0.011***	0.005
$\ln Q * \ln \omega$	0.011*	0.002	t*ATT*1977	- 0.029*	0.008
lnQ*lnk	- 0.033	0.040			

Note: The estimation method is Zellner's seemingly unrelated regression technique. Q refers to local output, P_L to wage rate, P_M to material price, P_K to user-cost of capital, t to time trend, ω to labor quality index, and k to capital quality index. ATT = 1 for AT&T Long Lines and = 0 for LECs. 1977 = 1 for the period 1977-1991 and = 0 for the period 1951-1976. The estimation is carried out with pooled cross-sectional time series data for AT&T Long Lines Department and 8 LECs. The number of observations is 357. *,**, and *** imply that the coefficient is statistically significant at the 1%, 5%, and 10% level. SE stands for standard error.

Table A.4 : Parameter Estimates of Single-Product Cost Functions Including the ATT-First Order Interaction Terms

(1) Model 3-2

variable	paramete r estimate	SE	variable	paramete r estimate	SE
intercept	0.970*	0.034	$\ln P_L * \ln P_K$	- 0.004	0.017
lnQ	1.049*	0.015	$\ln P_L * \ln CI$	0.003	0.004
$\ln P_L$	0.428*	0.010	$\ln P_L * \ln\omega$	- 0.066*	0.023
$\ln P_M$	0.193*	0.009	$\ln P_L * \ln k$	- 0.258*	0.019
$\ln P_K$	0.379*	0.009	$\ln P_M * \ln P_K$	0.071*	0.012
lnCI	- 0.338*	0.022	$\ln P_M * \ln CI$	0.008**	0.003
$\ln\omega$	- 0.428*	0.010	$\ln P_M * \ln\omega$	0.061*	0.014
lnk	- 0.172	0.111	$\ln P_M * \ln k$	0.098*	0.018
$0.5*(\ln Q)^2$	0.019*	0.005	$\ln P_K * \ln CI$	- 0.011*	0.003
$0.5*(\ln P_L)^2$	0.066*	0.023	$\ln P_K * \ln\omega$	0.004	0.017
$0.5*(\ln P_M)^2$	- 0.010	0.014	$\ln P_K * \ln k$	0.160*	0.019
$0.5*(\ln P_K)^2$	- 0.067*	0.018	$\ln CI * \ln\omega$	- 0.003	0.004
$0.5*(\ln CI)^2$	0.070*	0.007	lnCI*lnk	- 0.122*	0.038
$0.5*(\ln\omega)^2$	0.066*	0.023	$\ln\omega * \ln k$	0.258*	0.019
$0.5*(\ln k)^2$	1.017*	0.313	ATT	- 0.359*	0.047
$\ln Q * \ln P_L$	- 0.006*	0.002	lnQ*ATT	- 0.239*	0.034
$\ln Q * \ln P_M$	0.017*	0.002	$\ln P_L$*ATT	- 0.068*	0.009
$\ln Q * \ln P_K$	- 0.011*	0.002	$\ln P_M$*ATT	0.069*	0.008
lnQ*nCI	- 0.018*	0.004	$\ln P_K$*ATT	- 0.001	0.009
$\ln Q * \ln\omega$	0.006*	0.002	$\ln\omega$*ATT	0.068*	0.009
lnQ*lnk	- 0.073***	0.039	lnk*ATT	0.535*	0.131
$\ln P_L * \ln P_M$	- 0.061*	0.014	lnCI*ATT	- 0.091*	0.019

Note: The estimation method is Zellner's seemingly unrelated regression technique. Q refers to local output, P_L to wage rate, P_M to material price, P_K to user-cost of capital, CI to competition index, ω to labor quality index, and k to capital quality index. ATT = 1 for AT&T Long Lines and = 0 for LECs. The estimation is carried out with pooled cross-sectional time series data for AT&T Long Lines Department and 8 LECs. The number of observations is 357. *,**, and *** imply that the coefficient is statistically significant at the 1%, 5%, and 10% level. SE stands for standard error.

(2) Model 4-2

variable	parameter estimate	SE	variable	parameter estimate	SE
intercept	1.538*	0.089	lnP_L*lnω	0.110*	0.014
lnQ	1.096*	0.025	lnP_L*lnk	- 0.050*	0.012
lnP_L	0.727*	0.010	lnP_M*lnP_K	- 0.038*	0.012
lnP_M	0.084*	0.017	t*lnP_M	0.005*	0.001
lnP_K	0.190*	0.016	lnP_M*lnω	- 0.089*	0.008
t	- 0.053*	0.008	lnP_M*lnk	0.022	0.020
lnω	- 0.727*	0.010	t*lnP_K	0.006*	0.001
lnk	0.919*	0.245	lnP_K*lnω	- 0.020	0.014
0.5*$(lnQ)^2$	0.035*	0.005	lnP_K*lnk	0.028	0.020
0.5*$(lnP_L)^2$	- 0.110*	0.014	t*lnω	0.012*	0.000
0.5*$(lnP_M)^2$	- 0.051*	0.013	t*lnk	- 0.040*	0.010
0.5*$(lnP_K)^2$	0.018	0.019	lnω*lnk	0.050*	0.012
0.5*t^2	0.001*	0.000	ATT	- 0.154	0.150
0.5*$(ln\omega)^2$	- 0.110*	0.014	lnQ*ATT	- 0.041	0.092
0.5*$(lnk)^2$	2.050*	0.493	lnP_L*ATT	- 0.095*	0.005
lnQ*lnP_L	- 0.001	0.001	lnP_M*ATT	0.070*	0.008
lnQ*lnP_M	0.015*	0.002	lnP_K*ATT	0.025*	0.008
lnQ*lnP_K	- 0.014*	0.002	t*ATT	- 0.026*	0.009
lnQ*t	- 0.003*	0.001	lnω*ATT	0.095*	0.005
lnQ*lnω	0.001	0.001	lnk*ATT	0.510*	0.139
lnQ*lnk	- 0.150*	0.043	1977	0.200	0.135
lnP_L*lnP_M	0.089*	0.008	ATT*1977	0.757**	0.306
lnP_L*lnP_K	0.020	0.014	t*1977	- 0.013**	0.005
t*lnP_L	- 0.012*	0.000	t*ATT*1977	- 0.028*	0.010

Note: The estimation method is Zellner's seemingly unrelated regression technique. Q refers to local output, P_L to wage rate. P_M to material price, P_K to user-cost of capital, t to time trend, ω to labor quality index, and k to capital quality index. ATT = 1 for AT&T Long Lines and = 0 for LECs. 1977 = 1 for the period 1977-1991 and = 0 for the period 1951-1976. The estimation is carried out with pooled cross-sectional time series data for AT&T Long Lines Department and 8 LECs. The number of observations is 357. *,**, and *** imply that the coefficient is statistically significant at the 1%, 5%, and 10% level. SE stands for standard error.

Table A.5 : Parameter Estimates of Single-Product Cost Functions Considering Serially Correlated Errors

(1) Model 1

variable	parameter estimate	SE	variable	parameter estimate	SE
intercept	1.528*	0.085	lnQ*lnPM	0.037*	0.005
lnQ	1.168*	0.034	lnQ*lnPK	- 0.019*	0.003
$\ln P_L$	0.568*	0.020	t*lnQ	- 0.006*	0.001
$\ln P_M$	0.153*	0.030	lnPL*lnPM	- 0.006	0.013
$\ln P_K$	0.279*	0.019	lnPL*lnPK	- 0.061*	0.011
t	- 0.043*	0.006	lnPM*lnPK	0.006	0.012
$0.5*(\ln Q)^2$	0.037*	0.009	t*lnPL	- 0.007*	0.001
$0.5*(\ln P_L)^2$	0.067*	0.013	t*lnPM	0.004*	0.001
$0.5*(\ln P_M)^2$	0.000	0.019	t*lnPK	0.003*	0.001
$0.5*(\ln P_K)^2$	0.055*	0.014	ATT	- 0.427*	0.084
$0.5*t^2$	0.000	0.000	t*ATT	- 0.018*	0.004
$\ln Q*\ln P_L$	- 0.018*	0.004	ρ_c	0.780	
			ρ_s	0.856	

Note: The estimation method is a two-step Cochrane-Orcutt procedure and Zellner's seemingly unrelated regression technique. Q refers to local output, P_L to wage rate, P_M to material price, P_K to user-cost of capital, and t to time. ATT = 1 for AT&T Long Lines and = 0 for LECs. ρ_c and ρ_s are the first-order autocorrelation coefficient in the cost function and the share equations, respectively. The estimation is carried out with pooled cross-sectional time series data for AT&T Long Lines Department and 8 LECs. The number of observations is 348. *,**, and *** imply that the coefficient is statistically significant at the 1%, 5%, and 10% level. SE stands for standard error.

(2) Model 2

variable	parameter estimate	SE	variable	parameter estimate	SE
intercept	1.661*	0.103	$lnQ*lnP_K$	- 0.019*	0.003
lnQ	1.168*	0.035	t*lnQ	- 0.007*	0.001
lnP_L	0.569*	0.020	lnP_L*lnP_M	- 0.005	0.013
lnP_M	0.148*	0.030	lnP_L*lnP_K	- 0.061*	0.011
lnP_K	0.283*	0.019	lnP_M*lnP_K	0.009	0.012
t	- 0.057*	0.009	$t*lnP_L$	- 0.007*	0.001
$0.5*(lnQ)^2$	0.032*	0.009	$t*lnP_M$	0.004*	0.001
$0.5*(lnP_L)^2$	0.066*	0.013	$t*lnP_K$	0.003*	0.001
$0.5*(lnP_M)^2$	- 0.004	0.019	ATT	- 0.783*	0.045
$0.5*(lnP_K)^2$	0.052*	0.014	1977	0.207	0.240
$0.5*t^2$	0.001***	0.001	ATT*1977	0.478**	0.224
$lnQ*lnP_L$	- 0.019*	0.004	t*1977	- 0.010	0.009
$lnQ*lnP_M$	0.037*	0.005	t*ATT*1977	- 0.019*	0.008
			ρ_c	0.790	
			ρ_s	0.856	

Note: The estimation method is a two-step Cochrane-Orcutt procedure and Zellner's seemingly unrelated regression technique. Q refers to local output, P_L to wage rate, P_M to material price, P_K to user-cost of capital, and t to time. ATT = 1 for AT&T Long Lines and = 0 for LECs. 1977 = 1 for the period 1977-1991 and = 0 for the period 1951-1976. ρ_c and ρ_s are the first-order autocorrelation coefficient in the cost function and the share equations, respectively. The estimation is carried out with pooled cross-sectional time series data for AT&T Long Lines Department and 8 LECs. The number of observations is 348. *,**, and *** imply that the coefficient is statistically significant at the 1%, 5%, and 10% level. SE stands for standard error.

(3) Model 3

variable	parameter estimate	SE	variable	parameter estimate	SE
intercept	0.811*	0.043	lnQ*lnω	0.012**	0.006
lnQ	1.000*	0.026	lnQ*lnk	0.016	0.045
lnP_L	0.356*	0.016	lnP_L*lnP_M	- 0.022**	0.013
lnP_M	0.227*	0.017	lnP_L*lnP_K	- 0.067*	0.010
lnP_K	0.418*	0.012	lnP_L*lnCI	- 0.009*	0.003
lnCI	- 0.205*	0.026	$lnP_L*ln\omega$	- 0.089*	0.012
lnω	- 0.356*	0.016	lnP_L*lnk	- 0.010	0.018
lnk	- 0.129	0.119	lnP_M*lnP_K	- 0.007	0.012
$0.5*(lnQ)^2$	0.005	0.011	lnP_M*lnCI	0.013*	0.005
$0.5*(lnP_L)^2$	0.089*	0.012	$lnP_M*ln\omega$	0.022**	0.013
$0.5*(lnP_M)^2$	0.029***	0.019	lnP_M*lnk	- 0.059**	0.026
$0.5*(lnP_K)^2$	0.074*	0.013	lnP_K*lnCI	- 0.004***	0.003
$0.5*(lnCI)^2$	0.013**	0.007	$lnP_K*ln\omega$	0.067*	0.010
$0.5*(ln\omega)^2$	0.089*	0.012	lnP_K*lnk	0.069*	0.016
$0.5*(lnk)^2$	- 1.612*	0.371	lnCI*lnω	0.010*	0.003
$lnQ*lnP_L$	- 0.012**	0.006	lnCI*lnk	0.100*	0.038
$lnQ*lnP_M$	0.041*	0.006	lnω*lnk	0.016	0.019
$lnQ*lnP_K$	- 0.030*	0.004	ATT	- 0.563*	0.068
lnQ*nCI	- 0.019*	0.006	lnCI*ATT	- 0.047**	0.021
			ρ_c	0.831	
			ρ_s	0.914	

Note: The estimation method is a two-step Cochrane-Orcutt procedure and Zellner's seemingly unrelated regression technique. Q refers to local output, P_L to wage rate, P_M to material price, P_K to user-cost of capital, CI to competition index, ω to labor quality index, and k to capital quality index. ATT = 1 for AT&T Long Lines and = 0 for LECs. ρ_c and ρ_s are the first-order autocorrelation coefficient in the cost function and the share equations, respectively. The estimation is carried out with pooled cross-sectional time series data for AT&T Long Lines Department and 8 LECs. The number of observations is 348. *,**, and *** imply that the coefficient is statistically significant at the 1%, 5%, and 10% level. SE stands for standard error.

(4) Model 4

variable	parameter estimate	SE	variable	parameter estimate	SE
intercept	1.612*	0.107	$\ln P_L * \ln P_M$	0.004	0.013
lnQ	1.166*	0.033	$\ln P_L * \ln P_K$	- 0.069*	0.010
$\ln P_L$	0.560*	0.024	$t * \ln P_L$	- 0.007*	0.001
$\ln P_M$	0.093*	0.042	$\ln P_L * \ln \omega$	- 0.065*	0.013
$\ln P_K$	0.347*	0.028	$\ln P_L * \ln k$	- 0.026***	0.018
t	- 0.059*	0.009	$\ln P_M * \ln P_K$	0.008	0.012
$\ln \omega$	- 0.560*	0.024	$t * \ln P_M$	0.006*	0.001
lnk	0.380***	0.232	$\ln P_M * \ln \omega$	- 0.004	0.013
$0.5*(\ln Q)^2$	0.022*	0.008	$\ln P_M * \ln k$	- 0.027	0.025
$0.5*(\ln P_L)^2$	0.065*	0.013	$t * \ln P_K$	0.001	0.001
$0.5*(\ln P_M)^2$	- 0.012	0.019	$\ln P_K * \ln \omega$	0.069*	0.010
$0.5*(\ln P_K)^2$	0.061*	0.014	$\ln P_K * \ln k$	0.053*	0.016
$0.5*t^2$	0.001**	0.001	$t * \ln \omega$	0.006*	0.001
$0.5*(\ln \omega)^2$	0.065*	0.013	t*lnk	- 0.013***	0.009
$0.5*(\ln k)^2$	0.115	0.423	$\ln \omega * \ln k$	0.019	0.018
$\ln Q * \ln P_L$	- 0.019*	0.004	ATT	- 0.718*	0.038
$\ln Q * \ln P_M$	0.050*	0.007	1977	0.274	0.226
$\ln Q * \ln P_K$	- 0.031*	0.005	ATT*1977	0.176	0.244
lnQ*t	- 0.006*	0.001	t*1977	- 0.013***	0.009
$\ln Q * \ln \omega$	0.019*	0.004	t*ATT*1977	- 0.009	0.008
lnQ*lnk	- 0.046	0.041	ρ_c	0.736	
			ρ_s	0.889	

Note: The estimation method is a two-step Cochrane-Orcutt procedure and Zellner's seemingly unrelated regression technique. Q efers to local output, P_L to wage rate, P_M to material price, P_K to user-cost of capital, t to time trend, ω to labor quality index, and k to capital quality index. ATT = 1 for AT&T Long Lines and = 0 for LECs. 1977 = 1 for the period 1977-1991 and = 0 for the period 1951-1976. ρ_c and ρ_s are the first-order autocorrelation coefficient in the cost function and the share equations, respectively. The estimation is carried out with pooled cross-sectional time series data for AT&T Long Lines Department and 8 LECs. The number of observations is 348. *,**, and *** imply that the coefficient is statistically significant at the 1%, 5%, and 10% level. SE stands for standard error.

Appendix 2

Data Construction

This appendix describes the technical details on data construction. The major source of data is the *Statistics of Communications Common Carriers* (*SCCC*) published by the Federal Communications Commission (FCC) and the *Form M* reports in which reporting telephone companies file financial and operating data to the FCC. In particular, the *Form M* report provides detailed data for telephone plants or compensation and working hours. Annual data for AT&T Long Lines and 8 LECs were collected for the period 1951-1991. The sample LECs are composed of 2 large LECs, 5 midsize LECs, and 1 small LEC. 6 LECs are BOCs and 2 LECs are independent companies.

Output Variables

The quantity index for a type of telephone service is generated by deflating the revenues by an appropriate price index. The revenues of a telephone company are composed of local service revenues, toll service revenues, and miscellaneous revenues. The deflators are drawn from the Bell system price indexes for the period 1951-1971 and from the Producer Price Indexes (PPI) for the period 1972-1991.[1] The miscellaneous revenues are deflated by the price index of directory advertising, even though they include revenues from other sources. Constant dollar local, toll, and miscellaneous revenues are aggregated to create the Törnqvist index of aggregate output using revenue shares as weights. Similarly, constant dollar local and miscellaneous revenues are aggregated into local output.

Access revenues and expenses have been reported in the *SCCC* since 1984. Before the breakup, LECs collected all revenues, paid long distance companies their portion of toll revenues, and reported the remainder as their own toll revenues. This procedure was called 'separation and settlement.' To construct consistent output and cost data before and after 1984, I allocate access revenues of LECs to local or toll revenues and, also, subtract access expenses of AT&T Long Lines from both toll revenues and operating expenses. Specifically, end-user access revenues are added to local revenues and intrastate and inter-state access revenues to toll revenues.

Labor and Labor Quality

The *Form M* reports provide annual data for total number of employees, average weekly hours, and average hourly earnings of 34 occupational groups. Two types of total working hours are calculated; the simple sum and the weighted sum of working hours of each group. When the latter is computed, the ratio of each group's hourly earnings to average wage in the private sector is used as a weight. The quantity of labor input is the simple sum of working hours of every occupation group, while the index of labor quality is defined as the ratio of the weighted sum to the simple sum of working hours. The labor quality indexes are extrapolated for the period 1981-1991, because of lack of appropriate data.

Current-dollar wage rate which is actual compensation per working hour is divided by average wage rate in the private sector to derive constant-dollar wage rate. Real labor expenses are the quantity of labor times constant-dollar wage rate.

Materials

Material expenses are defined as a catch-all category which includes the part of operating expenses not attributable to labor or capital. Material expenses are computed by subtracting total compensations and depreciation charges from operating expenses. The PPI for intermediate inputs is used as a proxy for the price of material input. Material expenses are divided by the price index to derive constant-dollar material costs.

Capital, User-costs, and Capital Quality

I use two types of capital stock; net stock and gross stock. Net capital stock is constructed by the commonly used perpetual inventory method. The depreciation rates are obtained from Jorgenson, Gollop, and Fraumeni (1987).[2] I create gross capital stock series based on Gort, Bahk, and Wall (1993). My retirement pattern is based on the modified Winfrey S-3 distribution,[3] while Gort, Bahk, and Wall assume the simultaneous exit of investment goods of the same vintage.

Only the construction method of gross capital stock is reported because that of the net capital stock is the same. First, I classify telephone plants in service into five items: land and building, furniture and office equipment, vehicles and other work equipment, communication equipment, and cable and wire. Next, yearly dollar amount of each category of plants added is deflated by an appropriate price index to derive real gross investment in the specific year. Yearly real capital stock of each category of telephone plants is the sum of real gross investments cumulated over its service life. Aggregate capital stock is the weighted sum of five types of real capital stock in which capital expense shares are used as weights.

Price indexes and service lives were obtained from the Bureau of Labor Statistics and the Bureau of Economic Analysis.[4] The benchmark capital stock for each category of plants is constant-dollar gross plants in 1950. Transferred assets due to the breakup of the Bell system are deflated using the 1984 price indexes, which may cause some measurement errors. But experimenting with price indexes of different years produced nearly the same conclusions. The user-costs of each type of plant is the constant rate of return (0.035) plus its depreciation rate.[5]

I construct two capital quality indexes; one is average vintage and the other is (1+s) divided by average age where s is the ratio of telephones connected to the ESS to total telephones. In the case of AT&T Long Lines where the data are not available, s is approximated by the ratio of the dollar amount of ESS to that of total switching systems. The relationship between average vintage and average age is the following; average vintage=(year of measurement-base year)-average age.

Let me explain only the construction of average age. First, I compute the average age of each category of capital stock in a specific year as the weighted sum of ages of gross investments in which the weights are the ratio of yearly gross investment to the capital stock. The average ages of each category of capital stock are then averaged using capital expense shares as weights to derive the average age of aggregate capital stock. The age of each category of benchmark capital stock is assumed to be half of its service life. Transferred assets are omitted in the calculation of average age for AT&T Long Lines because no information on ages of transferred assets is available.

Total Costs

I sum constant-dollar labor, material, capital costs to derive constant 1972 dollar total costs.

Competition Index

A measure of competition in the U.S. telephone industry is developed by Chessler. Chessler's competition index is calculated as the ratio of the other common carriers' payments to LECs for connection plus the resale carriers' lease payment to the telephone industry's total toll revenues for the period 1971-1983. I extend the competition index using the market share of the OCCs in toll revenues for the post-1984 period.[6]

NOTES

[1] Even though the Consumer Price Indexes (CPI) for all telephone services are available for the period 1951-1971, the disaggregated Bell system price indexes are preferable. Using the Bell system price indexes may lead to a minor measurement problem in non-Bell independent companies.

[2] The depreciation ratio for each category of telephone plants is the following: building=0.033, furniture and office equipment=0.217, vehicles and other work equipment=0.277, communication equipment=0.118, and cable and wire=0.033.

[3] The modified Winfrey S-3 distribution is one of the bell-shaped retirement patterns. Young and Musgrave (1980) report the modified Winfrey retirement patterns used in the Bureau of Economic Analysis (BEA) capital stock study.

[4] The price index and service life for each category of telephone plants are in the following table.

Telephone plants	service life	price index
land	infinite	implicit price deflator for purchases of structure by public utilities
building	40 years	implicit price deflator for nonresidential building
furniture and office equipment	8 years	PPI for store & office machines and equipment
vehicles and other work equipment	10 years	PPI for motor vehicles and equipment
communications equipment	11 years	price for communication equipment
cable and wire	40 years	PPI for cable & wire

[5] This rate of return (3.5%) was obtained by averaging nominal rate of return (=after-tax net operating income/net plant) minus inflation rate (annual percentage change in CPI) for all observations of all sample firms.

[6] The FCC has reported the market share of telephone companies in toll service market in *SCCC* since 1984.

Bibliography

Bahk, B., and M. Gort. "Decomposing Learning by Doing in New Plants," *Journal of Political Economy*, 101, 1993, 561-583.

Baumol, W.J. "On the Proper Cost Tests for Natural Monopoly in a Multiproduct Industry," *American Economic Review*, 67, 1977, 809-822.

Baumol, W.J., and J.G. Sidak. *Toward Competition in Local Telephony*, The American Enterprise Institute Studies in Telecommunication Deregulation, Washington, D.C.: The MIT Press, 1994.

Berglas, E. "Investment and Technological Change," *Journal of Political Economy*, 73, 1965, 173-180.

Berndt, E.R., and M.A. Fuss. "Productivity Measurement with Adjustments for Variations in Capacity Utilization and Other Forms of Temporary Equilibrium," *Journal of Econometrics*, 33, 1986,7-29.

Bernstein, J.I. "An Examination of the Equilibrium Specification and Structure of Production for the Canadian Telecommunications," *Journal of Applied Econometrics*, 4, 1989, 265-282.

Boardman, A.E., and A.R. Vining. "Ownership and Performance in Competitive Environments: A Comparison of the Performance of Private, Mixed, and State-Owned Enterprises," *Journal of Law and Economics*, 32, 1989, 1-33.

Boddy, R., and M. Gort. "Obsolescence, Embodiment, and the Explanation of Productivity Change," *Southern Economic Journal*, 40, 1974, 553-562.

Brechling, F.P., and A. Surrey. "An International Comparison of Production Techniques: the Coal-Fired Electricity Generating Industry," *National Institute Economic Review*, May, 1966, 30-42.

Caves, D.W., L.R. Christensen, and J.A. Swanson. "Economic Performance in Regulated and Unregulated Environments: A

Comparison of U.S. and Canadian Railroads," *Quarterly Journal of Economics*, 96, 1981a, 559-581.

Caves, D.W., L.R. Christensen, and J.A. Swanson. "Productivity Growth, Scale Economies, and Capacity Utilization in U.S. Railroads, 1955-74," *American Economic Review*, 71, 1981b, 994-1002.

Caves, D.W., L.R. Christensen, and W.E. Diewert. "Multilateral Comparisons of Output, Input, and Productivity Using Superlative Index Numbers," *Economic Journal*, 92, 1982, 73-86.

Charnes, A., W.W. Cooper, and T. Sueyoshi. "A Goal Programming/ Constrained Regression: Review of the Bell System Breakup," *Management Science*, 34, 1988, 1-16.

Christensen, L.R., and D.W. Jorgenson. "The Measurement of U.S. Real Capital Input," *Review of Income and Wealth*, 15, 1969, 293-320.

Christensen, L.R., D. Cummings, and P.E. Schoech. "Econometric Estimation of Scale Economies in Telecommunications," In *Economic Analysis of Telecommunications*, ed. Courville, L., A. de Fontenay, and R. Dobell. Amsterdam: Elsevier Science Publishers B.V. (North-Holland), 1983.

Cohen, W.M., and R.C. Levin, "Empirical Studies of Innovation and Market Structure," In *Handbook of Industrial Organization*, ed. Schmalensee, R., and R.D. Willig. Amsterdam: North-Holland, 1992.

Cowing, T.G., and V.K. Smith. "The Estimation of a Production Technology: A Survey of Econometric Analyses of Steam-Electric Generation," *Land Economics*, 54, 1978, 156-186.

Courville, L., A. de Fontenay, and R. Dobell. ed. *Economic Analysis of Telecommunications*, Amsterdam: Elsevier Science Publishers B.V. (North-Holland), 1983.

Crandall, R.W. "Surprises from Telephone Deregulation and the AT&T Divestiture," *American Economic Review*, 78, 1988, 323-327.

Crandall, R.W. *After the Breakup*, Washington, D.C.: The Brookings Institution, 1991.

Denison, E.F. "The Unimportance of the Embodiment Question," *American Economic Review*, 54, 1964, 90-94.

Denny, M., M.A. Fuss, and L. Waverman. "The Measurement and Interpretation of Total Factor Productivity in Regulated Industries, with an Application to Canadian

Telecommunications," In *Productivity Measurement in Regulated Industries*, ed. Cowing, T.G., and R. Stevenson. New York: Academic Press, 1981.

Diewert, W.E. "The Theory of Total Factor Productivity Measurement in Regulated Industries," In *Productivity Measurement in Regulated Industries*, ed. Cowing, T.G., and R. Stevenson. New York: Academic Press, 1981.

Diewert, W.E., and T.J. Wales. "Multiproduct Cost Functions and Subadditivity Tests: A Critique of the Evans and Heckman Research on the US Bell Systems," Department of Economics, University of British Columbia Discussion Paper No: 91-21, 1991.

Dobell, A.R., L.D. Taylor, L. Waverman, T. Liu, and D.G. Copeland. "Telephone Communications in Canada: Demand, Production, and Investment Decision," *Bell Journal of Economics and Management Sciences*, 5, 1972, 175-219.

Ehrlich, I., G. Gallais-Hamonno, Z. Liu, and R. Lutter. "Productivity Growth and Firm Ownership: An Analytical and Empirical Investigation," *Journal of Political Economy*, 102, 1994, 1006-1038.

Evans, D.S., and J.J. Heckman. "Multiproduct Cost Function Estimates and Natural Monopoly Tests for the Bell System", In *Breaking Up Bell: Essays on Industry Organization and Regulation*, ed. Evans, D.S. Amsterdam: North-Holland, 1983a.

Evans, D.S., and J.J. Heckman. "Natural Monopoly", In *Breaking Up Bell: Essays on Industry Organization and Regulation*, ed. Evans, D.S. Amsterdam: North-Holland, 1983b.

Evans, D.S., and J.J. Heckman. "A Test for Subadditivity of the Cost Function with an Application to the Bell System", *American Economic Review*, 74, 1984, 615-623.

Fisher, F.M. "Embodied Technical Change and the Existence of an Aggregate Capital Stock," *Review of Economic Studies*, 32, 1965, 263-88.

Fuss, M.A. "A Survey of Recent Results in the Analysis of Production Conditions in Telecommunications", In *Economic Analysis of Telecommunications*, ed. Courville, L., A. de Fontenay, and R. Dobell. Amsterdam: Elsevier Science Publishers B.V. (North-Holland), 1983.

Fuss, M.A. "Economic Evidence on Scale, Scope and the Presence of Natural Monopoly: What Can It Tell Us About the Desirability of

Competition in Public Long-Distance Telephone Service?" Department of Economics and Institute for Policy Analysis, University of Toronto, Working Paper No. 9210, 1992.

Gentzoglanis, A. "Entry and Competition in the Canadian Telecommunications Industry: The Case of Alberta Government Telephones," *Atlantic Economic Journal*, 21, 1993, 38-46.

Gordon, R.J. *The Measurement of Durable Goods Prices*, Chicago: University of Chicago Press, 1990.

Gort, M., B.H. Bahk, and R.A. Wall. "Decomposing Technical Change," *Southern Economic Journal*, 60, 1993, 220-234.

Gort, M., and R.A. Wall. "Foresight and Public Utility Regulation," *Journal of Political Economy*, 96, 1988, 177-188.

Gort, M., and R.A. Wall. "Obsolescence, Input Augmentation, and Growth Accounting," working paper, 1993.

Greene, W.H. *Econometric Analysis*, New York: Macmillan Publishing Company, 1993.

Gregory, R.G., and D.W. James. "Do New Factories Embody Best Practice Technology," *Economic Journal*, 83, 1973, 1133-55.

Guilkey, D., and C.A. Knox Lovell. "On the Flexibility of the Translog Approximation," *International Economic Review*, 21, 1980, 137-147.

Hall, R.E. "Technical Change and Capital from the Point of View of the Dual," *Review of Economic Studies*, 35, 1968, 35-46.

Harper, M.J., E.R. Berndt, and D.O. Wood. "Rates of Return and Capital Aggregation Using Alternative Rental Prices," In *Technology and Capital Formation*, ed. Jorgenson, D.W., and R. Landau. Cambridge: MIT Press.

Hart, O.D. "The Market Mechanism as an Incentive Scheme," *Bell Journal of Economics*, 14, 1983, 366-382.

Hermalin, B.F. "The Effects of Competition on Executive Behavior," *Rand Journal of Economics*, 23, 1992, 350-365.

Hulten, C.R. "The Measurement of Capital," In *Fifty Years of Economic Measurement*, ed. Berndt, E.R., and J. Triplett. Studies in Income and Wealth, Vol. 54, Chicago: University of Chicago Press, 1990, 119-152.

Hulten, C.R. "Growth Accounting When Technical Change is Embodied in Capital," *American Economic Review*, 82, 1992, 964-980.

Hunt, L.C., and E.L. Lynk. "Divestiture of Telecommunications in the UK: A Time Series Analysis," *Oxford Bulletin of Economics and Statistics*, 52, 1990, 229-251.

Intriligator, M.D. "Embodied Technical Change and Productivity in the United States 1929-1958," *Review of Economics and Statistics*, 47, 1965, 65-70.

Jorgenson, D.W. "The Embodiment Hypothesis," *Journal of Political Economy*, 74, 1966, 1-17.

Jorgenson, D.W., and Z. Griliches. "The Explanation of Productivity Change," *Review of Economic Studies*, 34, 1967, 249-283.

Jorgenson, D.W., F.M. Gollop, and B.M. Fraumeni. *Productivity and U.S. Economic Growth*, Cambridge: Harvard University Press, 1987.

Jorgenson, D.W, and R. Landau. ed. *Technology and Capital Formation*, Cambridge: MIT Press, 1989.

Kiss, F. "Productivity Gains in Bell Canada," In *Economic Analysis of Telecommunications*, ed. Courville, L., A. de Fontenay, and R. Dobell. Amsterdam: Elsevier Science Publishers B.V. (North-Holland), 1983.

Kiss, F., S. Karabadjian, and B.J. Lefebvre. "Economies of Scale and Scope in Bell Canada," In *Economic Analysis of Telecommunications*, ed. Courville, L., A. de Fontenay, and R. Dobell. Amsterdam: Elsevier Science Publishers B.V. (North-Holland), 1983.

Lynk, E.L. "Telecommunications Divestiture in the UK: A Crossed Line?" *Applied Economics*, 23,1991, 379-384.

MacAvoy, P.W., and K. Robinson. "Winning by Losing: The AT&T Settlement and Its Impact on Telecommunications," *Yale Journal on Regulation*, 1, 1983, 1-42.

Mathios, A.D., and R.P. Rogers. "The Impact of Alternative Forms of State Regulation of AT&T on Direct-Dial, Long-Distance Telephone Rates," *Rand Journal of Economics*, 20, 1989, 437-453.

Morrison, C.J. "Productivity Measurement with Non-Static Expectations and Varying Capacity Utilization," Journal of Econometrics, 33, 1986, 51-74.

Morrison, C.J., and E.R. Berndt. "Assessing the Productivity of Information Technology Equipment in U.S. Manufacturing Industries," National Bureau of Economic Research, No. 3582, 1991.

Nadiri, M.I., and M.A. Schankerman. "The Structure of Production, Technological Change, and the Rate of Growth of Total Factor Productivity in the U.S. Bell System," In *Productivity Measurement in Regulated Industries*, ed. Cowing, T.G., and R. Stevenson. New York: Academic Press, 1981.

Nadiri, M.I., and I.R. Prucha. "Dynamic Factor Models, Productivity Measurement, and Rates of Returns: Theory and an Empirical Application to the U.S. Bell System," National Bureau of Economic Research, Working Paper No. 3041, 1989.

Nelson, R.A., and Wohar, E. "Regulation, Scale Economies, and Productivity in Steam-Electric Generation," *International Economic Review*, 24, 1983, 57-78.

Nerlove, M. "Returns to Scale in Electricity Supply," In *Measurement in Economics: Studies in Honor of Yehuda Grunfeld*, ed. Christ, C.F. Stanford: Stanford University Press, 1963, 167-198.

Olley, G.S., and A. Pakes. "The Dynamics of Productivity in the Telecommunication Equipment Industry," The Center for Economic Studies, CES 92-2, 1992.

Oum, T. "The Effects of Competition on the Productivity Growth in the U.S. Public Telephone Market," Unitel Evidence Part 11, 1990.

Oum, T., and Y. Zhang. "Competition and Allocative Efficiency: The Case of the U.S. Telephone Industry," UBC Working Paper, No. 93-TRA-004, 1993.

Palmer, K. "A Test for Cross Subsidies in Local Telephone Rates: Do Business Customers Subsidize Residential Customers?" *Rand Journal of Economics*, 23, 1992, 415-431.

Panzar, J.C. "Technological Determinants of Firm and Industry Structure," In *Handbook of Industrial Organization*, ed. Schmalensee, R., and R.D. Willig. Amsterdam: North-Holland, 1992.

Röller, L.H. "Proper Quadratic Cost Functions with an Application to the Bell System," *Review of Economics and Statistics*, 72, 1990a, 202-210

Röller, L.H. "Modelling Cost Structure: The Bell System Revisited," *Applied Economics*, 22, 1990b, 1661-1674.

Schankerman, M. and M.I. Nadiri. "A Test of Static Equilibrium Models and Rates of Return to Quasi-Fixed Factors, with an Application to the Bell System," *Journal of Econometrics*, 33, 1986, 97-118.

Scharestein, D. "Product-Market Competition and Managerial Slack," *Rand Journal of Economics*, 19, 1988, 147-155.

Shepherd, W.G. "The Competitive Margin in Communications," In *Technological Change in Regulated Industries*, ed. Capron, W.M. Washington: Brookings Institution, 1971.

Shin, R.T., and J.S. Ying. "Unnatural Monopolies in Local Telephone," *Rand Journal of Economics*, 23, 1992, 171-183.

Solow, R.M. "Technical Change and the Aggregate Production Function," *Review of Economics and Statistics*, 39, 1957, 312-320.

Solow, R.M. "Investment and Technical Change," In *Mathematical Methods in the Social Sciences*, ed. Arrow, K., S. Karlin, and P. Suppes. Stanford University Press, 1960, 89-104.

Solow, R.M. "Technical Progress, Capital Formation, and Economic Growth," *American Economic Review*, Papers and Proceedings, 52, 1962, 76-86.

Stanbury, W.T. "The Case for Competition in Public Long Distance Telephone Service in Canada," Unitel Evidence Part 9, 1990.

Stevenson, R. "Measuring Technological Bias," *American Economic Review*, 70, 1980, 162-173.

Stevenson, R. "X-Inefficiency and Interfirm Rivalry: Evidence from the Electric Utility Industry," *Land Economics*, 58, 1982, 52-66.

Stigler, G.J. "The Theory of Economic Regulation," Bell Journal of Economics, 2, 1971, 3-21.

Sueyoshi, T., and P.C. Anselmo. "The Evans and Heckman Subadditivity Test: Comment," *American Economic Review*, 76, 1986, 854-855.

Taylor, W.E., and L.D. Taylor. "Post-divestiture Long-Distance Competition in the U.S.," *American Economic Review*, 83, 1993, 185-190.

Waverman, L. "U.S. Interexchange Competition," In *Changing the Rules: Technological Change, International Competition, and Regulation in Communications*, ed. Crandall, R.W., and R. Flamm. Washington: Brookings Institution, 1989.

Westfield, F.M. "Technical Progress and Returns to Scale," *Review of Economics and Statistics*, 48, 1966, 432-441.

Wickens, M.R. "Estimation of the Vintage Cobb-Douglas Production Function for the United States, 1900-1960," *Review of Economics and Statistics*, 52, 1970, 187-193.

Willig, R.D. "Contestable Market Theory and Regulatory Reform," In *Telecommunications Deregulation*, ed. Allison, J.R., and D.L. Thomas. New York: Quorum Books, 1990.

Ying, J.S., and R.T. Shin. "Costly Gains to Breaking Up: LECs and the Baby Bells," *Review of Economics and Statistics*, 75, 1993, 357-361.

Young, A.H., and J.C. Musgrave. "Estimation of Capital Stock in the United States," In *The Measurement of Capital*, ed. Usher, D. National Bureau of Economic Research, Chicago: The University of Chicage Press, 1980.

Index

Tiv Economy

NORTHWESTERN UNIVERSITY AFRICAN STUDIES

Number Twenty

Tiv Economy

Paul and Laura Bohannan

NORTHWESTERN UNIVERSITY PRESS
Evanston
1968

Library of Congress Catalog Card Number: 66–17013
Manufactured in the United States of America

This volume has been published with the aid of a grant from the Program of African Studies, Northwestern University

Paul Bohannan is Stanley G. Harris Professor of Social Science at Northwestern University. Laura Bohannan is Associate Professor of Anthropology at the University of Illinois at Chicago Circle.

To the memory of
KARL POLANYI
who might have disagreed with much of
it and thereby enriched it

Foreword

Parts of this book have been previously published in other forms. The major portions of the descriptive passages of Chapters 2 to 6 appeared in *Tiv Farm and Settlement,* published as Colonial Studies No. 15 by Her Majesty's Stationery Office in 1954. Only five hundred copies of that report were printed, and it has been out of print for many years. It was done in a format that made library storage and use all but impossible—two columns on foolscap paper. Paul Bohannan wishes to thank Her Majesty's Stationery Office for permission to reprint shortened versions.

The second half of Chapter 7, on Tiv migration, was originally published as an article in *Africa* (Paul Bohannan, 1954). We wish to thank the International African Institute, and Daryll Forde, its director, for permission to republish that material.

Chapters 8, 9, and 10 are further analyses of material which appeared in rough form in *Three Source Notebooks in Tiv Ethnography,* which we did for the Human Relations Area Files. Aside from its inclusion in the files, only twenty-five copies of that book were printed. We are grateful to the HRAF for permission to reprint this material, albeit much of it is in a more highly worked form and most of it has been shortened.

Chapters 11 to 15 are detailed elaborations and further analysis of our article in the 1957 Transactions of the New York Academy of Sciences. The descriptive material is greatly enlarged in this book, and the analysis has been made more extensive; some of these detailed descriptions were adapted from Laura Bohannan's D.Phil. thesis, where the market material was first considered by us. We—particularly Paul Bohannan—wish to put on record the degree of profit that has arisen from association with Karl Polanyi and George Dalton.

The basic information of Chapters 16 and 17 has appeared in the form of two articles (Paul Bohannan, 1956, 1960). Both owe much to the late Franz Baermann Steiner.

This book on *Tiv Economy* supersedes all the abovementioned publications.

Foreword

Twenty-eight months' field work among the Tiv between 1949–1953 was financed by the Social Science Research Council, the Wenner-Gren Foundation, the Colonial Social Science Research Council, and the Government of Nigeria, all of whom we wish to thank.

PAUL AND LAURA BOHANNAN

Northwestern University
Evanston, Illinois

Contents

List of Figures

List of Tables

Tiv Economy

1

The Tiv and Their Economy

Tiv Country

THE COUNTRY INHABITED by the greater portion of the 800,000 people who call themselves Tiv is an undulating plain sweeping down from the peak of Koloishe, just south of Tiv country, through more than a hundred miles to the broad basin of the Benue: from wooded foothills whose summits exceed 4,000 feet in altitude to the bare flat sandbanks in the river, about 300 feet above sea level. North of the Benue, Tiv are well started on a creeping expansion up a hundred miles of a similar plain that ends at the abrupt escarpment of the Bauchi Plateau.

Tiv country today stretches from about 6°30′N. to 8°N. and from 8°E. to 10°E. Although Tiv settlements are found far beyond these meridians to the north and east, the Tiv Division of northern Nigeria, where over three-fourths of the Tiv live, is contained within them. The modern system of roads and railways, following the sweep of the two plains, forms a node where Makurdi, on the bluffs along the southernmost point of the great Benue loop, overlooks the impressive railway bridge across "the Munshi narrows." [1]

The southern and eastern portion of the country, which Tiv call the "top of the country" (*ityough ki taregh*), lies among the foothills of the Cameroon highlands and of the Sonkwalla Mountains of Obudu District (of which Koloishe is the highest peak). Here the hills are cut by swift-running streams with densely forested banks, where the light is always green. Even in the heaviest harmattan the air is soft and moist. The open park savannah on the rounded ridges between the streams is dotted with locust bean and silk cotton trees.

The ridges of southern Tivland form the divide between the drainages of the Aya River—one of the two main tributaries of the Cross River—and the Katsina Ala River and other tributaries of the Benue. The southern marches of Tiv country lie at the boundary between the Northern and Eastern Regions of Nigeria. In some places this boundary can actually be seen: the fertile soil of Uge and Udam country, to the south, supports a dense bush vegetation of high grass and small trees between the larger trees. On the Tiv side, the grass is short, and all the smaller trees have been cut down to make room for the fields where a population that rises in some areas to a density of over five hundred and fifty persons per square mile struggles to wrest food from an overcultivated, constantly deteriorating soil.

Farther to the east, where Tiv bound the Cameroon, the population density is as little as thirty or forty persons to the square mile, and there is (Tiv say) an abundance of farmland and game, firewood and water.

These south and southeastern areas of Tiv country also receive the heaviest annual rainfall: about sixty-five inches a year, most of it falling during the rainy season which lasts from late April to early October, with a short respite during August, the wettest month being September. November is usually cool, with morning mists, preceding the dust-bearing harmattan from the Sahara which arrives in late November or early December. In March and early April tornadoes and electrical storms herald the return of the rains.

As you "come down" (*sen*) toward the northwest, the ridges and hills become more rounded; the country changes to a rolling downland, some areas of which have been almost completely denuded of the original orchard bush. In these areas, Tiv say that people must cook over tiny

1. Munshi is the Hausa name for Tiv.

fires of guinea corn stalks and cannot build fires to warm themselves during the chilly nights of the harmattan. Although Tiv do not consider these areas overpopulated, the fertility of the soil is deteriorating, apparently because of Tiv methods of agriculture and their large-scale destruction of the bush. Schemes for at least partial reforestation have been thwarted by factors of climate and of Tiv social organization (McIntosh 1940; Briggs 1941; Betts 1941).

Outcroppings of large granite hills, singly or in threes and fours, break the monotony of the plain as the traveler proceeds north. Mkar Hill, one of the largest, is a huge, saddle-shaped hill beneath which Protestant mission headquarters are to be found. Nearby is the town of Gboko, the administrative center of Tiv Division. Here also is a large Roman Catholic mission. Gboko is on high ground where the temperature may be as much as 10° F. lower than in the south and perhaps as much as 15° lower than in Makurdi. At Yandev, the agricultural station five miles from Gboko, the average mean temperatures in 1950 varied between 76° in November and 85° in April, though diurnal variations of 40° are not uncommon.

The Benue River divides Tiv country, and the southern portion—by far the largest—is further bisected by the Katsina Ala River. Jukun have several riverside settlements in Tiv country (Abinsi is the largest). Both rivers flood during the rains and quite regularly inundate large territories along their banks. Although some Tiv in these areas have learned from Jukun and Chamba to make and use dugout canoes, Tiv are not at home on the water; where there are no large streams, as in the south, we estimate that over half the men cannot swim. Automobile ferries cross these rivers, but most Tiv cross in the canoes of Jukun and Chamba.

The country north of the Benue tends to be hilly, and is still covered for the most part with dense orchard bush in which antelope and other game are quite plentiful.

Tiv outside Tiv Division live intermingled with Hausa and pagan peoples in Nasarawa and Lafia divisions. Over half of the subjects of the Aka Uku of Wukari, the King of the Jukun to the east of Tiv country, are Tiv; Tiv are pressing in among and intermingling with the Chamba of Donga farther south and east. Tiv have settled as far north as Shendam Division. Many Tiv have migrated south into the Eastern Region and set up their homesteads among the various small tribes of Ogoja (known to Tiv collectively as the Udam), usually marrying Udam women, although they will not allow their women to marry Udam men. A few Tiv settlements are found among the Idoma, to the west, and even beyond it in Igalla country. This expansion is one of the most notable features of Tiv society (see Chap. 7).

Tiv Economy

THE TIV ARE FARMERS. They produce and trade what they eat and wear; they build their own houses and make most of their own weapons, including guns, for which they import only the barrels. Thus, they practice what is commonly called a subsistence economy.

The plains of Tivland are comparatively fertile, and in most parts of the country about an acre per person yields adequate subsistence for the entire year. Their farms—cleared and tilled with the short hoe, the matchet, and the digging stick (see Chap. 8)—are in production for two or, in some areas, three years. The first year the fields are mounded to produce a crop of yams, with various side crops such as okra, corn, peanuts, and greens grown along the sides of the mounds. The second year the same field is leveled and produces a crop of early millet and another of sorghum. It may be planted to beniseed (*sesamum indicum*), the chief cash crop, in the early part of the third growing season, although the beniseed may be interplanted with millet, or even replace the millet in the second year. After such a rotation the land is allowed to lie fallow until the wild plant life and earthworms it sustains show that it is rejuvenated. In some areas, notably in the south, land shortage has made necessary an increase in the number of years of cropping and has reduced the number of years of fallow.

From 1949 to 1953, when the data for this study were gathered, Tiv were not dependent to any appreciable degree on foreign goods, although European cloth and crockery, kerosene and matches, "Essence de geranium, non-alcoholic, made in Hackney, England," and an occasional pencil and exercise book were fairly common luxuries. Tiv were, however, by that time thoroughly accustomed to money as a medium of exchange and as a standard of value even in market transactions in which it was not a medium of exchange. They paid their tax and their bridewealth in money. They expressed disquiet about general-purpose money in an unformulated distrust for money or men who sought money (see Chap. 16).

There are two important indicators of the differences between "modern" industrial economies and subsistence economies. First, even where (as is the case with Tiv) there is an active output market, subsistence economies have no factor markets—that is to say, land, labor, capital, and entrepreneurial activity do not enter the market. Second, in subsistence economies, production is the task of the general institutions of the society, such as family and lineage, rather than of specialized organizations such as "firms."

The second point, despite its importance, requires only passing comment. To Tiv (and other peoples with subsistence economies) the activities which lead to production are part of the duties that form that complex network of rights and obligations which is the kinship, family, and sometimes the religious and political structure. The result is that the market place has the same immediacy to both producer and consumer that it has only to the ultimate consumer in our own system. Just as one goes to market to buy what one needs to use, so one goes oneself to market to sell what one has oneself produced.

It is, however, the lack of factor markets that is the more crucial point. Management is not an important factor. Tiv "capital" is meager. Producer's capital such as agricultural implements differs little from short-lived consumer goods. These implements may be made by the person who uses them (or by a kinsman), or they may be bought on the market from part-time blacksmiths. Tiv cannot amass consumer's capital because, given their climate and their technology, nothing lasts long enough. Prior to establishment of colonial government, Tiv had no coinage and no precise credit mechanism—let alone more elaborate financial institutions. Although they of course consider the natural increase of animals and plants important to livelihood, they do not invest any sort of capital in order to increase it, nor is there any institutionalized means by which they might do so.

Until the middle 1940's, after which a few Tiv became professional traders, there were no entrepreneurs. Economic activities, like any others, were organized by the father of the family, by the compound head, or in some cases by religious practitioners or men of political influence in the course of their familial and political duties. The institutionalization of land and labor was more highly developed.

Land

Tiv are primarily interested in land as the spatial dimension of social relationships; its part in any "property" system is secondary. Every Tiv has a right to a farm within the territory of his agnatic lineage, of whatever genealogical depth (see Chap. 6). This right is a condition of his agnation—a right of citizenship. The location of the farm is delimited not geographically, but socially.

Moreover, every man has a right not merely to a farm but to *sufficient* farmland. Every man has a responsibility to supply each of his wives with a farm; every compound head must see to it that each man of his compound can fulfill that obligation. In the Tiv form of rotational agriculture, a man does not always—or even often, except in areas of land shortage—put his farms in precisely the spot they occupied during a former rotation. Because of the short period during which a single farm is used, there is a constant shift in the actual position of a man's farms. His neighbors, however, do not change. As one man's family grows and another's shrinks, the plots are adjusted to fit the needs of each man. The actual mechanics of this adjustment do not always work smoothly; in areas in which there is even a slight land shortage, boundary litigation takes a disproportionate amount of time of both the farmers and the men who judge their disputes.

This system of land allotment might be summed up by saying that Tiv have "farm tenure" for the life of a single crop cycle, that is, for two or three years. They do not have "land tenure." Their "security of tenure" lies in the social organization, with its promise of sufficient "means of production."

Tiv land practice leads to a specific type of migration (see Chap. 7): as every small lineage area expands, so must each inclusive lineage area expand. Therefore, the whole of Tivland is expanding with every demand for more land. Toward the edges of the country, each small lineage not only expands but has a specific direction of migration. Until this system collided with enforceable British ideas of fixed boundaries, the lineages on the outside were proceeding four or five miles a generation against the Udam peoples in the south, and very much more rapidly than that in the north and east where little opposition confronted them. Tiv subdivide internally in a cell-like growth which necessarily pushes Tiv boundaries out in all directions. It is this outward spread which prevents the sort of land fragmentation so acute in some parts of Africa. While this practice solves the internal problem, on a wider canvas it leads to great disruption.

Land is not property among the Tiv. It cannot be sold—that would be tantamount to selling a genealogical position. It cannot be rented, for one's right to it depends on kinship status and residence. Tiv have precise rules about inheritance of cloth, weapons, and wives; they do not need such rules for land because all have rights already.

Land is a factor of production in Tivland as it is everywhere else. But that is, for Tiv, not its most obvious characteristic. It is, for them, the dimensional aspect of social organization. Status gives the right to exploit the land on which one finds oneself, living in one's social group.

Labor

Labor, like land, is an aspect of social organization. Production of food is an activity in which the whole family takes part. The rights and duties involved in work relationships are seen in terms of family and lineage values, not, as in the West, in terms of "labor" values and sanctions.

Within the family, labor in farm tasks is assigned on the basis of sex, age, prestige, and other considerations. Tiv do not, as a matter of fact, have a word that can properly be translated as "labor." The word *tom,* usually rendered "work," actually applies to fulfilling the duties of one's role.

A labor market is an institution whose primary value system is in terms of production and of roles which are linked by rights and duties bearing on the production process, the *quid pro quo* usually being money. Tiv have no such institution; rather, production is carried out in institutions whose primary value system is that of family, kinship, and neighborhood. Tiv work hard because they work as fathers, sons, brothers, and husbands, and so retain a right of the reciprocal work of their families. Today some Tiv go south to work for wages for a few months, although Tiv do not have a serious "labor migration" problem and only a very few young men are gone from the community at any one time. They are thus entering the labor market of the wider economy, but they have no labor market among themselves.

Markets

The absence of a factor market, and hence of a "market economy," means only that in subsistence economies a major part of subsistence does not enter markets and that the market is not responsible for the allocation of factors of production. It certainly does not mean that market places are absent, or that exchange according to market principles is unknown.

Tiv support an active "peripheral market"—a term that is meant to imply that it is recognizably a market, with all the characteristics of a market, but one that is not central to either economic or social integrity and prosperity.

The Tiv market is called *kasua*,[2] a word derived ultimately from Arabic and immediately from Hausa *kasuwa* (although the Hausa accent the first syllable, Tiv accent the second). Hausa are important to Tiv markets today, and even in the last century sometimes traded in Tivland—apparently the only foreigners who did. Still it seems doubtful that Tiv imported the institution of the market from the Hausa, as organizational and ideational details of Tiv markets resemble those of their southern neighbors more than they resemble those of the Hausa.

Kasua in Tiv is a word with several meanings. *Market* in English also has several meanings. To eliminate some of the ambiguity of translation we shall place the meanings of the two words in parallel columns. The Tiv examples are all taken from our notes; the English examples are from dictionaries or from economics texts.

Kasua	*Market*
1. A meeting of people in a place where goods are bought and sold. (*U za kasua?* "Are you going to market?" is a standard greeting on market day.)	1. A meeting of people. (Sometimes expressed as a meeting of buyers and sellers.)
2. The open space where goods are exposed for sale—the market place. (*Nyian m zende kasua tsegher.* "Today I am only walking about the market place" [instead of selling].)	2. An open space or covered building in which cattle, provisions, and the like are exposed for sale.
3. A right—that can be owned and inherited—to establish and operate such a place or	3. "The privilege granted to the lord of the manor, municipality, or other body, to establish

2. We hear this word *kasoa*, and have published it that way in all previous publications. However, D. W. Arnott of the School of Oriental and African Studies points out that to be in accordance with Tiv phonetic pattern, the *ua* is uniform whereas the *oa* would be unique. Since our clerks wrote *ua* almost as often as *oa*, and since the dictionaries write *ua*, we are falling into line.

	Kasua		*Market*
	hold such a meeting. (*Ka kasua u Kyagba.* "It is Kyagba's market.")		a meeting of persons to buy and sell" (*Shorter Oxford English Dictionary*).
4.	An act of exchange. (*Ka kasua ga.* "I am not selling it" [it is a gift].)	4.	The action or business of buying and selling; a purchase or sale (SOED).
5.	Ability or luck in selling and bargaining. (*Kasua wam doo.* "My market [luck] was good.")	5.	A bargain. Obsolete except in some set phrases (SOED).
6.	An item for sale. (*M ngu a kasua.* "I have something to sell.")	6.	(No parallel meaning.)
7.	A day in the five-day "week" cycle, or the five-day period itself. (*Nyian ka kasua nana?* "What market is today?" *M ve kasua uhar.* "I came two markets ago.")	7.	(No parallel meaning.)
8.	Price or exchange value. (*Kasua na nana?* "What is its market price?")	8.	Price in the market, market value (SOED).
9.	(No parallel meaning.)	9.	Sale as controlled by supply and demand; hence, demand (SOED).
10.	(No parallel meaning.)	10.	Opportunity of buying and selling.

In Western economics the term *market* has been given special meanings, illustrated by definitions 9 and 10 above. Or, more precisely, formal economics has discovered characteristics of markets which other peoples do not recognize. On the other hand, Tiv have extended their word *kasua* so that it plays an important part in their cosmological concepts (definition 7) and religious beliefs.

Whereas the price of foodstuffs and other subsistence (relatively little of which enters the market) follows familiar patterns, with a market-regulated price and a flow of goods from areas of low price to areas of high price, the absence of factor markets and the peripheral quality of output markets among Tiv is vividly demonstrated by the major importance in vast areas of their lives of non-market principles of exchange. Tiv exchange "prestige goods" and rights in women according to non-market principles. These two spheres of exchange were, in the traditional system, kept distinct from market-exchange of subsistence. The institutional overlapping which took place is analyzed in Chapters 11 and 12. Whereas most modern industrial economies are made "unicentric" by the presence of general-purpose money and the organizing force of the market, or a political equivalent thereof, Tiv economy is of a type that is manifestly "multicentric." The various spheres surrounding these multiple centers require a type of institutional correlation that can best be called "internal conversion": that is, within a single multicentric economy there is a type of activity which resembles conversion between two currency-defined economies in the Western world.

Economies based on conversion cannot be sufficiently described by mere negative definition. Their complexities require to be analyzed in their own terms.

2

Householding: The Domestic Unit (*Ya*)

"DOMESTIC ECONOMY" might indeed be defined as the art or science of consumption. Domestic groups everywhere are the "ultimate consumers." In a Western or Soviet type of economy the family is linked to the economy by two devices: First, the provender of the domestic economy comes from the wider economy either through the market or via more direct political allocation. Second, at least one member of each domestic unit leaves it specifically in order to acquire the means to pay for, or otherwise to obtain, the provender. Production is linked to consumption in an institutionally complex way.

In subsistence economies, however, the bulk of the product is consumed by the producers. The domestic unit becomes a general-purpose institution, its economic activities what Karl Polanyi, in his analysis of Dahomean economy (1966), called "householding." Householding is a mode of production, comparable at one level with peasant production for a market, with industrial firms producing for a market or for government quotas, and so forth. Modes of production may be conjoined in any specific instance—householding with cottage industry, for example.

In subsistence economies like that of the Tiv, the organization of the household is directly related to the total economy: production is planned and achieved for immediate consumption; allocation is planned and achieved within the same group, or between groups of the same sort.

To the economist concerned with modern Western or Soviet types of societies, the question of family organization is peripheral. Where, as among the Tiv, the primary institutionalized means by which people obtain their subsistence is the domestic unit, the description and analysis of the "householding" group is central to any understanding of the organizational principles of that economy.

The Compound

TIV LIVE in circular huts having what are said to be the thinnest mud walls in Northern Nigeria, pounded mud floors, built-in fireplaces, and sometimes built-in storage and drying platforms of hard wood. The hut is thatched with sword grass (*imperata cylindrica*) braided into long strips and wound around the conical roof from bottom to top. Though some huts may have a few slits in the walls, and others neck-sherds built in as chicken entrances, most have but a single opening: the door, usually placed a foot or two above the level of the ground. The overhanging eaves cover a narrow veranda where children play and a wife may set up her large granite grinding slab.

Every woman who is a senior wife, or who has daughters-in-law, has a granary. In most parts of Tivland, every married woman ideally has her own hut, at least after she has borne her husband a child, although in some areas a junior daughter-in-law continues to share a hut with her mother-in-law until a new wife is taken and, in turn, becomes junior daughter-in-law. The ideal is often not realized; a man never refuses to build his wife a hut, but he may procrastinate. Other men build a hut for each wife and still another for themselves. Sometimes a young man builds a rectangular house or a very large round hut containing two or more rooms, each wife having a room of her own.

In addition, elders and some important women have reception huts (*ate*) with two or more doors and somewhat lower walls than a sleeping hut; the roof is mounted on posts to allow more light to enter. Here Tiv eat, lounge, receive guests, and carry on business.

Huts and granaries are placed in an oval or circle—perhaps an arc if the compound is a small one—with the reception huts inside the oval (Fig. 1). This group of huts is called a "compound" (*ya;* plural: *uya*); it may be composed of as few as two or three or as many as forty huts. Usually in the south there are seven or eight huts in a *ya*, in northern

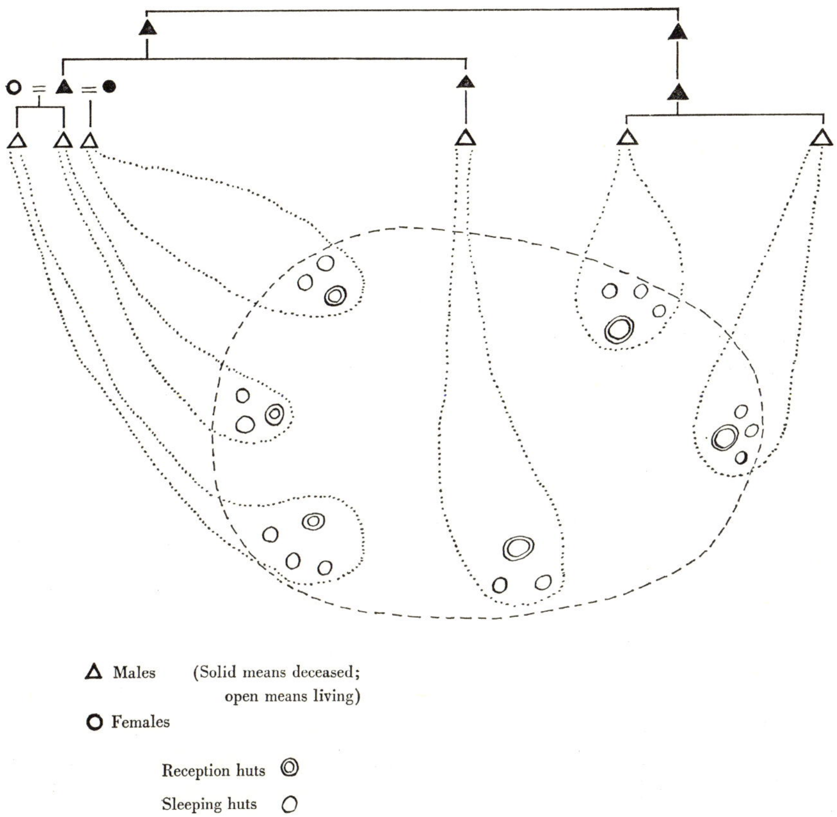

Fig. 1.—"Genealogical map" of a Tiv compound.

areas a few more. A cleared space is left in the middle of the ring of huts; this central open space (*tembe*) is the center of Tiv family life.

Beyond the huts and granaries is a circle of pawpaw trees, mangoes or bamboo. This area around the outskirts of the compound (the *ikongo*) is not only the garden where tobacco, vegetables, and useful medicines are grown but also provides the Tiv with a symbol of privacy.

Most of the myriad paths which crisscross Tiv country lead from one compound to another, although there are a few main paths which people avoid when choosing sites for their compounds. Seen from the air, the paths form a vast network covering the whole of Tiv country, with compounds forming the nodes. This arrangement, in which each compound appears as a hub in communication patterns, is unique in this part of Nigeria.

Members of the Compound

THE OLDEST MAN in the compound is called the *or u ya*. This means literally "the man (or person) of the compound" and will be translated "compound head." The compound head is always the oldest man in the compound. Ideally the recognized compound head is the man who actually runs the compound, but such is not always the case.

The compound head has a certain amount of authority within the compound. He must, for example, be consulted before a new hut is built; he is unlikely to object, however, if everyone else in the compound agrees. The compound head usually (but not in every case) coordinates the farming activities of the people of his compound, although every male member has a voice in the matter. If a man hoes an extra patch of beniseed or cassava for himself, the compound head is informed and agrees to the location. Tiv say this is necessary to prevent accusations of the man's working for himself rather than for the unit of which he is a part.

The compound head usually, but not always, controls most of the magical forces (*akombo*) that are considered important to farm prosperity and fertility. He is the usual arbiter of disputes within the compound; so long as his decisions are considered just, they are enforced. His most powerful weapon is influence; without it he finds himself deserted.

Tiv have no phrase for "compound member." Any man who lives in a compound can be spoken of, in compound context but not necessarily in any other, as the "youngster" (*wanye*) or sometimes as the "child" (*wan*) of the compound head. All the women married to men of the compound are referred to as "my wife" (*kwase wam*) both by their husbands and by the compound head, and as "our wife" (*kwase wase*) by the other members. Children can be called "my child" (*wan wam*) by the compound head and by most of the adults in the compound. All Tiv are, however, generally referred to and addressed by name. The people of a compound can be called the "people" (*iyor*) of the compound head, and he sometimes refers to them as "my people."

The individuals in any given compound may change rapidly and often. A few young men spend several years doing part-time trading, being at home perhaps half of the time. Children are encouraged to visit their kinsmen in other areas. Men move from their agnatic lineages to their mothers' agnatic lineages and on to their fathers' mothers' agnatic lineages quite readily. There is a constant shift in the population of the compound.

The people living in a compound usually form an extended patrilineal polygynous family consisting of an elder (the compound head), his wives, minor children, unmarried adult daughters, his adult sons and their wives and children. As often as not, there are full or half-brothers of the compound head living there with their wives and children. In some cases, there may be a sister's son, with his family, temporarily or permanently resident with his mother's people.

Wives come to live with their husbands in whatever compound the men are domiciled at the time of the marriage.

There may also be "guests" or "strangers" (*or va inya*) living in the compound: friends or age-mates of the male members, whose temporary residence endures a few days, weeks, months, or years. These strangers seldom (Tiv say never) marry into the compound; Tiv say that a disciplined wife and uxorilocality are mutually exclusive. A stranger must have permission of the compound head to live there, and at least the tacit consent of all compound members. He must do his share of farm work, build his own huts if he stays for long, and keep his part of the compound clean. He has only a few rights and a few duties.

Although Tiv say that the people of a compound are close kin (*angbianev*) and cite genealogical evidence, it is most important to emphasize that Tiv describe a compound not in terms of family or kinship organization, but in terms of status within the compound: its head, its "youngsters," its women (or wives), its children, and its strangers. To Tiv, a compound is a group of persons living together, who have certain rights and duties to others within the group. Sanctions and rationalizations for many of the rights and duties of compound members are phrased in terms of family and kinship, others in the values of production and consumption, witchcraft and religion. One may call a compound a "kinship group" or a "production group" only when emphasizing one quality of many.

All compound members are there "because their hearts tell them." The residential sanctions and rules elicited by the inquirer are statistical—not normative.

There are two related meanings of the word *ya*, here translated as "compound." It means the site and the huts and the cleared space in the center. It also means the people who live in it, the relationships between them, and the activities that make up not only the substance of their relationships but also the economy, the polity, and the family.

The Size and Plan of the Compound

WE GATHERED population data for 274 compounds from southern Tivland (see Table 1), in the lineage segments of MbaItyo of MbaDuku (221 compounds) and of Iyon of Shangev MbaShaya (53 compounds). A discussion of these data is followed by comparative statements from northwestern Tivland.

The largest compound in the southern sample contains 80 persons, and the smallest 2 persons (though one "mad man" was living entirely alone). The average number of persons per compound is 17, the median compound having 14. However, more than half of them live in compounds of 25 persons or more.

In this sample, the greatest number of adult (tax-paying) males in any compound is 21, although a number of compounds have only one adult male. The average number of adult males per compound is a little over 5, the median compound containing 4, though more than half of the total of 1,448 adult men in the sample live in compounds which contain 7 or more men.

The largest compound also contains 34 adult women—the largest number found. Several compounds have only one woman, and two have none at all.[1] The average number of women per compound is a little

1. One of these is a "mad man" who lives alone. The other is a compound of three huts in which an old man lives alone with his son of about twenty. He is jovial and well liked; he told us he had had several wives, but was weary of women, and that he would be getting his son a wife soon.

TABLE 1
Genealogical Chart of Tiv Lineages

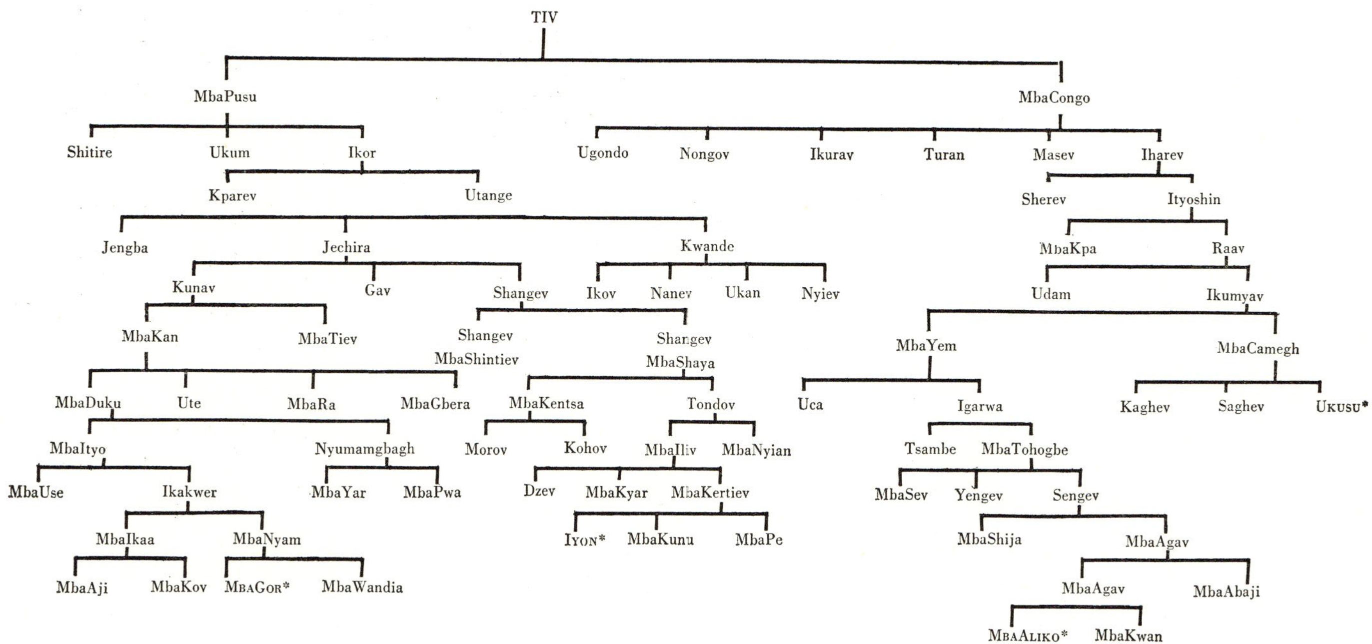

* Those lineages with which this report is primarily concerned.

over 6, the median compound having 5, but over half of the women in the sample live in compounds with 9 women or more.

Compounds in the northern and northwestern parts of Tivland are much larger than those in the south. Of 17 compounds of two different lineage segments of Iharev/Raav (see Table 1), the largest contain 139 persons, the smallest 9. The people of Iharev attribute the larger size of compounds in their area to the absence of land shortage and of witchcraft.

Tiv refer to one end of their compound as the "top" or "head" (*ityough*), to the other end as the "bottom" or "base" (*ityo*), and to the portion between as the "sides" (*mbusa*). The compound head most often has his reception hut at the base of the compound, while other members fill in the arcs of a circle or oval along the sides.

Tiv describe compound topography in terms of social relationships. Any compound containing people that extend beyond the nuclear family has within it sections associated with recognized genealogical cleavages. The relationships of the adult males determine hut placement. Reduced to its simplest terms, the compound is a spatial reflection of principles associating kinship and domestic grouping which may serve for exclusion as well as for inclusion. "We have one father but different mothers" is a reason for living together but also a reason for living apart. Having the same mother can serve as the sanction for full brothers' living together. Per se it cannot be used as a justification for their living apart—for that purpose emphasis is changed to "Our mother bore us [at] two [different times]" (*Ngo mar se uhar*). If two men's fathers had one mother, that fact can be their charter for living together if they so desire. But with different emphasis, that the fathers were "two men" is sufficient reason for their sons' establishing separate compounds.

The internal organization of the compound is seen by the people themselves as a spatial arrangement of the identifications and distinctions inherent in the agnatic polygynous family. Outsiders are attached to their nearest kinsmen for residence. Sisters' sons place their huts beside their mothers' full brothers (if there are none, by their mothers' half-brothers). A man whose mother was a member of some compound member's agnatic lineage will build next to that kinsman. A non-kinsman living in a compound builds his hut beside an age-mate.

A good example of these arrangements is to be found in Figure 2, which gives details for Kyagba's compound. In Figure 2, the numbers on the genealogical chart correspond with those on the sketches of com-

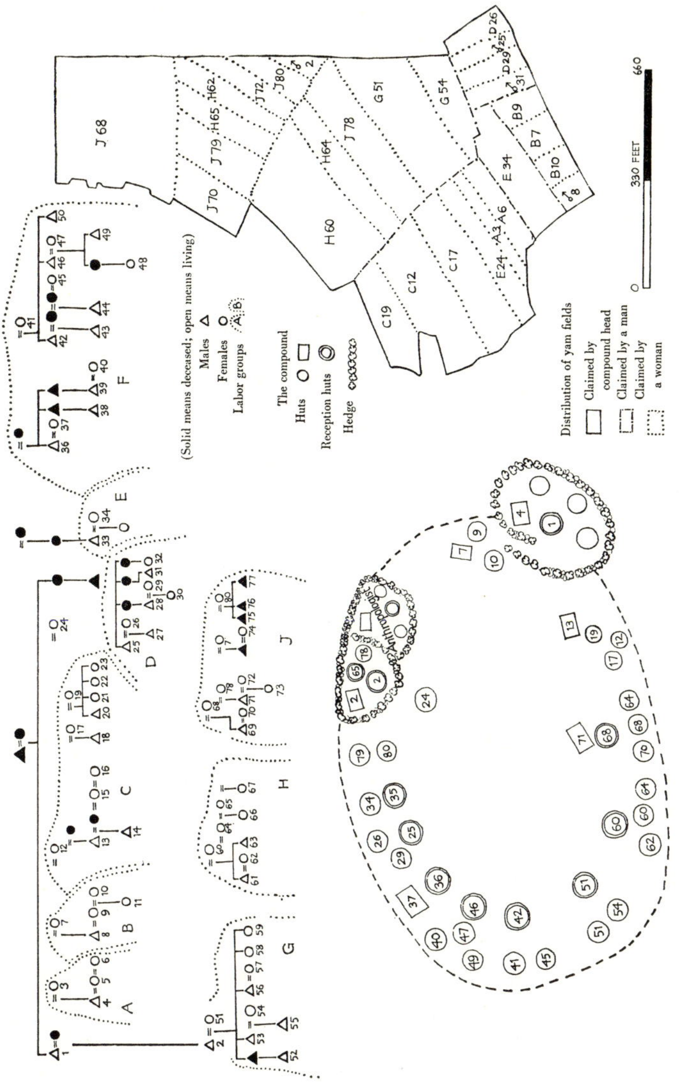

Fig. 2.—Kyagba's compound in MbaGor.

pound and farmland. Four specimen compounds from MbaAliko in northern Tivland are given in Figure 3. (A map of the farmland appears in Fig. 13.)

The genealogy, spreading backward in time, supplies not only the organizational basis for the compound but also the means for relating one compound to another. The compound therefore naturally reflects changes in relationships of the people occupying it. New huts are added for new wives. The death of a compound head also is reflected in compound topography. In one case where a son succeeded his father, he began to use the reception hut of his dead father very soon, but he allowed most of the houses of his father's wives to fall into ruin. Then, at the end of about two years' time, new huts were built for these wives among the clusters of huts of their levirs. The compound head himself repaired the reception hut of his father—it was now called his—and built new huts surrounding it for all his wives. In the rebuilding, the positions of huts were readjusted so that the ground plan corresponded precisely with the new genealogical make-up of the compound.

A suitable length of time is allowed to expire (usually about a year) before the new compound head takes on all of the rights and duties of his predecessor, but when he does so, in the course of keeping the compound in repair, the arrangement of the people within it is readjusted to conform with the new situation.

Fission of the Compound

ALL TIV EXPRESS and enact the process of social fission in terms of land for living and farming. Compound fission occurs in two contingencies: in the process of that particular kind of migration called "going to the farm" (see Chap. 7), and in the readjustment following the death of a compound head. We shall consider the latter first.

Tiv usually deny that there is in principle any connection between the death of a compound head and the fission of the compound, yet in any specific instance they stress the coincidence. By Tiv values, social groups, including compounds, should cohere; their fission is considered a poor reflection on their members and a sure indication that someone (in this case the new compound head) failed through inability or ill-will. This feeling has such force that it is unusual for a compound to split during the first year after the death of its head. During this year, everyone in the

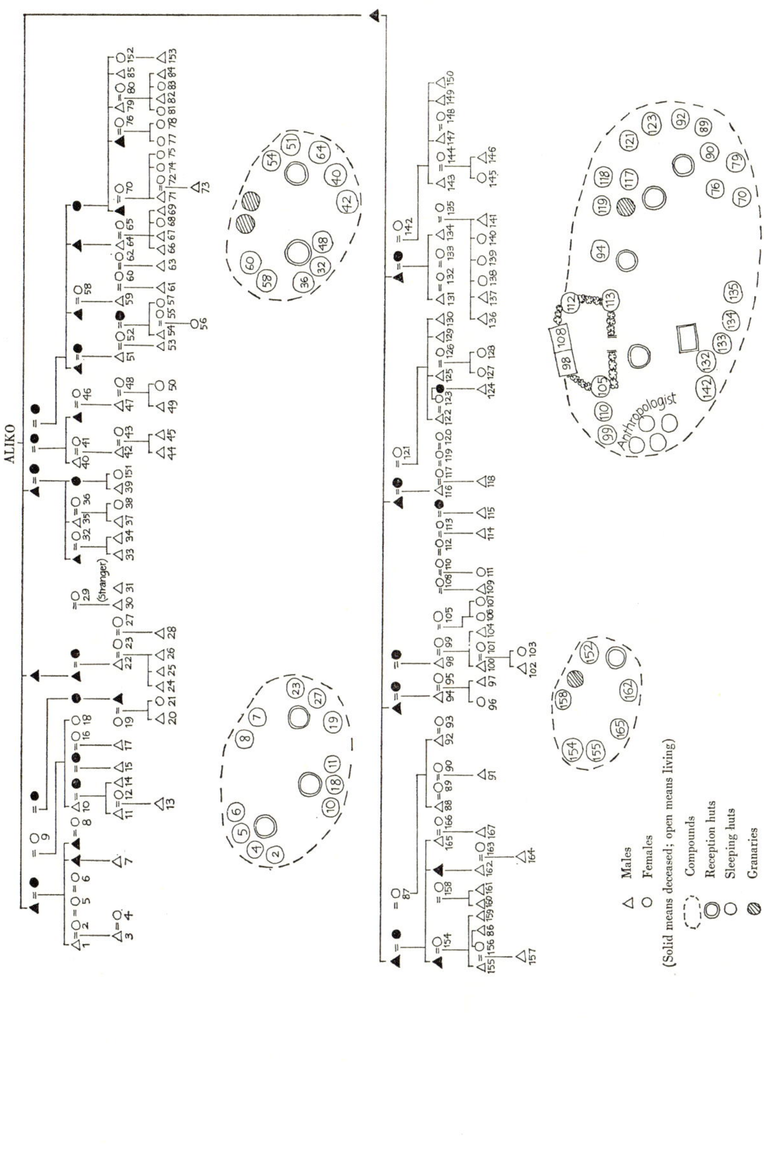

Fig. 3.—Compounds of MbaAliko.

compound and even in the community gives great verbal stress to the unity of the compound and tries to ignore or adjudicate any tensions that arise. After about a year, in the next appropriate building season (January to April), the compound is either repaired and remains a single unit or it splits at that time.

There are two possible courses of action in the splitting of a compound after the death of its head. The most common and amicable way is for all to continue to refer to the compound as that of the dead man and to tell the inquirer that the compound has no head, that he died. The new compound head, who is the oldest remaining male, and one or two other mature men in the compound begin to attend meetings of elders and to speak in moots (Paul Bohannan 1957). Thus, before a split has actually taken place, two or three incipient compound heads, each publicly controlling a particular segment of the compound, join in saying that there is no head at all.

But if witchcraft accusations and counteraccusations at the death of the old compound head have so embittered the members that an immediate split is unavoidable, each group (usually sets of full brothers and their families) goes to live with its mother's lineage for a year or two before returning to build new compounds.

Unless one of the points of contention which led to separation was land, or a matter dealing with farm work, it is unlikely that the two compounds will divide their fields. If both are small they may farm as a unit. They are "sitting separately" but they still have "one field." It may be as much as a generation or two later that the people speak of having separate fields as well as separate compounds.

When a compound splits, that part of it which moves out is said to be "going to the farm." The term indicates a general value among all Tiv and it includes in its meaning a need for new, better, and sufficient farmland, the difficulties of adjustment following the death of a compound head, and escape from a despot.

3

The *Tar*

To THE CASUAL OBSERVER there is no order in the scatter of Tiv compounds spreading as far as the eye can see. One can walk for miles through many, particularly southern, parts of Tiv country without ever being out of sight of at least one compound. Tiv, however, see their country as divided up in terms of lineage segmentation. Genealogies that connect and divide people into compounds are extended to divide and connect them into larger geographical units. There is, Tiv believe, a genealogy continuing back for fifteen or eighteen generations to "Tiv," the original ancestor who begot them all. The truth or falsity of the genealogy is contextual, for the groups and institutions associated with it and expressed in terms of it are not all of the same order. The more recent or "lower reaches" of the genealogy concern domestic grouping, which we considered in the last chapter. The "middle reaches" concern groups which are activated for certain ritual and marital purposes. The "upper reaches" of the genealogy form the basis for the territorial grouping of lineages.

Tiv always discuss matters of this sort in terms of actual examples;

indeed, to discuss them without actual examples is difficult. The example taken here is MbaGor/MbaDuku, which contains the compound of Kyagba, the oldest living male in his lineage, represented with some of his ascendants in Table 2 (for his descendants see Fig. 2). Kyagba

TABLE 2
DESCENDANTS OF GOR

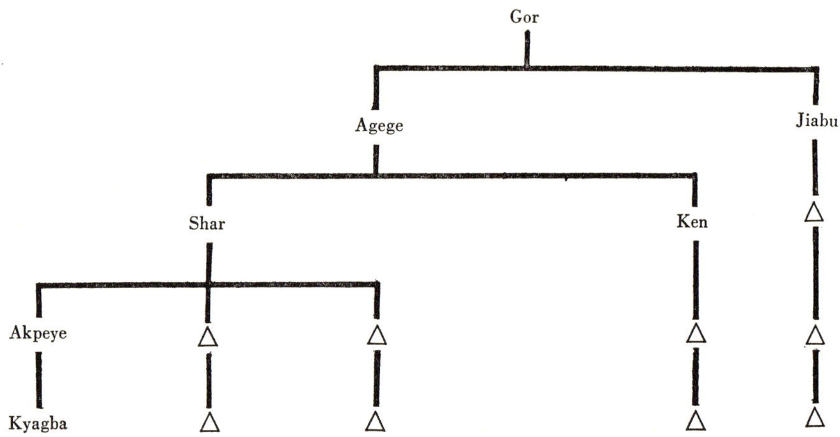

could, and often did, name 15 lineal ancestors stretching from himself back to "Tiv." All of the male descendants of Kyagba's father, Akpeye, happen to live in Kyagba's compound. But the other descendants of his grandfather, Shar, are numerous and scattered in several compounds nearby. The descendants of Shar's brother, Ken, are also living in the neighborhood. Their common father, Agege, had a brother named Jiabu whose descendants are also nearby. Agege and Jiabu had a common father, Gor. Those of Gor's descendants through males now living in the area of Kyagba's compound are divided into twenty-two compounds (Fig. 4). The members of each of these compounds, comprising the agnatic descendants of Gor, their wives, a few of their matrilateral kinsmen, and their "strangers," all have farms within a mile or so of their compounds.

Tiv call the collective descendants of Gor (including daughters, but not children of these daughters) "those of Gor" or "the people of Gor" —MbaGor. The territory within which these people live and have their

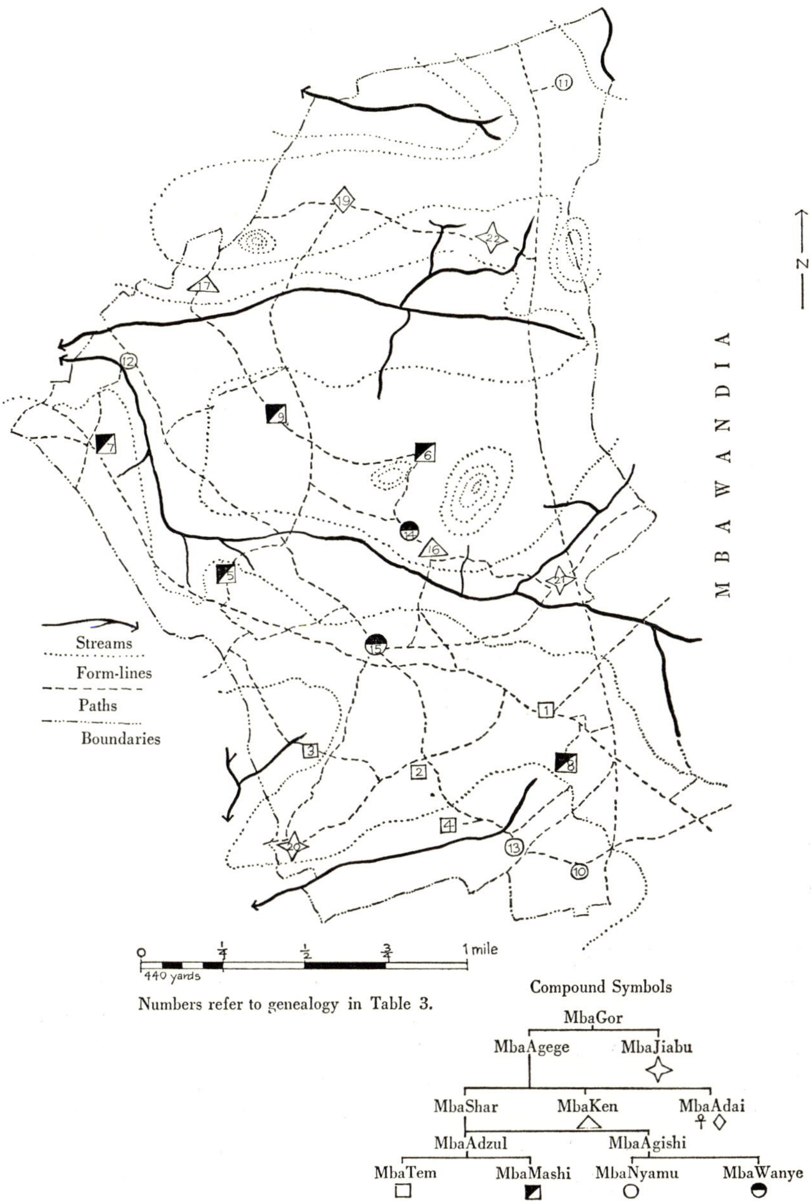

Fig. 4.—MbaGor of Kunav/MbaDuku.

farms is also called MbaGor. MbaGor is a discrete territorial unit. It is possible to draw a line around it at any given moment. Within the territory of MbaGor all the farms belong to descendants of Gor or to their kinsmen or to "their strangers." MbaGor is, however, the smallest lineage to be associated with a territory; none of the farms and compounds of the descendants of any single descendant of Gor forms a discrete territorial block.

Gor's father is said to have been a man named Nyam. Besides Gor, Nyam begot Wandia. The descendants of Wandia, with their kinsmen and strangers, live and farm in a discrete territorial block beside MbaGor. Both people and territory are called MbaWandia (Fig. 5). MbaGor and MbaWandia, taken together, are called MbaNyam after their common ancestor. Their combined territories are also called MbaNyam.

Nyam's father was named Ikakwer. Besides Nyam, Ikakwer had a child named Ikaa. The descendants of Ikaa live in a discrete territory (MbaIkaa) adjacent to MbaNyam. Both the combined descendants and the combined territories of MbaNyam and MbaIkaa are called Ikakwer.

Such description can be continued through another eight or ten generations. The collective agnatic descendants of any *ancestor* of Gor can be called by his name, and live as a discrete territory called by the same name. (Exceptions will be dealt with below.) Tiv may refer to a single patch of land as MbaGor, MbaNyam, Ikakwer, and so on, following up the genealogy to Tiv, the original ancestor (see Table 1). But each time one takes a further step back in the genealogy, the size and population of the relevant area are increased until all Tivland is included and all Tiv with it.

The territory of MbaGor has no name *except* MbaGor; within it only the streams and hills are named. No territory has a name other than a set of lineage names. Tiv place names are entirely dependent on social and political organization.

Tiv have a word for the territory occupied by the descendants of a single ancestor: *tar*, plural *utar*. Therefore, the territory of MbaGor can be called *tar u MbaGor*. The *tar* of MbaGor and the *tar* of MbaWandia, taken together, then, form the *tar* of MbaNyam. The *tar* of MbaNyam and that of MbaIkaa bound one another, and taken together they can be called the *tar* Ikakwer. The genealogy continues. Each increasingly large *tar* is called by the name of the agnatic ancestor of the people living in it until we reach *Tar Tiv*, as Tiv call their country (see Fig. 6, Table 1).

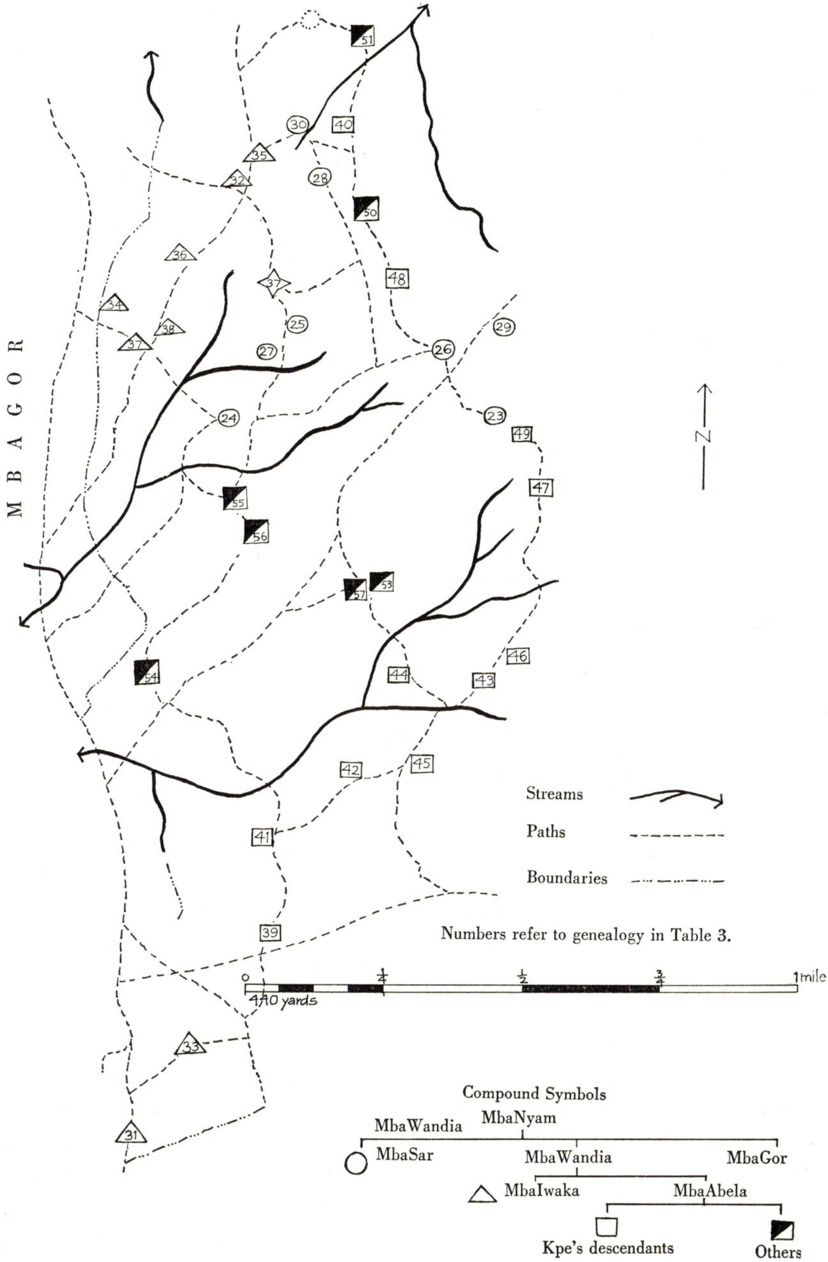

Fig. 5.—MbaWandia of Kunav/MbaDuku.

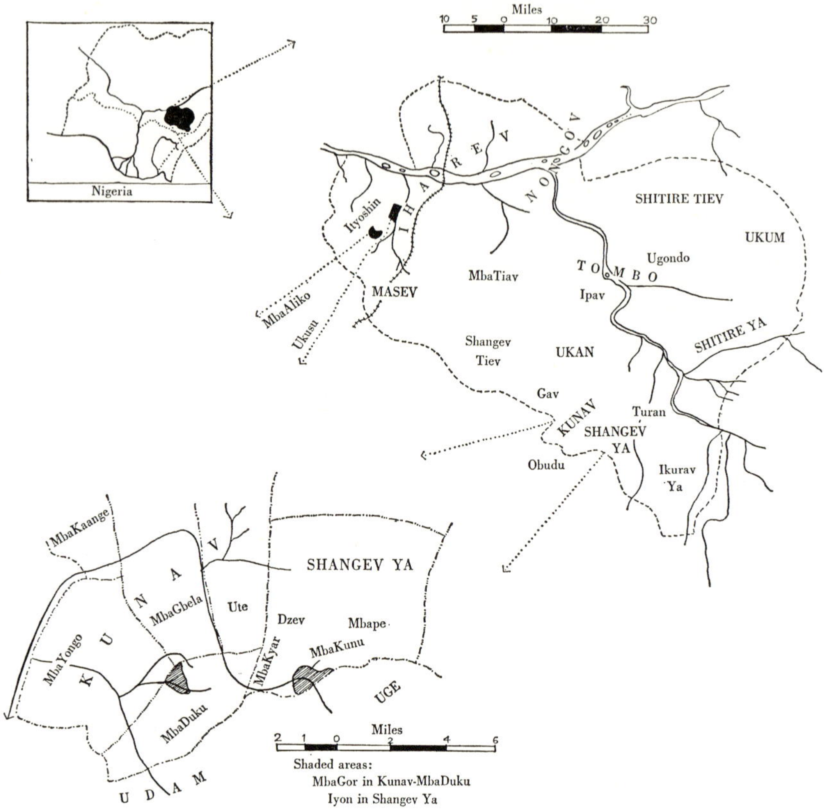

Fig. 6.—Tiv Division.

The Tiv speak of a lineage which is associated with a *tar* as an *ipaven* or "segment." *Ipaven* comes from the verb *pav*, "to split or divide." To split a piece of wood or to tear a piece of cloth is to *pav* it. When a compound splits, the word is *pav*. To divide a kola nut or an orange into its component segments is to *pav* it, and each of the resultant sections or segments is an *ipaven*. In the same way, when all of the descendants of Nyam are divided into two groups—the descendants of Gor and the descendants of Wandia—then MbaNyam is said to have the *ipaven* MbaGor and MbaWandia. MbaNyam itself is, however, an *ipaven* of Ikakwer.

Like the *utar,* the *ipaven* go to the top of the genealogy. The land of the Tiv is called *Tar Tiv,* but Tiv do not consider themselves to be a "segment" of anything. The term is not applied at the very top level, but is used for any segment springing from a descendant of Tiv, so long as that segment is associated with a *tar.*

There is no ancestor of Gor remembered today who does not serve as the eponymous designation of a *tar* (although there are one or two for which this association is practically irrelevant). MbaGor, however, is the smallest *tar.* No lineage comprising the group of descendants of any one of Gor's *descendants* is associated with a *tar.* The descendants of Shar and Jiabu, for example, do not live and farm on territorially discrete blocks, and therefore they have no *tar.*

For purposes of analysis it is sometimes necessary to refer to the smallest *tar,* the lowest point in the system of *utar.* We have called the smallest *tar*—of which the *tar* of MbaGor is an example—the minimal *tar.* We shall call the segment connected with it the "minimal segment." This is not to say that the minimal segment is the smallest lineage, for it is not. Segmentation continues below this level, but its use in territorial organization does not. The lineage system continues into groups smaller than those associated with *utar.*

No lineage which springs from a descendant of Gor (Table 3) is associated with a *tar;* none can be called a segment. Tiv are adamant on this point. Not only will they never call a subsumed lineage of MbaGor an *ipaven,* but they will not allow the ethnographer or anyone else to do so. Tiv stubbornly insist, "MbaGor does not segment (*pav*); it is one man (*or mom*)." If MbaJiabu and MbaShar are to be referred to by any terms other than proper names, they must be called "segments-within-the-hut" (*ipaven ken iyou*).[1]

Whereas the "segment" is associated with a territorial unit, a *tar,* and is *ipso facto* a political unit, the segment-within-the-hut is not a territorial unit (as reference to Figs. 4, 5, 7, 8, and 9 clearly indicates).

Both segments and segments-within-the-hut are sometimes called *nongo*

1. This term "segment-within-the-hut" is in current use only in parts of southern Tivland and is not common even there. Tiv generally do not need a term for these small lineages, and many phrases, of which this is the most vivid, are used when the need is felt.

TABLE 3

GENEALOGY OF COMPOUND HEADS OF MBAGOR AND MBAWANDIA

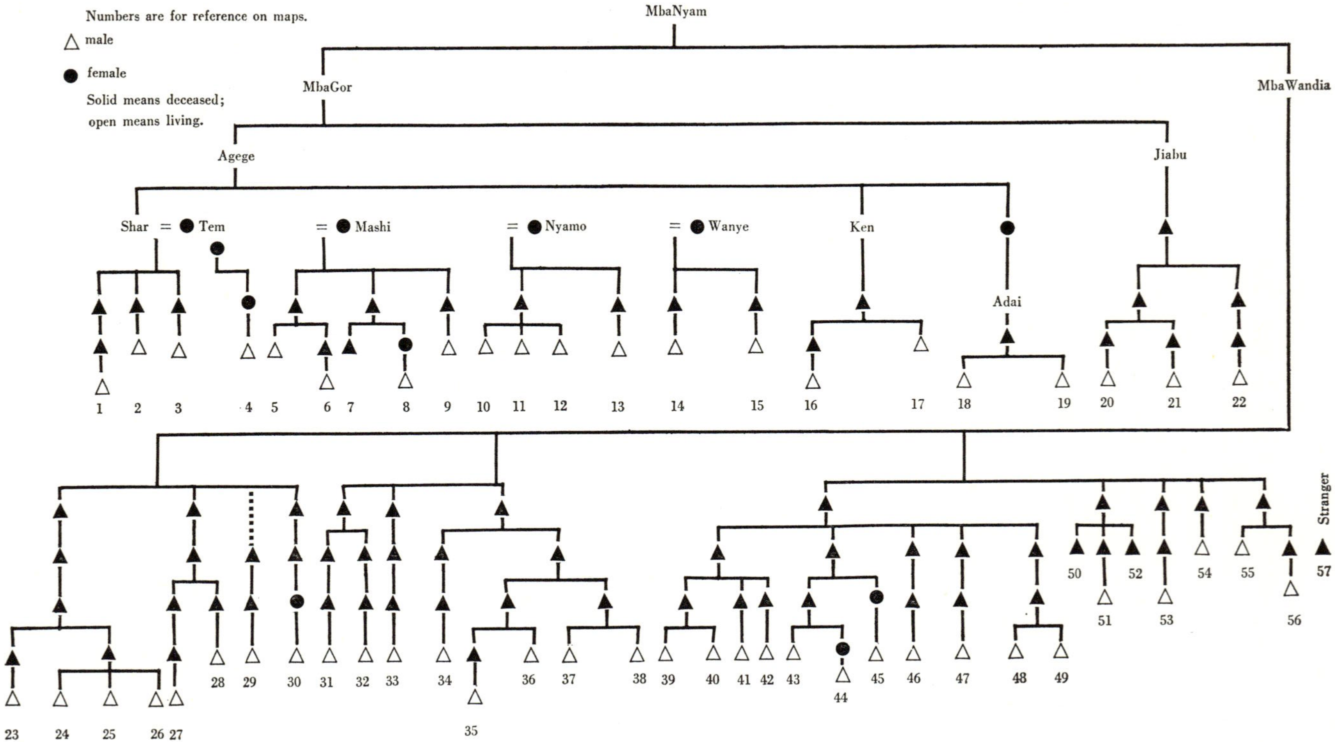

by Tiv. We translate *nongo* as "lineage" and say that each *nongo* is a lineage in a segmentary lineage system. An *ipaven* is a *nongo* associated with a specific area, called a *tar*. A lineage is an enduring group, composed of all agnatic descendants of a single ancestor. A lineage *system* is "the segmentary system of permanent, unilineal descent groups" (Fortes and Evans-Pritchard 1940:6). Lineages are means of classifying people; lineage systems are means of classifying social groups.

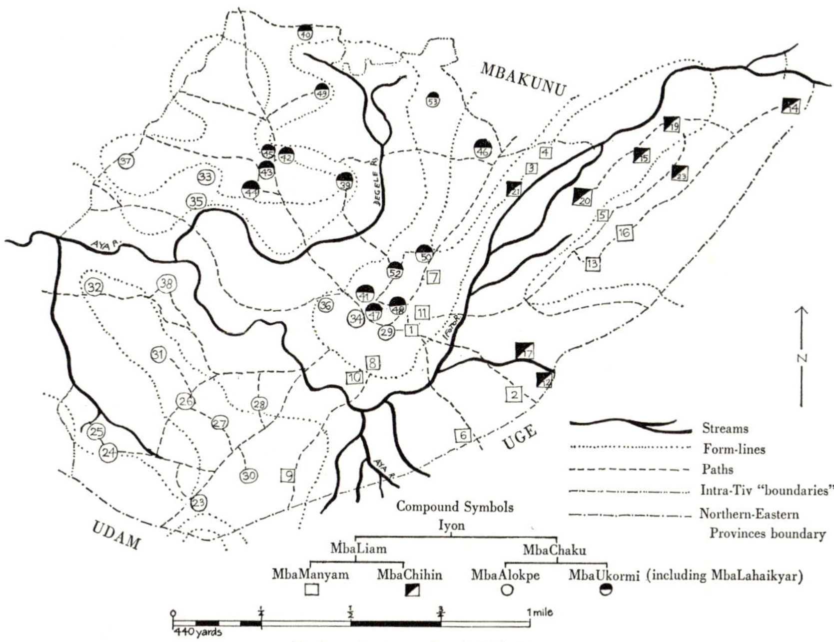

Fig. 7.—Iyon of Shangev Ya.

Balanced opposition between segments within a system of segments is the basis for the political organization of the Tiv; it is the basis of the geography of Tivland as well. Balanced opposition between segments-within-the-hut is a basis for many aspects of Tiv ritual life. Both are aspects of the segmentary lineage system, and each is validated by what Tiv believe to be one continuous, true (*vough*) genealogy.

TABLE 4
Genealogy of Compound Heads of Iyon

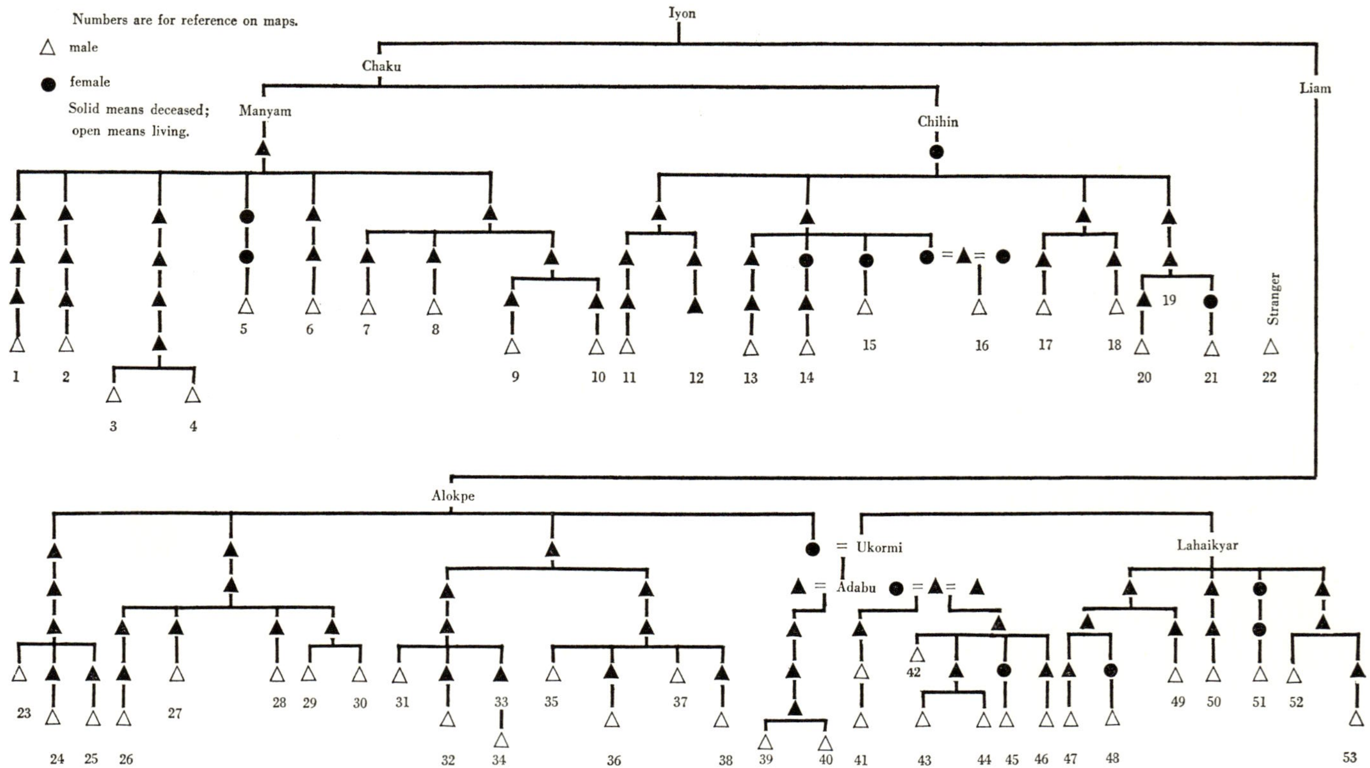

This is not the place to examine in detail the size and actual composition of Tiv *utar* (Paul Bohannan 1954:9–13). The minimal *utar* we studied varied in membership from 169 to 1,047, although it seems likely that those over about 500 would have split into two minimal *utar* if colonial political power and severe land shortage could have been overcome. Iyon, a lineage of Shangev MbaShaya, illustrates these points clearly. If the genealogy (Table 4) is coordinated with the map (Fig. 7), it immediately becomes apparent that the various segments-within-the-hut in Iyon have already spread more or less into separate units so far as the location of compounds is concerned. Iyon remains a single "minimal *tar*" only because farmland is not generally available and the adjustments of migration cannot be made. Figure 8, however, indicates that there is still considerable intermixture of farms within the *tar*.

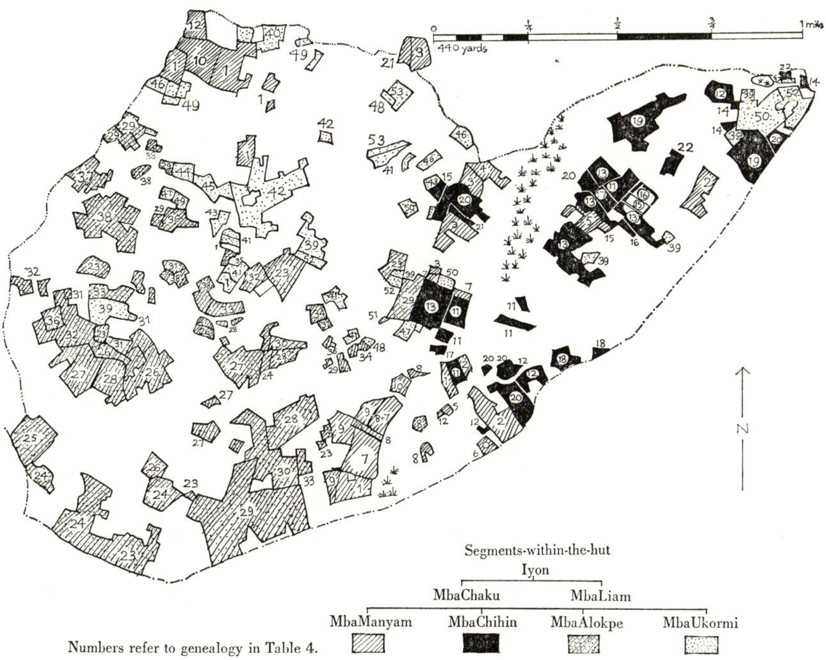

Fig. 8.—Iyon of Shangev Ya—land under cultivation (June, 1950).

Residence Rules and Settlement Patterns

EVERY MAN has a right to live in the *tar* associated with the smallest segment to which he is filiated by birth—normally that of his father. This right of residence cannot be lost unless a man's kinship with his agnates is ritually severed, which in the past amounted to his being sold into slavery.

In addition, a man can be *given* the right to live in his mother's agnatic *tar* or with other kinsmen. It is possible to live among compatible non-kinsmen if they agree and extend an invitation, though comparatively few men do so. In almost all cases, the reason a man lives with non-agnatic kinsmen is said to be that of greater compatibility. There are other reasons which show themselves, however: a man who can't get along with his compound head will move to his mother's agnates for a few years if it would be scandalous to split the compound; some men live with their mother's agnates because of better market facilities or because there is an influential man among her close agnates; others do so because they fear they will be bewitched among their agnates.

Compatible with these "residence rules," which affect individuals, is the settlement pattern made up of compounds and groups of compounds. As we have seen, the genealogies account for the spacing of huts within compounds but do not determine who lives in a particular compound. Brothers live together, but need not do so. Within the minimal *tar,* on the other hand, there is no necessity for brothers, classificatory or real, who live in different compounds to live in adjoining compounds (though many Tiv say it is desirable). A glance at Figures 4 and 9 indicates that the settlement pattern within MbaGor is almost random. MbaWandia is only a little less so (Fig. 5).

It is possible, then, to view the settlement pattern of Tivland as a series of small blocks of agnatic kin, held together and separated by the lineage system. Each *tar* is combined with another, next to it, to form a larger, more inclusive *tar.* All *utar* at every level are named identically with the lineage segments which inhabit them. The process goes from the minimal segment to the whole of Tivland: the span and population of the segment, the depth of the genealogy which validates it, and the size of the *tar* with which it is associated—all increase in association to one another.

It is possible to look at a representation of Tivland not only as a map but also as a genealogy. And, indeed, Tiv themselves think of their

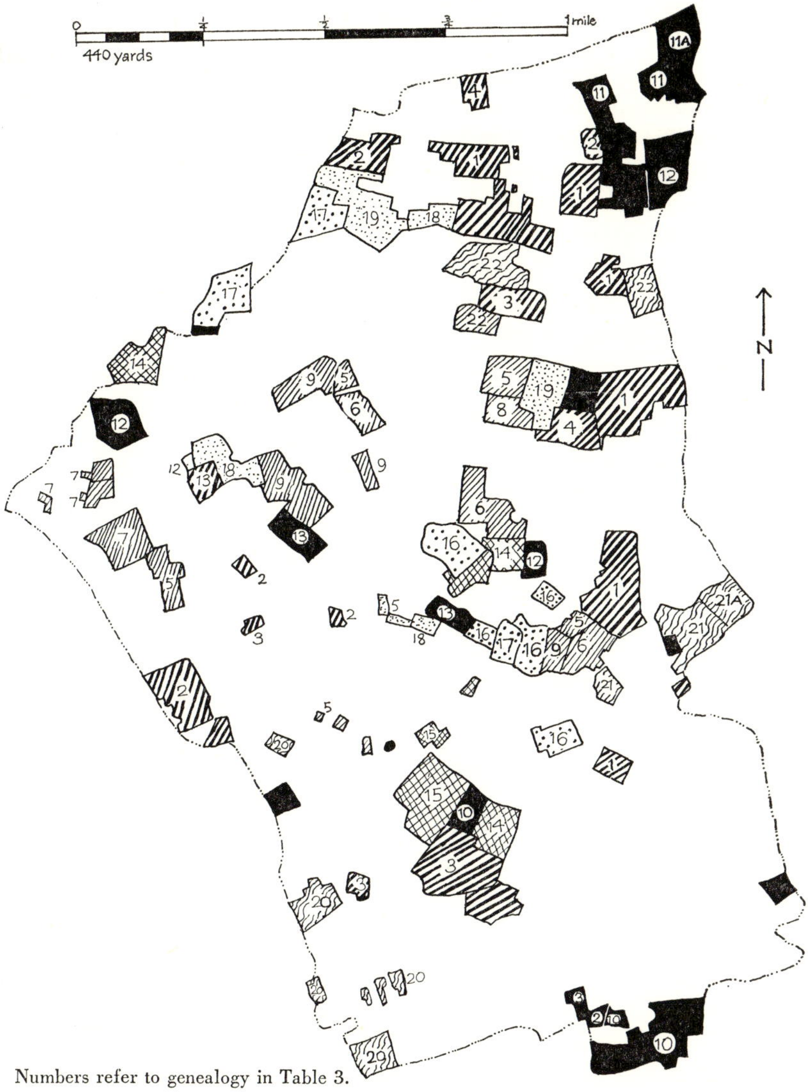

Numbers refer to genealogy in Table 3.

Segments-within-the-hut

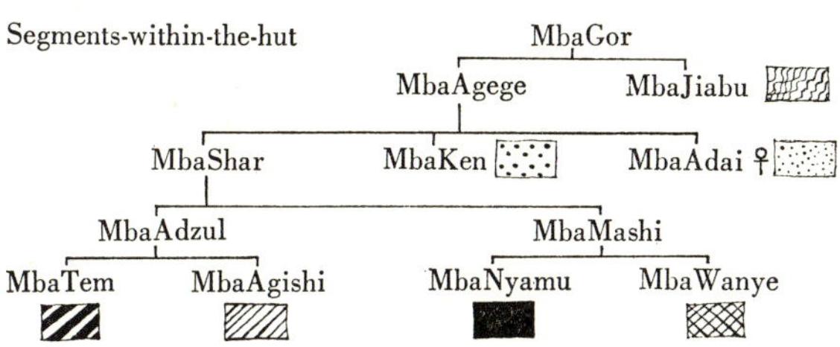

Fig. 9.—MbaGor of Kunav/MbaDuku—land under cultivation (May, 1950).

country as genealogically segmented. The only place names are the names of compound heads and of ancestors who are the genealogical apexes of the lineages associated with *utar*. All non-Tiv areas are called "outside" (*sha won*), or by the names of whatever people live in them.

Tiv geography is a paradigm of Tiv social structure: the lineage system and the genealogy which validates it. Economic activity, as well as all other activity, takes place within the overwhelming simplicity of the same system. We will, in Chapters 4–7, investigate the institutionalization of the factors of production within the confines of the Tiv social system.

4

Farms and Produce (*Yiagh*)

ONE OF THE PIVOTAL WORDS in the Tiv language is *yiagh;* in it their ideas of well-being and economy and work and society meet. *Yiagh* means subsistence in its broadest sense: food in the field, in the granary, in the pot, and in the belly; the farms that produce the food; the work that is required to plant, tend, and harvest it; and the whole range of activities that center around physical nourishment and comfort.

Crops and Crop Rotation

TIV COUNTRY lies between the yam-growing areas of the Guinea Coast and the grain-growing areas of the sudan. They are in the "middle belt" which is fortunate enough to have both roots and grains as staples. The staple crops are yams (*discorea* spp), millet (*pennisetum* spp), and guinea corn (*sorghum* spp). Tiv attitudes toward these and lesser crops are given in detail in *Akiga's Story* (1939:65–82).

Some parts of Tivland practice a three-year cropping cycle, other parts a two-year cycle. In the southern areas for which data is available (MbaDuku and Shangev MbaShaya), the third year is devoted to the important cash crop, beniseed. However, in the northwest (Iharev/ Ityoshin), beniseed is interplanted with guinea corn, and a two-year cycle results. In both areas plots of peanuts, sweet potatoes, and cassava are planted on fallow land especially cleared for the purpose; these crops are not part of the crop rotation. The quantity of such subordinate crops in proportion to the "staples" rises decidedly with increase in density of population.

To explain fallow periods and rotations merely in terms of years is to misrepresent the Tiv viewpoint. Southerners consistently say that land should lie fallow for at least six years and bewail the fact that such long periods of fallow are no longer possible. In the northwest, where there is no land shortage, Tiv do not think of fallow in terms of years at all, but actual practice seems to have varied from two years to sixteen years. Generally Tiv do not compute fallow periods in years; rather, they say that land should lie fallow until it is rejuvenated, a state which can be judged by grasses, trees, and earthworms. Tiv have told us that different types of soil take varying periods of years to rejuvenate, but none was able to give proportionate times. Since one can always determine the condition of any given piece of soil by examining it, there is no need to calculate time.

When discussing crop rotations, Tiv seldom talk in terms of years; instead they refer to the phases a particular farm goes through in a complete rotation cycle. Examination of the words for these phases provides a key to Tiv farming and land usage. They are summarized in Table 5.

All land which has obviously once been farmed is called "fallow" (*tsa*). It continues to be called that between the time that the final crops of one rotation are removed from it and the time it is again prepared for yams.

When the grass is pulled preparatory to making a new yam farm, the land is "cleared ground" (*ihyande*). As soon as the mounds for the new yams are well started, the land is called "mounds" (*avom*). After seed yams have been planted, it becomes a "yam field" (*sule*). Only a yam field can be called a "field"; all cultivated land can, however, be referred to as "farm" (*tiev*). As soon as the yams are harvested the farm becomes a "second-year field" (*akuur*). When the last crop (usually guinea corn) of the second-year farm has been harvested, the land

TABLE 5

WORK CALENDAR FOR KUNAV/MBADUKU

	Aug.	Sept.	Oct.	Nov.	Dec.	Jan.	Feb.	Mar.	Apr.	May	June	July
IHYANDE	———	———	———	———	→ ← AVOM	———	→ ← SULE	———	———	———	———	———
SULE	———	———	———	———	———	———→	← AKUUR	———	———	———	———	———
AKUUR	———	———	———	———	→ ← TSA	———	———	———	———	———	———	———
MEN'S WORK CALENDAR: SOUTHERN TIV												
CLEARING FIELDS			———	———	———	———	———	———	———	———→	←———	Millet harvest
		HOEING MOUNDS		———	———	———	———	———	———	———→		Sweet potato mounds
				Guinea corn harvest		Corn → Plant yams		———	———	———→	Plant guinea corn	
									Plant millet and beniseed			Pluck beniseed
WOMEN'S WORK: SOUTHERN TIV												
CLEARING FIELDS			———	———	———	———→		PLANT YAMS	———	———→	Weed yams	Millet and beniseed harvest
Second weeding			Weeding sweet potatoes									
YAMS DUG AS NEEDED			———	———	———	→ YAM HARVEST		Weed and *akuur*	level Peanuts	———	———→	Clear and plant sweet potatoes
				Rice harvest	Guinea corn and sweet potato harvest	Dry cassava and vegetables		Weeding in *pagh*				

becomes fallow again. Any crops of a third year of rotation are said to be planted on fallow (Figs. 8, 9, and 10).

Progression from one of these phases to another is marked by activities in the agricultural cycle. Pulling grass turns land from "fallow" into "cleared ground"; harvesting yams changes land from a "field" into an "old field." Tiv thus see their farming and their farmland in terms of these key activities.

Cleared Ground (*Ihyande*)

A FEW DAYS before the time to pull grass for new fields—about the last half of August in southern Tivland, November in the north—the compound head calls together all the other adult males in the compound to discuss selecting sites for new farms (*nema sule*). (We have never seen such a meeting, and we believe that Tiv over-formalize it in the telling.) Following conversations over several months, it is generally known and agreed where new farms are to be made: in the land-hungry areas of southern Tivland, on that piece of land which has been longest fallow; in the swampy northwest, on the piece of land which is not submerged in the normal year and which is close enough to the compound to allow easy access, but far enough not to be menaced by goats.

Tiv say that the compound head has final and complete control of the size and location of the farms of each man in the compound. Actually, practice varies markedly in different compounds: in some the head makes no attempt to correlate agricultural activities, each subordinate group within the compound farming where it sees fit. However, agreement is usually reached quite informally among all the adult males of the compound. Tiv insist that, should there be a dispute, the compound head has the final word, but a domineering and unreasonable compound head would soon find himself "sitting alone"—the worst fate that can overtake a Tiv.

When asked what sort of requests a compound head will refuse, Tiv answer in terms of labor. A man who wants to make a larger yam farm than he can control without stinting his obligations to the group will be refused permission by the compound head. If he wants to plant many more yams than his wives and children need, so that he can sell them to get more money and goods than others in the compound, he is likely to be refused, first on the ground that he will shirk communal work,

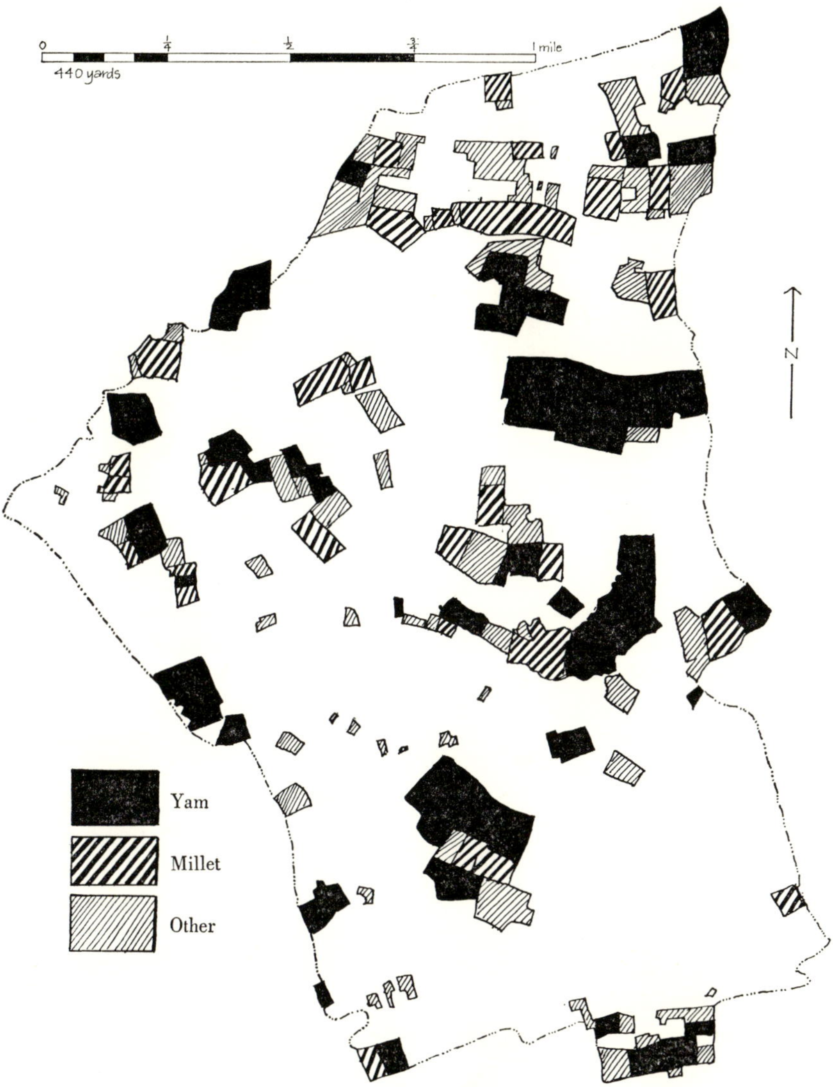

Fig. 10.—Land under cultivation, by crops—MbaGor.

second on the ground that if any person in the compound besides the compound head has a great deal of money the compound "spoils" (*vihi*). However, no reasonable compound head will refuse to let a man make an extra patch of ground nuts, beniseed, or cassava for sale, because the amount to be realized on them is small, and the work minimal.

On the day that grass-pulling begins, the compound is left in the care of a sick person or a child or two—in case of small compounds, often only the goats—while all of the others go out to pull grass. Women bring food to cook there; young women and girls bring pot after pot of water from the streams for cooking, for drinking, and for washing before eating. "Nurses"—children between seven and eleven—bring their infant charges out to be with their mothers who are cooking or pulling grass. All children above the age of eleven are working. But there is a definitely festive air; grass-pulling is a noisy time, with plenty of laughter and singing as well as plenty of work.

The people of the compound do not divide up into small labor groups to pull grass as they do for most other work in the fields. Younger men cut down the saplings with matchets and lop the lower branches off the large trees; the compound head, especially if he is an elder, may take little part in the work he oversees. But everybody—men and women, boys and girls—pulls grass. They form a long line, each person two or three yards from those nearest him; they bend from the waist, grabbing the tufts of grass in the left hand and pulling and cutting into the roots with hoes and matchets. The line keeps pretty well abreast as it goes from the top to the bottom of the field. The gay, picnic atmosphere lasts only for a day or two. After this, the women divide their time between the new clearings and their yam fields, and the work goes more slowly.

After a considerable portion of the land has been cleared, the compound head, with the advice and assistance of his brothers and sons, stakes out the new farms. Laying out a farm is done differently in the south than in the north.

In the southern lineages, the compound head first of all makes a large yam heap on which he will then erect an *akombo,* or magical emblem.[1] This mound, with the *akombo* set either on it or just in front of it, is the center point of the "top" (*ityough*) side of the field of the senior woman of the compound, usually the mother or wife of the compound

1. For discussion of *akombo,* see Akiga 1939:176–234; Abraham 1940:84–101; Downes 1933:63–67; L. and P. Bohannan 1953:85–90.

head. He then takes 58 (more or less) paces in a straight line to the bottom (*ityo*) of the field and marks the spot.

Most Tiv say that yam mounds should be one pace (32 in.–33 in.) apart and that there should be 58 mounds in a row (*iba*); hence, the compound head should take 58 steps. There is some variation in the number of mounds said to compose a row; although the most common number is 58, the range is from 50 to 80. There is even greater variation in the numbers of heaps actually found in rows.

Dechi

The line that the compound head paces off becomes the *dechi,* or center row of the field of the senior woman. Since it is important as a reference point, Tiv say it should be straight. In one case the compound head, after pacing off about 60 paces, pushed a pointed stick into the ground, returned to his mound, and sent three of his sons to different points along the line: one to the stake at the bottom of the field, the other two at interim points. He got them into a straight line by shouting instructions to each boy in turn to move a bit to left or right. The boys were not lined up directly behind one another; rather each was just visible to the left of the one in front of him. When the row was straight, each boy drove a stake into the ground as a guide for the men who were to hoe mounds of the *dechi.*

Southerners inspect and check the straightness of this line conscientiously. They hail every passer-by to inspect the line of the markers and if one demands a change of a few inches in one of the stakes, the change is made in the full knowledge that the next passer-by will probably move it back. The compound head sights down the stakes again and makes his own alteration, then finally announces that he is satisfied. He then paces off, at approximate right angles, the full width of the field, going in each direction from the *dechi* mound. If the senior woman of the compound is to have 80 rows of yams, he sets a stake 40 paces from each side of the original mound (actually 39 paces, for Tiv count the rows, not the paces). The *dechi* itself is counted twice: they always count the rows of yams in a field from the center to one edge, including the *dechi,* then multiply by two. There are thus 79 rows of yams in a field said to contain 80. If you point this out to them, they tell you that what you say is quite true, but the mounds in the *dechi* are bigger than the mounds in the other rows.

Only after the field of the senior women has been staked can the limits of the other fields be staked, usually with somewhat less care.

Southerners count yam rows to determine the size of a man's field or the status of one of his wives. One hears such comments as, "I see that WanGajo has 20 more rows of yams this year than last—is she pregnant?" Or, "Dzungwe has hoed 10 fewer rows for his senior wife this year—yes, her daughter was married and went to her husband." These are indicative of the fact that the size of a woman's fields is usually in direct proportion to her needs. A woman with many children or one who has many attached junior wives, daughters-in-law, or grandchildren may have 100 or more rows of yams. A new wife, who has borne only one child or none, will have only 20 or 30 rows. The favorite wife of an important man may have a larger farm than her co-wives, but all the persons in the compound, including her co-wives, say that this is because she supplies the major portion of the food which her husband shares with friends and guests.

The fields in the northwest have no *dechi.* They are laid out from one side to the other and "squared" on the "path row" (*iba-ndoro,* sometimes called *gwendor* in Iharev dialect). Iharev/Ityoshin people count all the rows in a farm; all the farms are larger, and the size of a woman's farm tends to be associated with the number of seed yams she can get together rather than with her status or the number of her dependents.

Mounds (*Avom*)

The day after the field of the senior wife has been laid out, the young men—including her sons, if she has any of working age—begin hoeing mounds, while the other men and the women continue to pull grass. The center row of mounds is usually completed before any other mounds are begun. Mounds of the *dechi,* like all yam mounds, are a pace apart, but *dechi* mounds are taller than others: they are usually about 30 inches in height, while the ordinary mound is about 18 or 20 inches.

The "small *dechi*"[2] are made next: rows of ordinary-sized yam heaps, one on each side of the *dechi.* The small *dechi* need not be finished before the other rows are made.

The mounds in a row of yams are set off from those in the next row, as shown in Figure 11. A well-laid-out field forms rows in both directions,

2. Translated "*dechi's* children" (*wan dechi*) by East in Akiga 1939, passim. As Akiga says, only Kparev make yam fields in this way.

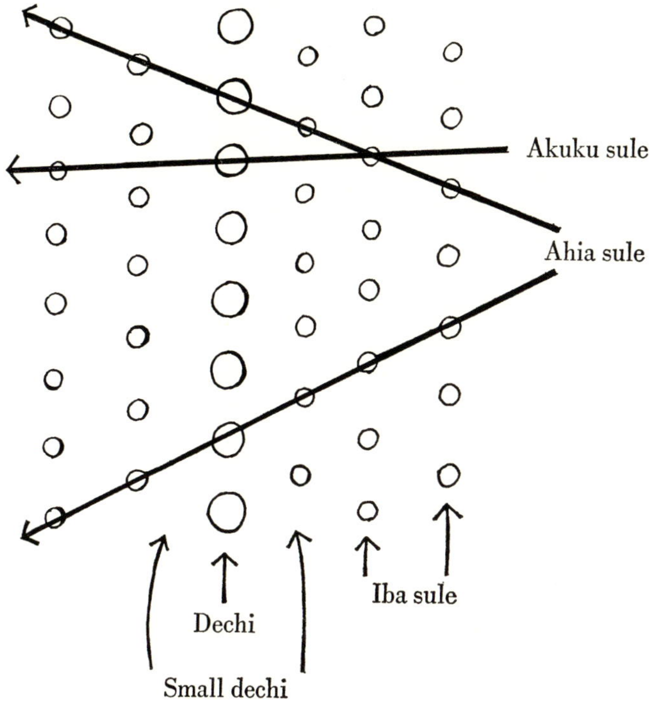

Fig. 11—Mounding of yam field.

as well as diagonally. Tiv describe this by reference to the top, bottom, and sides of the field. The rows which run from top to bottom are called *iba sule* in Kunav dialect; the rows from side to side are called *akuku sule;* while diagonal rows are called *ahia* (interstices) *sule.*[3]

Though the top of a field is usually higher in altitude than the bottom, it need not be so. Nor are "top and bottom" necessarily correlated with

3. Iharev dialect is somewhat different. An *iba sule* is a row running in any direction; the row from top to bottom is called *lyembe sule. Akuku sule* means diagonal row instead of cross row, while cross rows are called *gwambe sule.*

direction. In Ityoshin there is a cleared space some five feet wide around all yam fields to keep wild animals out. We have never seen this cleared space (*kpi sule*) in the south.

The work of making mounds is done by men with a heavy hoe. The actual operation of constructing the mound seems a simple one, but it is not. The legs are placed well apart, then the hoe is dug into the soil between the feet in such a way that a clod of about eight inches wide by two or three inches thick is scooped on to the hoe. This clod is then flipped, moist side out, into position on the mound which is being built up.

When the mound has been finished, dried grass and leaves—left on the field after clearing—are gathered into small bundles and set on top of the mound, weighted down with a hoeful of soil. This may be done by the man who digs the mound, or it may be done by younger boys. The purpose of this bit of grass is to prevent erosion of the mound and to afford protection from the sun for the sprouting yam.

Soon after the mounds have been made, often on the same day, a piece of cassava stalk, freshly cut, is placed rather low on the side of the mound. The cassava, planted thus several weeks or months before the yams, is given a chance to grow up high enough to provide a trellis for the yam vine to climb. Tiv point out that more than one stalk of cassava per mound stunts the yams planted there.

Clearing the grass is usually finished by the end of November (the end of December in Iharev/Ityoshin). Digging the mounds takes longer. Kparev say it is best to finish them a month or so after the dry season sets in; after that the ground is hard and mounds made from dry soil will crumble. This ideal is seldom realized. A light rain during the dry season (there are usually three or four) is followed by a few days of feverish activity in the new fields, before the soil again gets too hard to work. Although the best and biggest yams are those which are planted in February, many people do not finish digging their mounds until the new rainy season is well under way—sometimes the middle of May. This latter practice is the norm in Iharev/Ityoshin.

When a man has finished his yam fields he cuts a branch of a plant known as "man's ikula" (*ikula i nomsu;* in Iharev, *kwakula*) if he is living with his agnatic lineage, or "woman's ikula" (*ikula i kase—sarcocephalus esculentis*) if he is living with his mother's agnatic lineage. This branch is placed across the most important patch which borders or traverses the field; the symbol of an *akombo* may be put beside it. The branch is covered with a hoeful or two of soil to keep it in place.

Tiv say that this notifies persons going down the path that a farm has been finished, so that they will examine it and remark on it. The man who puts up such a symbol will himself first scrutinize his fields closely. He will look down each row to see that it is straight; he asks advice of passers-by. We have seen a man change the position of two mounds slightly because he was convinced they were a bit out of position on the diagonal, thereby spoiling the symmetry of his field. A well-laid-out field in which all the rows—vertical, horizontal, and diagonal—are straight is a source of aesthetic satisfaction. We have been told, "Nothing so gladdens the heart as a field in which all the rows are straight." We have seen Kparev Tiv spend an hour or so admiring their own fields, if they were well done, and almost as long admiring those of a kinsman. If one walks in a strange area with Tiv at any time during the growing season, one is given a continuous commentary which can be summarized, "That field is a good one; this one is not."

The Field (*Sule*)

THE ACTIVITY which turns the mounds into a field (*sule*) is the planting of the seed yams. Seed yams are sometimes stored in the compound but more often in special trenches or in last year's mounds. They are carried to the new field by women, who sort them by types and lay them out, one seed yam or section on top of each mound. Both men and women do the actual planting—men with the heavy hoe, women with a digging stick. The grass crown is removed and the hoe is sunk into the top of the mound just behind the seed yam (only the inept first remove it; more skillful men jeer at those who do so, or those who injure a seed yam with the hoe). The hoe is then held in the left hand and pulled forward to leave a gap; the seed yam is slipped into the gap with the right hand and the hoe removed easily, allowing the soil to fall back gently. The most skillful men can perform this operation without losing any of the soil from the top of the mound.

When a woman plants seed yams, she takes her digging stick in both hands, and with a sharp downward stroke forces it into the top of the mound. Then by twisting and working it about, she creates a hole into which the yam will fit, inserts it, and closes the top by hand. In all cases the grass cap is replaced and again weighted with soil.

In May, after the rains are well under way and all the yams should have sprouted, those which prove infertile are removed from the mounds and replaced.

Planting completes men's work with yams. The major part of women's agricultural work is with yams. Yams should be weeded two or three times—most Tiv agree that three times is best, but many compounds manage only two weedings. Women perform this work with the small hoe (see Chap. 8).

Many side crops are grown in yam fields. We have counted as many as eight crops in a single mounded field: yams, cassava, peanuts, okra, sorrel (both red and white), *ahuma* beans, *adenge* beans, and pepper. Cotton, corn, gourds and cucurbits of several kinds, and a great many "soup plants" may be grown as well. Most of these side crops (cassava is the obvious exception) are planted low on the sides of the mounds by women during and after the first weeding. They are harvested as they ripen or are needed.

Harvest

The first yams are usually ready to be dug and eaten in early July in Kparev; in Ityoshin, not until early September. First yams are said to be associated with various rites in different parts of the country (East 1939:90–92). These rites are casual; they affect only adult males and are commonly omitted.

After the "first yams" are dug (and often before), women harvest those they need for food. The general harvest takes place in late December when the dry season is well advanced, after the guinea-corn harvest is completed. It is carried out by the basic labor group of women: a senior woman with attached co-wives and daughters-in-law. Most of the digging is done by young women with digging sticks. The stick is held in both hands and driven into the top of the dry mound; then with a forward motion one whole side of the mound is peeled off, leaving the yam exposed. The dry soil can be loosened the rest of the way round the yam and the whole thing lifted out undamaged. For a few weeks after the yams have been removed, the field looks like a series of half-domes. The younger women bring the yams to one of several collection points in the field, where the senior woman separates them carefully by types. Eating yams are separated from seed yams, then dry yams (*iyough*) from water yams (*agbo*), and there is a further separation into red and white yams. There are well over fifty types of yams, with names known only to women; except for a half dozen of the simplest and most common, few women agree on identifications.

In Kparev, after separating and grading, the yams are cached about the field in trenches, covered with a matting of dried grass. They are not simply dumped into the trenches, but are carefully and symmetrically laid in, graded by type and size.

In Kparev, seed yams are carried either to the compound or to the cache in the new fields, whichever is nearest. Women divide their time between old farms and new farms at this time of the year. The men, who are probably still working at hoeing mounds in the new fields, build field shelters (singular, *tumbe*) for their wives, and a woman and her young children often spend many nights in her fields watching over the harvest, with only an occasional trip back to the compound. These shelters are about ten feet in diameter, comprised of six or eight uprights with a thatched conical roof. They have built-up, pounded floors; the sides may be open or mats may be tied along one side. A fire is kept going in these shelters and cooking is done here. There is usually a single shelter for all the women in the basic labor group.

Even in Ityoshin, where all other crops are safely stored in the fields, it is risky to leave seed yams unattended in a cache. People say that they are short almost every year and that yams are likely to be stolen no matter how strong a magical emblem protects them. They add, however, that when yams are laid out on mounds before planting, they will not be stolen—though they may be eaten by cattle or monkeys.

The yams for eating are gradually, over a period of weeks, transported by women to the compound, a headload at a time. The ordinary headload of yams varies between fifty and seventy pounds in weight; it is sometimes over a mile from farm to compound but usually half a mile.

Second-Year Fields (*Akuur*)

As soon as the yams and cassava have been removed, the land on which they grew becomes a second-year field (*akuur*). In March or April, usually after the harmattan has stopped but before the rains, what remains of the old mounds is leveled by the men. Then, after the first big rain which is considered to be the start of the rainy season proper (there may be two or even three heavy rains before this time, perhaps separated by dry periods of a fortnight or more), bulrush millet (*amine*) is planted by broadcasting the seed; the soil is then turned and worked quickly to bury the seed. This is men's work, though women often assist.

When the millet has been planted, a man again puts *ikula* leaves on the path to indicate that he has finished a farm. It is not so important an occasion as finishing a yam field. Tiv appreciate a neat millet field, but not with the same zeal with which they appreciate a well-made yam field.

Millet is a tricky crop, Tiv say, because minor changes in the weather affect it. Millet must be planted after a soaking rain, and should then have about five days of cloudy weather, but no rain. Without these conditions for germination, the yield is poor and the grains—if they form—are brown and bitter. Tiv are always in doubt as to just when millet should be planted, and they emphasize the element of luck involved. What was conceded by all to be the best millet crop in MbaGor—the only really good one in 1950—was planted almost two weeks before any of the others. Almost everybody else in MbaGor told the compound head who had planted so early that he would get no yield at all. Iyon planted millet even later than MbaGor, and the yield was even poorer.

The crop most commonly interplanted with bulrush millet is guinea corn (*wua*), which is planted a month or six weeks after the millet, being broadcast into the field, usually with no subsequent turning of the soil. Kparev people broadcast guinea-corn seed from their hands, walking through the field. Iharev people take a mouthful of mixed guinea-corn seed and beniseed and walk along the rows blowing it out in a fine spray. Most people say that millet should be weeded immediately prior to the planting of the guinea corn, but it often is not done. Tiv in the north do not plant guinea corn until the millet has been harvested.

Other crops found interplanted with millet include cassava (usually to be found growing in the field before the millet itself is planted) and beniseed. Planting beniseed and guinea corn together is somewhat more common in the south and is standard practice in the north, where beniseed is planted late in July (if planted earlier in this swampy, fertile soil, it goes to stem and does not bear well).

At millet harvest, men break down the stalks with an arrangement of rope and pole (*aga*), cut them with matchets, and then lay the stalks in rows running from top to bottom of the field, about eight to ten feet apart. Women then go down the rows, cut off the heads, tie them into bundles, and carry them back to the compound.

Millet is harvested about two months or slightly longer after planting—usually in June. It is valued by all Tiv because it makes the best beer; it is especially valued by southern Tiv because it is a hunger-breaker, ready between the time that the guinea corn is exhausted and the new yams are dug. It shares this role of hunger-breaker with corn, which is grown as a side crop in the main fields, in the kitchen gardens, or in stream beds.

In those parts of southern Tivland which we know, millet stalks are used for fuel. In the areas of Ityoshin where we worked, however, and in all those we walked through, there is plenty of firewood; there millet stalks are laid in rows in the fields, and the soil is piled over them, making ridges in which the next crop, guinea corn, is planted. Iharev people recognize that this makes the soil more fertile but say the effects are not noticeable the year it is done.

Beniseed is harvested by plucking each spike, then tying several together and hanging the bundle over rails to dry. It is, when thoroughly dry, pounded in a mortar and winnowed.

Guinea corn is harvested in the same way as millet; it is trod down with the *aga,* and the heads cut off with a knife. People in the south say that guinea-corn harvest begins the morning after the first harmattan arrives; this was true on December 1, 1949. Northerners do not make this correlation, and in late 1951 the guinea-corn harvest followed the arrival of the harmattan by more than three weeks.

Fallow (*Tsa*)

As soon as the guinea corn has been cut, the land is referred to as "fallow." In the south, wet season beniseed (that is, planted in April and harvested in June) is grown on "fallow" in the third year of a rotation. Peanuts, sweet potatoes, Bambara groundnuts, and cassava may be grown on "fallow" without necessitating a lengthened fallow period, Kparev people say. Iharev, however, say that Kparev grow these crops in quantity only because of land shortage, and that they themselves like these foods only for occasional snacks.

In the south, when fallow land is cleared for planting fallow crops, grass and small trees are allowed to dry and are burned. The ashes are mixed with the soil when mounds or ridges are made. Some Tiv say this increases the yield; others say that burning the grass and brush is the easiest way to get rid of it, adding that if you wanted it for fertilizer you would bury it rather than burn it.

Clearing for fallow crops is usually done by women, though men assist by making the mounds. Sweet potatoes (*atsaka*) are planted from runners in round, flat-topped mounds about eight or ten inches high. Southerners plant sweet potatoes late in June; northerners plant the few they want at any time during the agricultural year. Weeding and harvesting sweet potatoes is woman's work.

Peanuts are planted by women on three or four inch high ridges made by men; Bambara groundnuts are planted by women on small rounded mounds about ten inches high. Both are harvested by women. Cassava, when grown alone, is planted by either sex in ridges a foot or so high made by men.

These fallow crops are of great importance in the diet of southern Tiv; they are not highly regarded, however, and much less care is given to them than to yams, millet, guinea corn, or the "swamp crops."

One of the ways to tell that a piece of fallow is ready for recultivation is by the type of grass it produces. Tiv say that after a piece of land has been cropped, the grass which it grows is predominantly sword grass (*ihila*). The next year, *agom* grass begins to replace it. For the next few years *agom* clumps get larger and heavier until they are about a foot in diameter at the base and there is no sword grass left; then it is ready to be farmed again. Another grass which is said to be indicative of good soil is *aper*.

In central parts of Tivland, according to informants in Jengba/Ipav, trees and earthworms are used as the indicators of rich soil. Large undergrowth trees and large-size grass clumps with plenty of earthworm mounds mark fallow which is ready for recultivation.

Reclaimed Swamp (*Pagh*)

In Iyon there are fifteen or twenty acres of irrigated land in stream beds, which do much to ameliorate conditions of severe land shortage. This land (see Figs. 9 and 12) is in production for ten months or more of the year, affording as much as three crops of quick-maturing corn and one of rice. During the rains, proper care of this reclaimed land insures fresh supplies of silt, and the quick decay of luxuriant vegetation, constantly being cut back and left to rot, has made a deep, thick soil.

Tiv call the irrigated land along a stream bed *pagh*. *Pagh* also means to prepare land in such a way that water is retained or redirected. The heavy work of keeping such land clear, and the almost constant work on ditches and the building of ridges and beds, falls on the men. Most of it is done with the heavy hoe and the matchet. Corn—the most important crop grown in such reclaimed swamp—is planted in slightly raised ridges by stamping a hole in the soft moist soil with the heel of one foot, dropping in two grains of corn, then brushing soil back into the hole with the toes. Rice is planted by broadcasting it among the stalks of the second or third crop of corn.

There are no large areas of such irrigated land in MbaGor, although we believe all compounds have small corn and rice fields along the streams. The first crop of corn is planted as early as February, maturing in April.[4] Corn ears are sold in local markets during the season, to be eaten on the spot. Rice, in 1949, was primarily a woman's cash crop which eventually found its way to the Ibo and Cross River peoples in the Eastern Provinces. By 1952, men had taken over the rice, and acreage was vastly increased.

Reclaimed swamp land of this sort is highly prized by southern Tiv; they say that a good piece of such land need never be left fallow for more than a few months at a time. It is also highly important because the corn it produces ripens when food is short, and because the rice which can be grown on it is a ready source of cash. We found no reclaimed swamp land in the northwest of Tivland.

Kitchen Gardens (*Ikongo*)

THE GARDENS which surround southern compounds—and are in turn often surrounded by a ring of trees—provide Tiv not only with most of their green vegetables, tobacco, indigo, cotton, and fruit, but with small amounts of staples, particularly millet and corn. The *ikongo* also symbolizes privacy and discreteness. The kitchen gardens, forming a ring round the compound, are a buffer between the people of the compound and their neighbors. Encroaching on someone else's land to make a new

4. During the time the map of MbaGor was made, there was no activity in the stream farms, and most were fallow. We did not map them, since pacing these muddy paddies is almost impossible. Those in Iyon (see Fig. 12) are all indicated, but are in such tiny blocks that we did not have time to get exact distribution.

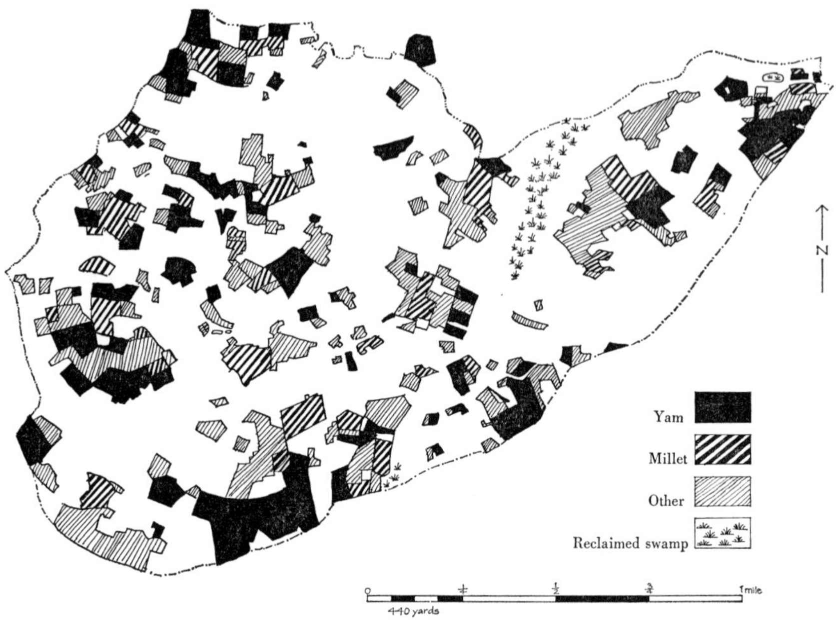

Fig. 12.—Land under cultivation by crops—Iyon.

farm is common; but encroaching on kitchen gardens is a serious breach of etiquette.

Every man is entitled to a plot just behind his wives' huts for his kitchen gardens. He may give each wife a patch or he may keep it all for himself, telling the wives what work to perform.

Kitchen gardens and the trees that surround them are important in some rituals; they are the scene of assignations; they symbolize the unity and discreteness of the compound; they are kitchen middens and latrines.

Ikongo, in Iharev dialect, means only the area surrounding the compound. It is seldom used for kitchen gardens because the goats (more numerous than in the south) eat any crops planted close to the compound. Crops which Kparev people plant in their kitchen gardens are planted by Iharev in their fields—except tobacco, which would be stolen.

Tobacco has a high sale value, whereas food crops in that area do not. A few men in Iharev/Ityoshin fence small areas for planting tobacco.

The Size of Tiv Farms

SOUTHERN TIV say that land shortage reduces the size of their fields and shortens periods of fallow. This situation can be investigated quantitatively, even though the quantities which an anthropologist can collect readily are limited, since he must do his own measuring and counting. In Tivland, an investigator, after several months' residence and a thorough acquaintance with genealogies, can take a census that is accurate for the few days he is taking it. He can also measure farms. It is possible to measure the total acreage within a minimal *tar* with less accuracy, since boundaries, which are necessary to measuring acreages, can only be "roughed in."

Facts it would be desirable to know, but which an anthropologist cannot reasonably be expected to collect, make a formidable list. He has no way to measure differences in soil fertility or the percentage of cultivable land. It is impossible for him—in Tivland, at least—to obtain accurate data on crop yields. Tiv do not have a general harvest in which all of any single crop is put in one place. Millet and guinea corn, which are cut all at once, are seldom stored in such a way that the investigator can do more than estimate weight roughly. Yams are dug as needed until the middle or late dry season. Tiv do not measure the yield of their farms. Therefore, any attempt to do so meets misunderstanding, mistrust, active opposition, or resentment. The agricultural department of the colonial government could supply yield statistics from its own experimental farms, but conditions on the farms for which these figures are available do not correspond with those of an ordinary Tiv farm, and the statistics are useless for our purposes. The 1951 agricultural census was much too general. In short, obtaining accurate yield statistics would require one assistant per woman whose farm was included in the survey. He would have to follow her about from morning to night to weigh and record what she picked or harvested to eat or cook. Needless to say, Tiv women would not put up with this sort of thing, even if the assistants could be found.

In order to investigate Tiv statements about land shortage, it has been necessary to adopt different methods. We can ask, "What percentage of a total area of a minimal *tar* is under cultivation each year, and what is the relationship between this percentage and the density of population?"

Table 6 shows quantitatively that as the density of population rises, the percentage of land in cultivation also rises. If the same proportion of land per person were used in Iyon as in MbaAliko, 62 per cent of Iyon would be under cultivation constantly. Since it is impossible, if fertility is to be maintained by fallowing, for 62 per cent of the land to be under cultivation at a given time, the land shortage is obviously severe. The actual figure is about half that: 35 per cent. Hence, some adjustment must be made. Several things can be done: (1) The cultivation period can be shortened; (2) the size of the fields can be decreased; (3) a change in crops can be introduced; (4) the length of the fallow period can be reduced; and (5) new farming methods can be introduced.

TABLE 6
Cultivated Acreage and Population Density

Location	Acres in Cultivation	Total Acreage	Percentage of Land in Cultivation	Density of Population per Square Mile
Southern Tivland				
MbaGor	327.75	1,145	28	267
Iyon	426.05	1,217	35	552
Northern Tivland				
MbaAliko	125.27	c. 1,700	7	65

In both MbaGor and Iyon the cultivation period has been reduced. In this part of the country, where beniseed is normally planted as the third-year crop in the rotation, many farmers plant beniseed on second-year fields in place of millet, rather than on fallow in the third year. When these two crops are harvested in late June or early July, the entire acreage is put into guinea corn. This practice definitely reduces the amount of food available, especially millet, a hunger-breaker. Moreover, not all land that is planted to yams in the first year of a rotation cycle is used during the second year. Some is allowed to revert to fallow after a single year, thus further reducing food supplies. All these practices lead to greater reduction of grain than of yams in land-hungry areas.

The size of fields has also been reduced, especially in Iyon. The yam acreage per person is given in Table 7.

TABLE 7
YAM ACREAGE

Location	Acres in Yams	Population	Acres of Yams per Person
Southern Tivland			
MbaGor	152.35	477	.30
Iyon	163.50	1,047	.16
Northern Tivland			
MbaAliko	57.56	167	.33

Though Iyon have reduced the size of yam fields to about half that of the other areas, MbaGor have not. MbaGor people say that they would increase the size of their yam fields if they had sufficient land to do so. If Iyon yam fields were the same size per person as those of MbaGor, and all other factors were equal, Iyon would require almost three times the total amount of land available at present.

Much the same picture emerges from an examination of the total cultivated acres per person (see Table 8).

TABLE 8
CULTIVATED ACREAGE PER PERSON

Location	Total Cultivated Acreage	Population	Acres per Person
Southern Tivland			
MbaGor	327.75	477	.68
Iyon	426.05	1,047	.42
Northern Tivland			
MbaAliko	125.27	167	.75

MbaGor have cut the size of their fields very little. Iyon, who have only about half the amount of yam acreage per person as MbaGor, have two-thirds the amount of total cultivated acreage per person. This leads to the next point: the possibility of a change of crops.

If, by change of crops, new and different crops are meant, there has been no change. Iyon grow the same crops as do the rest of the Tiv. However, a change of proportion in the crops has taken place. The data in Table 9, as well as Table 10 and Figures 12 and 13, give some indication of the extent of the change.

The most obvious difference is the decreased acreage in millet and guinea corn in the land-hungry areas, and the concomitant increase in fallow crops, such as beniseed and cassava. These figures support our

TABLE 9
ACREAGE IN VARIOUS CROPS
(percentages)

Location	Yams	Millet and Guinea Corn	Beniseed (3d year)	Cassava	Fallow Crops
Southern Tivland					
MbaGor	46.5	30.3	10.0	3.1	10.1
Iyon	38.4	29.9	17.5	4.4	9.8
Northern Tivland					
MbaAliko	46.0	51.0	—	—	3.0

TABLE 10
BENISEED ACREAGE PER ADULT MALE

MbaGor	.24
Iyon	.27

observations that in the north guinea corn lasts until the new yams are ready to eat, leaving almost all the millet for brewing, whereas southerners are lucky if the guinea corn lasts until the millet is harvested and, further, if the millet lasts until the new yams are ready to eat.

Northerners usually have no hungry season. This fact is largely attributable to a much greater grain supply. Although the yam supply is somewhat larger, yams are staple for fewer months of the year.

The increase in fallow crops in the southern areas is in all probability the obverse of this comparative shortage of grain. Peanuts, Bambara groundnuts, and sweet potatoes are all grown extensively in the south. Although some northern women plant one or all of these crops, they seldom plant more than can comfortably be grown as side-crops on their yam mounds. They are used in the north as flavoring or snacks; in the south they form an important part of the diet. Northerners say they do not like them, that they are not proper food. Southerners, however, say that they do like them and that they are good solid food.

The increase in cassava proportionate to the shortage of land does not show up as clearly in these figures as it might if there were a more detailed breakdown. Cassava can be and is interplanted with all the other crops. In collecting the data on which these figures were based, it was necessary, therefore, to eliminate cassava when it was intermixed with other crops, lest the number of categories become utterly unwieldy; only those patches in which cassava was planted alone appear

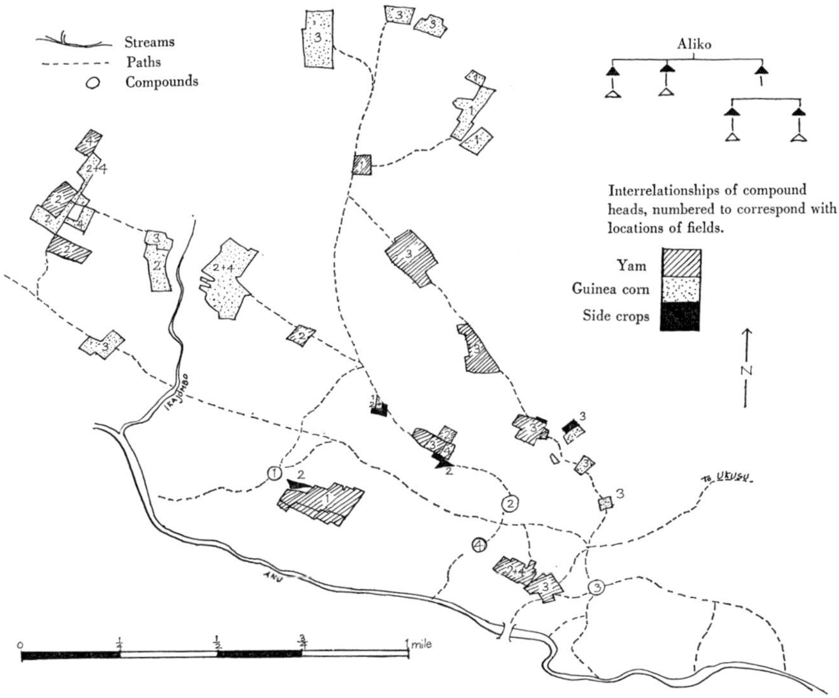

Fig. 13.—Land and compounds of MbaAliko.

in the figures. What is more apparent is that there is some cassava grown in Ityoshin, but only that grown on yam mounds, and by no means in all yam fields. In MbaGor there is much more cassava grown in yam fields than in Ityoshin, both in amount per field and in the number of fields, for in MbaGor almost all yam fields had cassava. Sometimes cassava was interplanted with kitchen garden crops; it was almost always done in Iyon. MbaGor seldom or never interplanted cassava with guinea corn; it was standard practice in Iyon. In almost no place in MbaGor was cassava interplanted with beniseed; it was common in Iyon. One man in Iyon knowingly took a chance of reducing the yield of his yam field by planting two cassava roots instead of one in each mound.

Tiv do not like cassava. Southerners say that a small portion of it can be added to yam porridge (*luam*) without hurting it; northerners say that the yam porridge containing cassava and sweet potatoes made by the people of central and southern Tivland gives them headaches and stomach-aches. No Tiv like cassava alone; all say that if you eat too much of it, stomach-ache and indigestion result. This is due in part to the fact that Tiv do not leach cassava to make flour of it as do Ibo, Hausa, and other Nigerian peoples. They usually eat it roasted. They know the techniques used by these other peoples, but do not employ them.

Another important, but not so obvious, divergence which comes to light in Table 9 is the beniseed acreage. It would seem unusual—at least it would demand explanation—that in an area as hard pressed for land as Iyon, almost one-fifth of the land is devoted to a "cash crop." This can be explained, however, by reference to another set of figures (see Table 10). Throughout Tivland there is a direct relationship between beniseed and tax. Beniseed is called "tax" (*kpandegh*) and not "food" (*yiagh*). This is the one crop for which the acreage per male is nearly the same in the two lineage areas. Men in both areas say they plant only enough beniseed for tax.

In the 1950 season the price of beniseed was 20*s.* per hundredweight, and the agricultural census figures made that year indicated that the average yield in Tiv Division was 200 pounds per acre. This figure was said to be below average because of unseasonal rains. We were in MbaGor and Iyon at the time and recorded no statements that a bad crop was harvested. If these figures are correct, each male received about 10*s.* for his beniseed, while the tax was 8*s.* 6*d.* per adult male that year. We left the field before beniseed sales began on a large scale, and before taxes were due, so we did not learn whether the amount of beniseed actually did cover the tax. We did learn, however, that when the tax was announced—it had been 6*s.* 6*d.* the year before—increases were made in the amount of land allotted to beniseed.

Beniseed acreage in the north is not strictly comparable because it is interplanted with guinea corn and ripens at a different time of the year. During the 1951 season, when everyone said the beniseed crop was poor because of heavy rains in October, the dozen or so men we asked about sales reported that they got about 15*s.* for their beniseed—enough to pay their 10*s.* head tax in addition to a cloth for the wife who carried it to town to sell it.

Any reduction which may have occurred in the length of the fallow period is not so easy to quantify. Tiv in MbaGor said that they had extensively reduced their fallow periods—too much to allow for proper revitalization of the soil. Iyon people said that they had nearly done away with fallow. One Iyon man whom we knew well planted millet and guinea corn in 1950 on a farm which had been cleared of yams the year before. Instead of planting beniseed there the following year, or letting the land revert to fallow, he intended to put a new crop of yams on it in 1951. They wouldn't be very big yams, he said, but by eating them mixed with cassava and sweet potatoes, he could leave some of his other fallow idle or in fallow crops for an additional year. He said this would make one year out of three more difficult. During that year he would have to sell cassava to the Hausa and Ibo in Obudu in order to buy yams from the Uge, but he would be better fed in the long run, because his land would have longer in fallow or fallow crops.

The last alternative for bringing into accordance agricultural practice and the amount of land available for cultivation—that of new and improved agricultural methods—needs careful consideration. Tivland is (except for a few places which have been cleared) tsetse-infested, and as a result animals in sufficient number to provide manure for fields cannot be kept. Most Tiv use no form of fertilization on their farms. Some of them do use human and animal excreta in their kitchen gardens when planting fruit trees, but they regard the idea with repugnance. The government agricultural service tried to introduce green manure crops and compost heaps. In Shangev and southern Kunav these methods were not successful. Tiv so far have not been made to see that if they plant part of their land with a crop which will be dug under to provide green manure, they will get much better yields from the land (though the northern practice with millet stalks indicates that the idea is not foreign to them). To Tiv, such a practice represents hard work with no visible gain, and—more important to them—it would (they say) reduce even further the amount of land available. Composting has been even less successful: Tiv are indifferent to green-manure crops. They are downright anti-compost: "Have *you* ever tried to carry such stuff on your head?"

When discussing the introduction of new agricultural methods, Tiv always come back to fallow. They insist that if there were enough land available, it could lie fallow until it grew the right grasses. It would then be as good as it ever was for growing food. They say they do not need green-manure crops or compost heaps but the abolition of the boundary between themselves and the Udam and Uge, their neighbors to the south, thus making it possible for them to take sufficient land to prolong the fallow periods.

Both points of view are correct: either longer fallow periods, with accompanying expansion, or fertilizers and improved farming methods will increase agricultural production in southern Tivland. Proponents of each view are not prepared to accept that the other has any virtue while the alternative exists. Here is a real clash of cultural values.

We have seen that there is a wide variation in the size of farms in various minimal *utar* and that the amount of land under cultivation per person and the proportions of crops are, generally speaking, directly correlated with the density of population and resulting land shortage. The variation in size of farms within the minimal *tar* is also greater in those areas in which land shortage is most severe (Table 11).

TABLE 11
TOTAL CULTIVATED ACREAGE PER PERSON

Location	Compound with Maximum	Compound with Median	Compound with Minimum	Ratio of Maximum to Minimum
MbaAliko (including a sample from Ukusu)	.77	.74	.58	1.33:1
MbaGor	1.79	.65	.40	4.48:1
Iyon	.92	.41	.14	6.57:1

In the northern sample, the compound with the most acreage per person has only 33 per cent more than that with the least. In the southern area of greatest land shortage, the compound with the greatest acreage per person has 650 per cent as much land as that with the least. Though a ratio of 6.5:1 between poorest and richest is truly very small, it is noteworthy that land shortage is one of the primary correlates of a growing economic differentiation.

Both median and minimum acreages per person decrease as the density of population increases. Yet the maximum acreage of cultivated land per person is higher in both areas of land shortage than in the area in which there is no land shortage. The cases discussed in Chapter 5 will help to explain this fact.

5

Work (*Tom*)

TIV SAY that work is a good thing. They spur one another to work with a chant: "Is work a bad thing? No, work won't kill a man."

Tom means work. But there is no Tiv word for the additional meanings of the English word "labor": "an instance of bodily or mental toil, especially when painful or compulsory" (*Shorter Oxford English Dictionary*). "Labor" implies a market on which work can be sold, and that portion of the population that actually sells its work on the market.

Tiv work hard, and they agree that "hoeing fields is the most important work of the Tiv." But they do not find their work painful (even when they get tired, or when they would prefer to do something else); they do not traditionally sell "labor" on the market, although they are beginning to do so. They do trade work for work.

The opposite of *tom* is not "play" among the Tiv, but rather "laziness" (*hwev*): shirking whatever responsibilities one must perform to assure one's place in a social group.

Yet "labor" is obviously as much a "factor of production" for Tiv subsistence as it is anywhere else. The question becomes, "How do Tiv organize and allocate their work for production in the absence of a simple 'economic' mechanism of allocation, such as the market?"

Sexual Division of Agricultural Labor

TIV ALL KNOW whether a specific task is man's work or woman's work. If one asks men why women don't do men's work, one is told that they aren't strong enough, that it would be immodest, or that they would run away from the man who required it. If one asks men why men don't do women's work, they say that such work does not require a man and that men have quite enough to do already.

The number of tasks which both men and women perform—especially those which they perform in the same way—is small. Spinning is one, so is pulling grass for new farms. Even in clearing farms there is a tendency toward complementary rather than identical tasks. Men break down the grass with the *aga,* fell small trees and lop the large ones, while women pull grass. When men pull grass, they do it the same way as women, but as soon as enough has been done to begin mounding, the grass-pulling is left to the women.

Men and women both plant yams, but use different methods. Occasionally a man and his wife work at this task as a team: one does the actual planting while the other takes care of rearranging the grass caps on the mounds, cleaning up, and leaving each mound in good shape. Harvest and husking of millet, guinea corn, beniseed, and rice is a joint effort between men and women, but each performs separate tasks.

Young and early middle-aged men do the heavy work of hoeing mounds and ridges for all crops. Most men of forty-five or so slacken their efforts because there are younger men to take the heavier portion of the work; the head of a compound of four or five adult males does little or no hoeing. Women keep mounds and ridges in repair by building them up again when crops are harvested, but this is a comparatively light task. Planting millet and guinea corn is primarily a task of men; although women may assist, it is described as man's work.

Women do all the weeding; it is a job no man will perform, except in his tobacco plot. Women weed with the small hoe, bending forward from the waist; they say it is a task that is never finished. Women plant and harvest side crops and harvest all yams. Drying and preserving vegetable food at the beginning of the dry season (December and January) is time-consuming but not hard work. Women carry crops and seed from field to compound or from one field to another.

TABLE 12
MEN'S WORK AND WOMEN'S WORK

	Men	Women	Both Sexes
Ihyande	Lay out farms.		Pull grass and clear land.
Avom	Hoe mounds.		
Sule		Plant and harvest side crops; weed; harvest.	Plant yams.
Akuur	Level mounds; plant millet and guinea corn.	Weed.	Harvest millet and guinea corn.
Tsa	Prepare for beniseed. Dig mounds for sweet potatoes, peanuts, etc.	Weed sweet potatoes, etc. Harvest.	Harvest, dry, and husk beniseed.
Pagh	Trench and hoe mounds and ridges.	Weed and plant.	Harvest and husk rice.
Ikongo	Make mounds, ridges, etc. Care of tobacco.	Weed, harvest, and prepare (by drying) fruit and vegetables.	

Table 12 explains men's and women's work by the type of field; Table 5 is a work calendar for Kunav/MbaDuku. These tables show that women perform a greater variety of tasks than men and are busy the year round; men have a free month between millet and beniseed harvest and the preparation of new yam fields. In Iharev/Ityoshin men do little farm work during the three months between preparing *akuur* for planting guinea corn in July and early August and the beniseed harvest in mid-November, after which they have the heavy work of hoeing mounds. Much of Ityoshin is flooded during these months; communication is difficult and working on the farms is impossible. Thus, there is little to do but brew, drink, and perform ceremonies.

When other tasks are considered, both sexes have work the year round. During the dry season men build houses and gather the thatch to roof them. Women pound the mud floors of the houses, construct their own fireplaces, and keep the environs of the hut cleaned and weeded. Men do all the weaving, both cloth and basketry, though both sexes spin cotton. Men gather materials for building, dying, basketry, and other tasks; women gather plants for sauces.

Some Tiv men do considerable hunting (see Chap. 9), but more in the dry season than in the wet, when luxuriant vegetation impairs visibility and swamps hinder mobility. However, hunting is more a prestige and recreational pursuit than a source of livelihood.

Women grind millet and guinea corn into flour and cook. Men clear the paths in August and September. Women tend their children (though after he is a few months old every child has a nurse, another child five to eight years his senior); men spend a great deal of time playing with small children and instructing their sons.

Men govern the country: in an egalitarian political system like that of the Tiv, this task weighs heavily on almost all men above the age of thirty-five. But unless a meeting concerning a political or jural matter is of utmost importance, it will not convene before 10:30 A.M. (Only the largest markets—and a very few with a highly specialized purpose—convene before noon. Tiv usually put in several hours of work on their farms before attending meetings or markets. Any market is "far" (*cha*) if one cannot put in a good day's work before or after attending it.)

Division of Labor on Other Bases

THERE IS SOME DIVISION of labor by age, although it is difficult to separate this from differentiation by status, because status is age-linked. A mature compound head continues to spend a great deal of time on his farms; he performs light tasks there, constructs *akombo,* and oversees the work of all the people of his compound. We know one case in which a compound head sent all of his people to their farms with admonitions to work with special speed and thoroughness because they were all behind in their agricultural activities. He then went off to a meeting which dealt with the welfare of one of the wives in the compound. His people smarted under his reproof, but acknowledged that he was right. When asked if he should not have shared their work, they said that his work was important and that only he could do it. Another compound head upbraided a woman for spending too much time with hair-dressing (for which she was paid by other women) and too little time on her farms: "You may have done that in the compound you left; we don't do it here!"

There are few farm activities peculiar to children. The main one is scaring birds away from the ripening millet, performed by boys between six and sixteen. We have seen adult men perform this task when there were no children in the compound, and occasionally girls, though people said it was too bad they had no brothers. Both boys and girls do a little farm work from the time they are about seven, but it is the same as that of adults of the same sex. Boys of seven or eight make some mounds—and most can do the job almost as well as their elders—but they do it because they want to. Almost no Tiv father has to insist that his young sons work; no Tiv father would demand that his sons overwork. A boy of fifteen can do and does a grown man's work—he is, in fact, a grown man.

There is some division of labor based on status alone. No government chief or "man of the *tar*" (*or taregh*) hoes mounds, although he oversees the work on his farms in the same way as any other compound head. He may call communal labor for such purposes as building a large reception hut in which court will be held. When we studied them, Tiv considered this not as work for the chief but for the British administration.

A woman's work decreases after she is old enough to have a junior attached wife or a daughter-in-law. Increase in the size of her group accentuates her managerial role, besides giving her younger people to perform the heaviest tasks. There is little difference in the type of work done by young and old women, but a woman's hardest work—including agricultural work—is done while she is of child-bearing age.

Division of farm labor on economic grounds is beginning. Successful part-time traders hire men to hoe mounds, most often younger brothers, who will take less pay. The trader may pay his compound head, who sees to it that all the young men of their compound hoe the trader's farms. A few people hire comparative strangers to mound their fields; more use the already existing forms of labor cooperation (described in the next sections), sometimes increasing the payment. One comparatively wealthy young trader in rice paid an organized hoeing party, which had come to work for another man, for two days' work after they had finished their previous task.

The Basic Work Group

DESCRIPTION OF GROUP FORMATION among Tiv is difficult. Large numbers of latent relationships can be activated. The casual, spontaneous quality of the original behavior is easily lost in the verbiage of principles and exceptions. Tiv fit easily into groups; they know which groups they "belong to." Describing group formation falsifies the pace of Tiv life. Work groups are no exception.

Communal work such as building a school or cleaning a market place may be done by the men of a lineage or an age-set who are accustomed to working together. However, men of adjoining compounds—almost always within the minimal *tar,* but not necessarily segments-within-the-hut—may band together to clear paths or to build a bridge which they all use.

Generally, the people of a compound form a basic work group. Many agricultural tasks—particularly those of men—are spoken of in terms of compounds and compound heads. However, a large compound breaks up into smaller groups in performing many agricultural tasks. The basis for forming smaller groups is inherent in the polygynous family: full siblings work together. This principle can be factored into two general rules: (1) brothers tend to work together; and (2) co-wives (unless they are attached) work separately.

All the men of Kyagba's compound (Fig. 2) form a single group to clear paths. When asked why this is so, they will probably answer with another question: "Aren't we brothers?" "Aren't we one man?" or "Isn't it our land?" If a man's sons by different wives work separately in hoeing mounds, the reason will be, "It is age-sets; my age-set will help me. But since my brother is younger his age-set will help him. Therefore we hoe separately," or an insistent, "We may hoe separately, but it is one field!" The investigator seldom gets beyond the stock answer, "Their hearts tell them to do so." A child may be more basic: "It is their mothers" (*Ka u-ngo ve*).

Work groups of women tend to be more circumscribed than those of men. Within a large compound like Kyagba's, there are several groups of women whose fields are close together, who work as a team, store their food together, and cook together or alternately. These groups are composed of a senior woman, those junior wives who are "put into her hut," and her daughters-in-law. There is no Tiv word for such a group of women, but we may call it a "basic work group." There is little cooperation between such groups. The three senior wives of the compound head maintained a friendly but fairly distant relationship. The only agricultural activity in which all the women of the compound worked together was clearing fields.

Not all polygynists attach junior wives to senior wives. It is rare in Iharev, where each woman, with her daughters-in-law (if she has any), has separate farms. Here, and in central Tivland as well, a woman and her most recent daughter-in-law share a farm. When another daughter-in-law joins the group, the older daughter-in-law is given her own hut and her own farm, though all usually continue to work and cook together.

The optimum day-to-day size of a working group of women is three or four. If there are more, they tend to form two groups; if fewer, they join to give mutual assistance. (Women who find themselves in an uncongenial group leave their husbands; Tiv say a man's success as a polygynist depends as much on the character of his mother and his wives as on his own.) It is this group of women who weed and harvest together, usually store their yams in a single granary, and share a single drying platform. One of the simplest ways for the investigator to determine such groups is to inquire about ownership of these platforms (see Chap. 10).

Men, especially during the lifetimes of their mothers, do most of their agricultural work as a complement to basic women's groups. However, there is a pull for half-brothers and father's brother's sons to work together. For example, the oldest living son of the head of the compound shown in Figure 2 was a man who could do prodigious amounts of work, but he did not like to work alone. He organized his brothers, and the group of six or eight hoed mounds noisily and with amazing speed for a day or two. Then he went off to market or on a visit; each of the others drifted back to the fields of his own wife and mother. Then the senior son returned and again whipped them up into a new fervor as a team. This was a personal attribute of this man, not a function of the senior son, although he might have found it more difficult had he not been the eldest.

As the link between "brothers" grows more distant (*cha*), the tendency to work together becomes weaker. In the compound already referred to, all of the compound head's sons except one had their yam fields adjoining. The head's half-brothers, however, had their farms in a different location. In no agricultural activities did these two groups work together, although they said that they were working separately only because the group was so large.

When full brothers have adult sons, they are likely to farm separately, although they continue to discuss the location of farms. Full brothers whose adult dependents continue to form a single production group may be accused of being lazy. A large group, Tiv say, encourages a man to feel that his share of the work is unimportant, whereas in a small group, he can more easily see that his efforts are vital in ensuring the food supply. Other men stated that fission of the labor groups of mature full brothers was connected with granaries: when a man's mother is alive, his wives and his brother's wives share her granary. After her death, each set of wives needs a separate granary.

This change in the size of male work groups is an aspect of the process of fission from the extreme identification of unmarried full brothers, through their growing apart with their different affines, separate huts, and ultimately separate farms. After their deaths, their sons may continue to live together, but in the next generation the compound is sure to split. No matter what permutations individual relationships may put on it, there is a tendency for brothers to work together and co-wives (except those who are attached) to work separately. With time and increase in size (two ideas which are almost indistinguishable to Tiv) the brother notion of unity grows weaker and the co-wife notion of disparateness grows stronger.

Cooperation among Women

There is one condition under which unattached co-wives work together. All the women of one compound, or of several nearby compounds, may band together two or three times a year for a hoeing party to weed the fields of each of them. Such a hoeing party is "called" by a senior woman, who supplies food and beer. The women meet early in the morning, accompanied by their daughters and unweaned sons. These groups are usually composed of fifteen to twenty women, but there may be as many as thirty-five. The senior woman directs the work of the others and, perhaps with the help of one or two other older women, acts as song leader. The group may break up into two or three parties, working close together. Each woman takes a row or two of yams, and, singing and dancing, they weed down the field. After a pause for breath and more dancing at the end of each set of rows, they begin the next set. Some of the songs are praise songs of the senior women of the group; others are songs and dances of a sexual nature. Men and women both say that no man would dare go too near one of these hoeing parties. Hoeing parties may convene for several successive days, although two days are usually sufficient for completely weeding the fields of all.

These parties serve two purposes. First, Tiv prefer to work very hard and at a terrific pace and then do almost nothing for a day or two. There is satisfaction in seeing a field completely weeded in an hour. Secondly, the parties also underscore the prestige of important women. If a woman calls a big hoeing party and supplies generous amounts of food and beer, she will be called "important woman" (*shagba kwase*) for months afterwards.

Other forms of agricultural cooperation among women are rare. Lineage sisters occasionally help one another. A young woman we know walked about two miles to help a very old lineage sister, whose daughters-in-law were away. Neither she nor the old woman touched the fields of the absent daughters-in-law. Since full sisters and even half-sisters are almost never married into the same minimal segment—they are, in fact, except in the case of twins, married as far away from their own lineage, in different directions, as possible—they cannot help one another. Occasionally all of a woman's daughters, some of them joined by attached co-wives, visit their mother and weed her fields. Any time a woman is visiting her own home, she works on her mother's field as she did when a child. If her mother is not present she works on the fields of her full brothers' wives, but not so diligently or so long.

Cooperation among Men

COOPERATION IN AGRICULTURAL TASKS is more formally developed among men and follows several kinds of relationships which have no equivalents among women. A man can call his best friend (*hur-or*), his age-set (*kwagh*), or a junior age-set to which he stands in a relationship known as *igba;*[1] he can call on individuals (usually not a group) from his mother's lineage. Women do not have best friends in a formal sense; they are not divided into age-sets; and their relationships with their mothers' lineages are more diffuse than those of men.

The biggest mounding party we saw was called by a man of about forty-five, Yilaun of MbaDuku/MbaYar. He called his best friend from the neighboring lineage of MbaYongo (a segment equivalent to MbaDuku), who gathered about twenty of his own agnatic kinsmen about him and came to MbaDuku late one evening. They brought drums and hoes, a hurricane lamp, and their best clothes. They danced at several com-

1. The primary meaning of *igba* is a person of the mother's agnatic lineage in any depth up to the level at which it is a segment equivalent with one's agnatic lineage. The term is reciprocal.

pounds along the way, to songs composed for the occasion and to well-known work songs. When they got to Yilaun's compound they danced until about 1:00 A.M. They began work the next morning at daybreak. This group was joined by Yilaun's agnatic kinsmen: all the agnatic descendants of his father's father, and a few youths with more attenuated agnatic links. Drawn by dancing and the promise of rich food, several young men of Yilaun's mother's lineage came, and two of his wives' brothers also arrived.

This group made yam mounds at a feverish pitch for about four hours. They were then given food, prepared and brought to the fields by all of the women of Yilaun's compound. They ate in groups of about half a dozen; each group was given a huge calabash of yam porridge with three different sauces. There was so much food they could not eat it all. Work was suspended during the heat of the day.

They began again about four in the afternoon, worked until dark, and were fed again. That night they danced until well after midnight. The next day the procedure was repeated, and all the farms of Yilaun's compound were finished. That night they had the biggest dance of all: Yilaun furnished pot after pot of millet beer and killed two small goats. They again danced far into the night, and the next day danced for almost two hours in the local market.

It was not until several weeks later that we learned what lay behind Yilaun's party. If a land dispute is decided against a man, few Tiv judges will make him give up yam mounds he has already dug on the disputed land; usually he is told that he can harvest this yam crop or the complete cycle of crops. Yilaun had driven deep into land claimed by a neighboring lineage, and—as he had predicted—the chief and judges told him that he had trespassed but that he could harvest the complete rotation of crops. If the mounds had not been completed, he would not have been allowed to hoe them.

In addition Yilaun reaped great prestige from his lavish treatment of the relatives of his best friend; during the dancing at the market place, his name was on the lips of all. The best friend also gained great prestige.

When hoeing mounds in large groups each man makes a diagonal row of mounds, but all work together in a single vertical row (see Fig. 10). This enables them to mound the field with regularity. As soon as a man's diagonal row reaches the side of the field, he goes to the other end of the line. Between forty and fifty men mounded about five acres of yams in two days.

Hoeing parties based on age-sets are usually not so large. Most men who call an age-set are likely to be younger than Yilaun, since men of Yilaun's age-set are no longer interested in or capable of great physical exertion. Young men of about thirty reckon their age-sets only to the limit of the minimal *tar,* at most the next larger *tar.* As a man gets older, the span of the lineage within which he recognizes age-mates increases, but the primary function of the age-set changes from mutual aid in farm work to mutual protection against witches. For men above sixty-five, the age-set may cover a lineage span relative to a *tar* of well over a hundred square miles, but it has almost no social function.

A young man who calls his age-set to mound his fields is expected to feed them, and his wife or mother brews beer for them. Usually, however, the age-set works for four or five hours, and does not reconvene.

Men of thirty-five or so sometimes call together the members of the youngest age-set (men of about twenty) and give them a feast of chicken and beer. Thereafter the men of the junior age-set call these elder men *igba.* In return the men of the junior age-set mound the fields of their *igba.* The relationship is a permanent one. The *igba* acts as adviser to the younger men in learning the difficult business of taking part in witchcraft moots and consulting diviners on behalf of age-mates. If he expects the junior age-set to mound his farms a second year he must reaffirm the relationship by another feast.

When a man "calls his age-set" (*loho kwagh*) or his junior *igba* age-set to work on his farm, he may not include all the age-set, even from his minimal *tar.* Sometimes no more than three or four age-mates work together. They are usually, but by no means always, given a meal. This sort of aid is usually reciprocated in the same year; calling a larger segment of the age-set is not.

Mature men may call their sons-in-law to mound their fields. We know of only one such case. A young man of about twenty-six was requested by his father-in-law, who lived fifteen miles away, to come with his kinsmen. Only three others would go with the young man—his younger full brother and two half-brothers. Three sons-in-law besides himself were also called. They worked for two days and were given food, but the father-in-law did not kill a goat or even a chicken. The young man considered such behavior niggardly.

Thus, Tiv activate many sorts of latent relationships for mutual aid in agriculture. The basic work group is connected with family structure and limited by residence, but for special tasks many other persons—

age-mates, kinsmen of best friends, one's own kinsmen or affines—can be called in. Most Tiv, especially young men, define age-sets and best friends primarily in terms of such mutual aid. Since farm work is the major economic activity among Tiv, the principles of organization are pulled into its orbit.

Amount of Work Performed

WE DID NOT COLLECT precise data on the amount of time Tiv spend in agricultural work. We made several attempts, and found that collecting such data was disproportionately time-consuming. Tiv have no unit for measuring labor. Because they do not make a connection between time and work, the quantitative idea of "man-hours" is impracticable; accurate data can be had only by direct observation. Some men spend much more time on their farms than others do, and Tiv make a direct correlation between agricultural work and sufficient food. A text collected in Iyon says, "I work hoeing my fields; I don't rest—hunger won't catch me. But some men run to markets and don't hoe or look after their food crops—they will be hungry."

Tiv also make a rough correlation between the size and appearance of a man's fields and the amount of work he has expended on them. One man said of a particularly large and fine field. "That is a lot of work!"

"Labor" is the factor of production in greatest supply. Land is sometimes short, capital is scarce, and tools are rough. Just as Americans who go to Africa notice the extreme "waste" of labor and prescribe labor-saving devices, so Africans who see Western production processes note the material waste and the dependence on capital and tools. To Tiv, work is a part of producing food directly, in a householding economy, and hence the very substance of family relationships. With such a view, they judge themselves adequate or inadequate as husbands and fathers, not merely as producers.

6

Land Rights: Social Relations in Terrestrial Space

IDEAS ABOUT SPACE reflected in English and related languages are so specialized—so refined to fit specific institutional situations—that it is difficult to understand other cultural perceptions of social ecology (Hall 1963; 1966). The concepts of "ownership" and "property" are highly specialized and are dependent on specific technological, social, and legal institutions which are by no means universal. Nevertheless, a large number of writers dealing with "land tenure" have assumed the English and American specifics and then tried to understand the general and the exotic in terms of them. Needless to say, it has not worked well.

The literature on land tenure is one of the largest—and one of the poorest—in all of social and legal science. Its poor quality is comparable only to that of the late nineteenth-century writings on "primitive religion," with the difference that "land tenure" had no Tylors, Robertson-Smiths, or Durkheims. Lévi-Strauss (1963) recently claimed that "totemism" could be understood in terms of nineteenth-century Western needs and philosophies, but scarcely in folk terms or in terms of twentieth-century social and cultural theory. In analogy, we claim here that "property" and "land tenure" and "ownership" studies in the ethnographic literature can be explained in terms of Western legal philosophy and economic history from the waning of the Middle Ages until about World War II. Only if we first consider Western history a specific example of a generalized phenomenon can we bring into focus the problems and data of essentially different systems of arrangement of society in terrestrial space, and exotic economic, political, and mythopoeic arrangements for linking men and space together.

Land Tenure

WHAT IS IT, then, that Westerners mean by "land tenure"? What do we mean by "land," and what by "tenure"?

Every people must have some view of the physical milieu—the "folk geography" of the world and their part of it. Westerners divide the earth's surface by use of an imaginary grid, itself subject to manipulations and redefinitions. We then plot the grid on paper or on a sphere, and correlate the physical features of the land and sea with this grid. We have developed instruments for locating ourselves on the earth's surface in relation to the positions of the stars. There are precise rules for symbolizing the information from the instruments. We have perfected a system of measurement which allows us to repeat precisely operations carried out in the past; thus we have been able to locate and measure pieces of the earth's surface and to record our computations on gridded maps. These measured pieces become, for some purposes at least, identifiable "things."

Land, whatever else it may be, is for Westerners a measurable entity divisible into thing-like "parcels" by means of mathematical and technical processes of surveying and cartography. This complex notion of "land," with its accompanying technology, is an essential of present-day Western land tenure, as well as of the Western market-oriented economy.

The ideas of "tenure" in our own system are even more tangled. First of all, "tenure" assumes a conception of "land" of the sort we have already described. Only if it is divisible and the divisions stable and measurable can "land" be "held." Only if land is cut up into definable units can it enter the market or, as the jurist sees the same phenomenon, be subject to contract. Contract and a land market constitute specific types of relationships between men and land.

However, holding a piece of "real property" is more than a mere relationship between a man and a thing-like piece of land. As the best jurisprudential opinion has long told us, it is a relationship among persons. "Tenure" has to do with rights *in* land *against* or *with* other persons. Thus, in addition to the man-land unit, usually called a "property system" by Westerners, we have a man-man unit, which is part of the social system in general and of the legal system in particular.

When Westerners discuss land in terms of the social or legal system, they do so in terms of "rights." "Rights" are attributes of persons against other persons. But in European languages, with the particular notions of land they reflect, "rights in land" can become attributes *of the land*. This "land right" links a person and a piece of land.

The equation of rights of people with rights in land, making one the obverse of the other, does not occur in most African societies. Therefore, when we turn from the Western to the Tiv situation, we must ask how we can classify Tiv attitudes toward "land" and Tiv versions of "tenure."

This matter can best be examined by generalizing the factors we have found so they can guide our work in exotic cultures. The three factors are: a concept of geography, a mode of correlating man with his physical environment, and a social system with a spatial dimension. These factors can be restated as axioms: (1) A people has a representation of the country in which it lives; that representation can be analyzed in analogy to the Western map. (2) Members of any society have a set of concepts for speaking about and dealing with the relationships between themselves and the earth they exploit. (3) The spatial aspect of any social organization has some overt expression in word and deed. The study of land in society is the way in which any people associates map, property, and spatial relationships.

Land Maps

It is, we believe, a fact that no African societies used indigenously an astrally based map such as our own, although a few peoples elsewhere, notably the Polynesians, did do so. Astral maps are, of course, seafarers' maps. A few African peoples did divide up the earth's surface into pieces by using terrestrial landmarks. The Kikuyu (Sluiter 1960) and the Haya (P. Reining: personal communication) are notable examples. In both cases, some form of land transfer similar to "sale" did exist under limited conditions; in at least the Kikuyu case, however, land "tenure" was regarded as part of the kinship system.

Other African societies have "tenure" relationship to land of a sort very different from "ownership" or "contract." Two of probably many types can be distinguished here. The first is represented by rain shrines among the Tonga (Colson 1954) or saints' graves among the Cyrenaican Bedouin Arabs (Evans-Pritchard 1949). Rather than pieces of land with boundaries, there are points on the land, marked by shrines, that form the center of activity of a social group; the "map" is a juxtaposition of geographical points intruded into the social system by a religious idiom.

In the second type of "non-property" system, of which the Tiv are an example, the social organization is conceived in terms of pure space, and is only incidentally linked with the physical environment for comparatively short periods of time by activities such as farming. Tiv see geography in the same terms in which they see social organization. As we have noted (Chap. 3), the idiom of descent and genealogy provides not only the basis for lineage grouping, but also for territorial grouping.

This "genealogical map" of Tivland moves about the surface of the earth in sensitive response to the demands of individual farmers as those demands change from year to year. Association of the genealogical map with specific pieces of ground is of brief duration only; a man or woman has what we have called "farm tenure," i.e., precise rights to a farm during the time it is in cultivation. But once the farm returns to fallow, the specific rights lapse. However, a man always has rights in the "genealogical map" of his agnatic lineage, wherever that lineage may happen to be on the terrain. These rights, which are part of his birthright, can never lapse. A mathematician has suggested that whereas the Western map, based on surveys, resembles geometry, the Tiv notion resembles topology—"geometry on a rubber sheet." The Western map must be rigid and precise if the principle of contract is to work; the Tiv map must constantly change if the principle of kinship grouping is to work.

Instead of seeing maps in terms of man-thing units such as property, Tiv axiomize the spatial aspect of their social groups. This "social map" leaves them free to question the ways in which social groups or individuals are attached to exploitational rights in the earth. This is the exact opposite of Westerners, who axiomize their maps in terms of their property norms and values and see the social system which results as fundamentally a series of contracts and hence open to change and to question. In short, Westerners can question the social group more easily than a concept of property, while Tiv can question their property-like concepts more easily than their social groups.

Land Rights

WITHIN THE MINIMAL *tar* every piece of cultivated ground, from the largest yam field to the most insignificant patch of cassava, can be referred to by somebody as "mine." Such a statement does not imply "ownership" of the land, but rather the "farm tenure" which gives that person some rights in the crops then growing on that farm. Such rights can be defined closely.

If the investigator is standing beside a farm in a minimal *tar*, A, and asks a man of a different minimal *tar*, B, "Whose farm is this?" he will be told, "It is the farm of minimal *tar* A," giving the patronymic. One is not likely to be successful in getting a more precise reply. If, however, one is standing beside the same farm in minimal *tar* A and asks a man of the segment associated with minimal *tar* A, "Whose farm is this?" the first reply will probably be, "It is ours." If the investigator asks which compound head the farm belongs to, the name will be given; the name of the woman who works the land and who will harvest and sell the crops or the name of her husband may be volunteered.

Again, if the ethnographer asks a woman whom he finds weeding a farm, "Whose farm is this?" she will say, "It is mine." In order to get further information one must ask, "Whose wife are you?" If, however, one asks this woman's husband, "Whose farm is this?" he will say, "It is mine." In order to get any further information one must ask, "For which wife did you hoe it?" or, "Who is your compound head?" If one further asks the compound head whose farm it is, he will reply, "It is mine." To get more precise information one must ask, "For which woman did you hoe it?" or "Which of your youngsters hoed his wife a farm here?"

In these answers are found the cluster of rights in a particular farm. A woman has rights in the farm: she owns and controls most of the produce which is grown on it; she has corresponding obligations to feed her husband and her children from the produce of her farm. Her husband also has his rights in the farm: he made it for his wife with the help of some of his kinsmen; millet or beniseed grown on it during its cultivation cycle belongs to him; he will eat from it and pay his tax from it. The compound head has other rights in the farm: he allotted it to the husband for him to make his wife a farm there; he has a right to a small portion of the crop for purposes of hospitality and ritual if he needs it; he has the final word in determining the size of the farm.

That farm is the symbol of the compound head's good faith in fulfilling the needs of his subordinates.

The map of Figure 2 shows the farms of most of the people of a large compound in MbaGor. It is meant to bring out the different portions claimed by different women, the juxtaposition of farms of the various women making up a basic labor group of attached wives or daughters-in-law, the larger portions claimed by the husbands of the women, and the whole claimed by the compound head.

Besides these rights in specific pieces of cultivated land, Tiv also claim rights to *sufficient* land. (1) Every wife or widow of a compound member has a right to a farm from which to feed herself and her dependents. (2) Every adult male who lives in the compound has a right to a farm sufficient to feed himself and his dependents. (3) Every compound head has a right to sufficient land to feed himself and all the members of his compound, including the rights of sufficient fallow land to allow for revitalization of exhausted soil.

Sanctions

BOTH SETS OF RIGHTS—to specific farms and to sufficient land—have easily recognized sanctions. Rights to specific farms are protected by ritual sanctions. Tiv erect magical emblems called *akombo* to protect farms against interlopers, thieves, and witches. This emblem consists of a bundle of sticks or a braided palm frond, a stone or a potsherd, placed on the ground or hung on a pole at the "top" of a field. The *akombo* are said to inflict disease or misfortune on anyone who meddles with the farm. The right to erect this magical emblem on a specific field is part of the basic cluster of rights in the field: a woman can erect an *akombo* on her own field; and her husband or the compound head can erect an *akombo* on this field. A woman may ask her husband to erect one on her farm, though more often he simply does it. A man can ask his compound head to erect an *akombo* on his farm; the compound head would not do so without being asked. No one else dares do so (unless he is paid by one of these three) lest the *akombo* attack the person who put it up.

In southern areas an *akombo* is usually found on any piece of cultivated ground, although some men do not erect them until late in the season. Except for *akombo* hung on economically valuable trees, they are rarely found on uncultivated ground. Fallow is not in need of *akombo* protection, although in some boundary disputes the emblems are brought into play.

Although an *akombo* would seize (*kor*) any person within the compound who took food he was not entitled to, it serves primarily as a protection against outsiders. Within the compound, moral sanctions of family (including co-wives) and co-residence are usually sufficient protection of one's rights to specific land.

Rights to sufficient land also have obvious sanctions and reciprocal obligations. A woman has a right to a farm sufficient to feed herself and her children. If she doesn't get it, she leaves. Her husband must either make her a farm or lose his wife and bridewealth. If a man is too ill to make his wife's farm, his brothers or age-mates hoe for her. Every adult male in the compound, then, must have a right to land on which to make farms for his wives.

An agnatic member of the lineage associated with the minimal *tar* has rights to sufficient land within that minimal *tar* because of his agnation. A man who is not an agnatic lineage member is given such rights for the period of his residence. If a compound head does not provide fields—and provide them willingly, with no fussing, nagging, or fighting—an aggrieved agnate follows one of two courses: he either moves out and builds a compound for himself, after which he can claim a part of "his father's fallow" and have the claim settled by the elders of the minimal segment; or he moves his family to his mother's people, where he is given fields. A non-agnate simply leaves. The sanction is the same as in the case of a wife: physical separation. The fear of "sitting alone" is acute among the Tiv. A man who wrangles with his sons and brothers soon finds himself "sitting alone," without labor, unable to work his fields. A bad temper or a surly, stingy nature "spoils the compound."

The right of every compound head to sufficient land to feed his dependents is of a basically different kind: a "right against the world." Generally speaking, the only sanction among Tiv to enforce a right against the world is self-help, traditionally in the form of fighting or "fighting by night" (the Tiv euphemism for witchcraft). Self-help calls into play all sorts of subsidiary rights and duties, the sanctions of which are found in lineage, kinship, religious, economic, and other values.

If a compound head does not have sufficient land to feed the members of his own compound, he exercises his right against the world and takes more land. His right to do so is limited only by the possibility of his being driven from it. The procedure for taking more land, however, is circumscribed. Except for migration—a very different matter—it is always accomplished by extension of "present holdings."

The following example of such an extension shows its main points. The boundaries of fallow are almost never marked. Markers, especially within the minimal *tar,* are universally deplored. It was, however, known throughout the minimal lineage that Andza had rights to land next to Ikpaor, Doga, and Iyorver (see Table 13). Andza, Doga, and Iyorver had a common father's father's father, but their agnatic link with Ikpaor was still another generation removed. When the time came for Andza to farm fallow he had used several years previously, he therefore extended his former holdings toward Ikpaor, not toward his closer agnates, Doga and Iyorver. In areas of land shortage Tiv say, "Would you cause hunger in the compound of your mother's son (*wanngo,* 'a close relative') if you can get land from your father's son (*wanter,* 'a more distant relative')?"

TABLE 13

LINEAGE RELATIONS IN A LAND DISPUTE

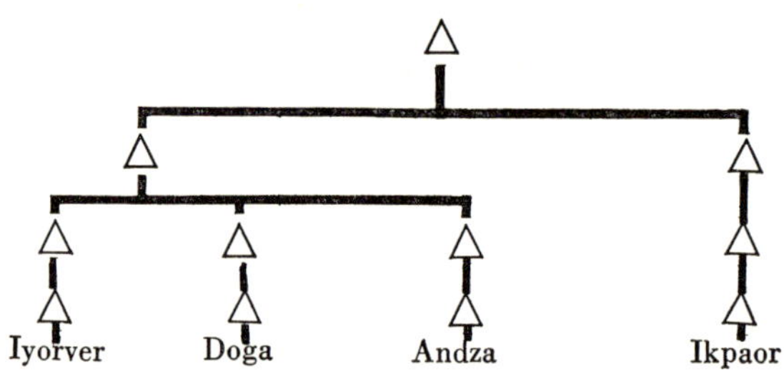

This pattern was vividly apparent when a group of Hausa established a village at a market site within the area of MbaDuku. The presence of resident Hausa is the mark of a big market. Most MbaDuku favored a Hausa settlement. After considerable negotiation, a site was selected in the *tar* of MbaKov (a lineage of MbaDuku/MbaIkaa; see Table 1). MbaKov people agreed to have the Hausa on this site, but they foresaw a problem sooner than did the others. At this site a market place and the Hausa village were laid out. About twenty acres were taken out of cultivation. No food crops were destroyed; all were harvested. But the fallow and fields covered by this area were claimed by six different compounds.

We asked one of these compound heads if the new settlement did not run him short of farm land. "Oh, no," he replied. "All MbaDuku has given land to the Hausa."

We didn't understand. "You *saw* us," he continued. "All MbaDuku met and decided that this was the place for the market and the village to be built."

"But it's *your* food land!" we insisted.

His reply stated the Tiv view precisely: "All MbaKov [the minimal segment] are the sons of one mother. When you need more land you go to your brother and you say, 'My brother, give me land on which to make a farm for my wife so that she may eat.' The brother says, 'Of course' (*Ka sha mi*), and if he in turn is short of land he goes to the brother at his back [i.e., the person who bounds his land but who is more distantly related than the first man, to whom he gave land] and says, 'My brother, will you give me land so that I may hoe a farm for my wife, so that she and her children may eat?' He says, 'Of course,' so the man hoes the farm. That man then goes to his brother in MbaAji [the companion minimal segment] and says, 'My brother [using the same term, *wanngo*], will you give me land . . . ?' " and so on.

Discussing this later we received a fuller explanation. Our informant drew a diagram on the ground. First, he said that MbaKov land looks like this (a circle), and it is surrounded by other *utar* (which he named; see the larger concentric circle of Fig. 14[A]). Then he explained again the process of extending holdings at the expense of the genealogically most distant kinsmen whose land adjoins your own. As he went on he changed the diagram until it had the appearance of that shown in Figure 14(B). "You see," he ended, "all MbaDuku—all Kunav [for the concentric circle surrounding MbaKov could represent either]—are giving land to the Hausa!"

While the Hausa community was being laid out, the men who were being dispossessed from part of their farms shouted and argued foot by foot. They were opposed by the government chief and a few of the more vocal elders; the Hausa very wisely stayed out of the dispute. On the surface one might have assumed that the chief had a right to move his people off a given piece of land or that he had some sort of superior right in it. No Tiv, including this chief, saw it that way. By shouting over this specific piece of land the farmers were impressing on all comers their

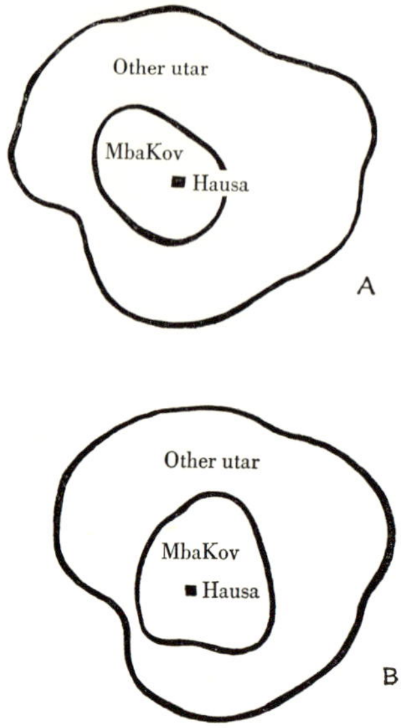

Fig. 14.—Sketch of land adjustment for a Hausa market.

rights to sufficient land. When the time came for them to take more land, they wanted their case to be remembered. The shouting and argumentation were, in a way, a mnemonic formality.

Rights of women always depend on residence and marriage. A woman has rights in her farms in the *tar* of her husband as long as she is resident there. She has a right to sufficient land in her husband's minimal *tar* as long as she is married to him (or her levir), whether the right is exercised or not. Women have no land rights in their natal *utar*, and few live there once they have married. If one does, she works on "the farms of her mother." Her brothers may hoe her a very small farm for a year or two, but she is still said to be "tending" the farm of her mother. No one has an obligation to her; such an obligation to an unmarried woman is incumbent only on sons and brothers.

Rights of men depend on agnation or matrilateral kinship and residence. A man has rights to sufficient land, as well as to specific cultivated land, in the *tar* of his agnatic minimal segment. Although these rights can be exercised only while he is resident in this *tar*, they never lapse. Rights of agnates can be revoked only by ritual exclusion from the agnatic lineage. This was formerly possible if a man changed lineages when the children of complementary exchange marriages were equalized (Akiga 1939:118–20), or if he were sold into slavery for his crimes. Today both these possibilities are prohibited by law. If a man is accepted as a resident by a non-agnatic kinsman or friend, he is given land. His rights lapse when he leaves that *tar*.

The right of an agnate to a specific piece of land within the minimal *tar* is often in question, but in the several score of land disputes we saw from 1949 to 1953, the right of an agnate to sufficient land was never questioned. No informants could discuss, even hypothetically, a situation in which an agnate was refused land in his minimal *tar*. None got beyond the platitude, "You can never refuse your father's son." To do so would be an act canceling agnation.

Thus land rights are a part of the birthright of every man; they are an aspect of his social and political status. To be without land is to be without kin. But a man has specific land only when he is in residence. On his return after an absence, he settles among his full brothers, beside his half-brothers. This agnatic birthright carries with it rights to land intermixed with, but distinct from, that of a man's full brothers. Their land is intermixed with, but distinct from, that of their half-brothers. The land of the entire group is intermixed with, but distinct from, that of other men of the compound. Land rights are a spatial aspect of agnatic filiation.

Land rights are simple among Tiv; they are, in fact, so simple that the anthropologist must state some of the things which they are *not* in order to avoid being misunderstood.

"Communal Tenure"

TIV LAND TENURE is *not* communal tenure. It is a matter of regret that this concept, which Malinowski (1935 1:319) called "the undying fallacy of anthropological work," has refused to die. In a fully developed, contractually oriented society like our own, communal ownership can and does exist. That is to say, the commune, whatever its nature, is regarded as a jural person—a corporation aggregate capable of legally owning property. The difficulty arises when this legal fiction is applied to African land systems. "Communal tenure" is an illusion that results from viewing the systematic exploitation by kinship groups of their environment through the distorting lens of Western market-oriented and contract-dominated institutions of property and ownership. Sir Henry Maine pointed out long since that in a community based on kinship, land is an aspect of the group, but *not* the basis for grouping.

Rights are attributes of persons. "Titles" are attributes of land. Tiv rights of persons in specific farms and to sufficient land are all individual rights. Tiv do not recognize any concept of "title." No matter how many other persons have another right to a piece of land in which X has a right, X's individual right still remains. A woman, her husband, and their compound head each has an individual right in a given farm. No one else has those particular rights. To change the frame of reference from the people to the farm by saying that the farm is the "community" property of some members of the compound is to distort the Tiv view so greatly as to make comparative study banal.

Tiv always use a *person's* frame of reference. They never use the *land* frame of reference with its insistence on some concept of "ownership." There are several references in this report to "our field" and "our fallow." This meaning of "our" is analogous to the commonly heard "our wife." "Our wife" means a woman who is the wife of a man with whom the speaker is identified, usually by lineage membership. It certainly does not indicate polyandry. Equally, "our farm" means one to which one or several individuals with whom I am associated can claim rights. It does not indicate "communal ownership" or even collective rights.

Inheritance, Rental, and Sale

LAND RIGHTS can be acquired in two ways: by status and by gift. "Inheritance" cannot give one greater rights than one already has. To make this point completely clear, we shall discuss inheritance briefly. Tiv do not have a word in their language which corresponds to the English word "inherit." Both Tiv dictionaries define *dyako* as "heritage"; nevertheless, it has this meaning only by extension. While a man is alive his *dyako* is his property: wives, spears, reception hut, clothing, crops, and livestock. When he dies, his heir is said to "eat" (*ya*) the *dyako*. One may also "eat the *dyako* of so-and-so" before the death of the owner. This sentence refers to major gifts from one's elders, and was said of gifts we made from our belongings before leaving the field. Basically, *dyako* means "personal property." It includes seed yams, but not land or farms.

To succeed a man to a status or position is to "sit on his head." This phrase is also used when a man's son, younger brother, or recognized agent deputizes for him. A new compound head either "sits on the head" of his deceased predecessor, or, most commonly, is said to "eat the compound" of his predecessor.

Land is relevant in the context of splitting a compound or splitting the field (see Chap. 2) but not in the succession to the status of compound head or inheritance of *dyako*. The Tiv state two "rules" which sound enough like "inheritance laws" that they should be examined carefully: (1) At the time the fields are "split" a man has a right to that portion of his father's land which was farmed by his mother, and (2) if full brothers split their land, they say it is split on the basis of what has been farmed by their respective wives.

We were discussing inheritance with a son of a polygynist who had three sets of sons. The young man assured us that each set of sons would take the land of their mother; he pointed out the actual yam fields of these women as the indication of where each group of sons would farm. We asked, "What will happen if your group gets larger than that of your half-brother? What if you become twenty, but they become only ten?" His immediate reply was "He will give me land; we would not split the field." But we asked what would happen if the half-brothers too were short of land; would he still get land from them? "That would spoil the compound," he said seriously. "I would get it from my father's son" (i.e., from a still more distant agnate).

What Tiv are in fact saying is that they take their position in the lineage and the rights that go with it from their fathers, and that their separate mothers account for the internal structure of the group. Although they see the lineage in spatial terms, they do not see any correlation between genealogy and specific farms. They see the genealogy as evidence of rights to sufficient land.

These concepts do not pretend to allow for equal sharing of farms among all of a man's sons and brothers; rather they allow for every man's right to land, in view of his genealogical position. Indeed given the migration patterns of the Tiv (Chap. 7), it would be impossible for a man actually to use the ground that his mother farmed. He does, however, "inherit" his position in the system through her and through his father.

A Tiv may be given temporary rights to reside and farm in areas where he has no agnatic rights to the land. He pays no one for this land; his kinsmen say that they "gave him a farm." However, his rights lapse when he leaves. In a few exceptional cases a farm is "given" for the length of a rotation cycle to non-residents. In the land-hungry areas of southern Tivland, a compound head who finds himself with many absent dependents—hence a nominal right to more land than he can use without making his fields too big to work (thereby leading to envy and witchcraft)—grants the right to make a farm to a man living in a neighboring minimal *tar*, and filiated to the lineage of the neighboring *tar*.

In MbaGor, Tsavkor found himself "sitting alone" because people were convinced that he had perpetrated a particularly heinous piece of witchcraft. He knew that if his farms lay fallow for more than the minimal time for bare recovery, he would lose them through occupation by his neighboring agnates. His problem, then, was to keep enough land in legitimate use to be able to retain it against the day when his people returned to him or to his successor. He solved his problem by giving land to four different individuals who were living in the companion minimal lineage, MbaWandia. Two of these men were his matrilateral kinsmen. The third recipient was a daughter's son of MbaWandia, whose position there was weak and who was not a strong enough personality to gain as much land from his host as he thought he needed. The fourth was an agnatic member of MbaWandia: a quiet, meek man who was the younger full brother of one of Tsavkor's most cherished adversaries.

These men hoed farms for their wives on Tsavkor's land. The trilogy of rights to these farms shows some important differences from the ordinary. Tsavkor had no right to any of the produce, and certainly he received no rent in money. He had no obligation to see that a non-resident farmer of his fields had *sufficient* land. The non-resident farmer had no obligation to take part in any land action brought by or against Tsavkor, although his interest might lead him to do so. The rights of the non-resident farmer are temporary, almost never exceeding a single crop rotation.

To call such persons "tenants" seems a marginal usage of the term. No question of residence is involved, no question of rent—even in kind or as token—enters into the matter. It is solely a device on the part of the "giver" to maintain his rights in a specific farm against the rights of his agnates to sufficient land.

Tiv are unequivocal (once you have made them understand the question) in saying that they cannot and do not rent or sell land. There is no word in Tiv which even remotely approximates the English word "rental," nor is there a word that can be used in respect to land that might be translated as "pledge" or "mortgage." Tiv of Kunav and Shangev are not unfamiliar with the idea of rental, however. They knew the Hausa colony in Obudu paid a small annual ground rent which went into the Native Treasury of the Obudu District. Tiv consider this procedure immoral. To them it smacks of taking money for hospitality, thereby nullifying the hospitality. One Tiv summed it up by saying, "Isn't a guest a guest?"

The Hausa settlers in MbaDuku certainly paid no land rent; they said so themselves, the Tiv all said so, and we believe that we obtained enough information on the money that changed hands among them to be sure that nothing even resembling a rental for the village site was ever considered. They did pay for labor performed and for building materials, but they did not pay for the land on which they lived or had their garden plots.

Whenever we asked too insistently about rentals, Tiv always asked, "Are *you* paying anything for permission to live here?" and we had to admit that we were not.

Though no cases were found in 1950 of land being farmed for money or produce rent, we were told that about three years before our study one Iyon man had offered the Uge £6 for the right to farm a complete crop cycle in their country. The particular people whom he approached agreed to the arrangement, but the elders of the Uge refused to give their permission, so nothing had ever come of it.

When we returned to MbaDuku in 1952, we discovered one case in which a man of MbaGor had accepted 3*s*. from a man of MbaWandia for permission to clear some swamp land and plant rice. The MbaGor man said it was for one year only, and his swamp land would be cleared for him.

We also found an account in old administrative files about an occasion on which Gav (companion lineage to Kunav) encroached on Ogoja Division tribes: they were fined heavily but were allowed to retain the farms they had planted. As soon as the people and administration of Ogoja Division discovered that the Tiv thought this an excellent bargain, they no longer allowed Tiv to retain the farms.

All Tiv deny that they sell land. When pressed they are likely to ask, "What would we sell?" To Tiv, such an answer is sufficient. Land is, for them, a unique category. Nothing is exchangeable for it. Tiv could not exchange land for wealth to use as bridewealth as do the Jaba Kwoi (Cole 1949) because land and such wealth (or women themselves) have, in the Tiv view, nothing in common to make them exchangeable. On the other hand, they do not exchange land for other land (Cole 1949 cites such cases among the Kagoro; Jones 1949 mentions it as an Ibo practice) because land is associated with a segment in a lineage system, and hence with agnation.

There is one other, and even more important, reason why Tiv cannot sell land. We have seen that rights to land follow residence rights and are of two kinds: permanent rights to land and temporary rights to farms. No one can *buy* rights to land in a *tar* in which he is living; he has a right to sufficient land already. No one can *buy* land in a *tar* in which he is not living, because he is not living there. No one can buy a farm because it is a part of his living arrangements.

If land runs short in one's own *tar,* one takes land from an adjoining *tar.* All Tiv see the *tar* as expanding with the synchronous processes of passage of time and increase of population (proved to Tiv satisfaction by reference to lineage genealogies). This has led to the doctrine, *not* of transfer of land, but of movement of people and social groups, as the next chapter will show.

Summary

THE TASK of this chapter has been to examine the ways in which "land" and territoriality are institutionalized by the Tiv in the absence of market allocation. In carrying out that task we saw that the relationship between people and things, which in English is translatable into a set of "ownership" ideas, backed by "property" law and deep regard for the property of others, is seen as a social relationship by Tiv. Such a task is one of the most difficult that anthropology presents: understanding phenomena as they are manifested in the absence of simplifying ideas. Yet only by so doing can we see the significance of the simplifying idea as well as the need for it.

7

Going to the Farm (*Udzan shin Tiev*)

IF ALL AGNATIC MEMBERS of the minimal segment, and their wives and the kinsmen living with them, have rights to sufficient land in the minimal *tar*, and if the technique for asserting these rights is to push one's genealogically most distant neighbors, two inferences can be made: most land disputes among Tiv are boundary disputes, and Tiv as a whole are expanding their territory. As a matter of fact, all land disputes are about boundaries, and Tiv are moving outward in every direction at the rate of from four to thirty miles a generation.

Boundaries

THERE IS NO precise equivalent in the Tiv language for the English word "boundary." There are, however, three separate "boundary" concepts.

(1) *Dondo* means "to be next to" or "to follow next after." My next younger full sibling "follows after me" but "I follow after" my next older sibling. The same word is used to say, "His compound is next to ours." The connotation is not of distance but of adjacence.

Dondo is best translated by the word "march," in such expressions as "His field marches with mine." In both English and Tiv, emphasis is on the fact of being next to one another rather than of limiting each other. This word "march" (*dondo*) is used in an emotively neutral sense.

The same general and neutral idea is expressed by the word *kighir,* "to come to an end." One often hears Tiv say, "My field ends here." This word ignores the person one bounds.

(2) The second idea is *ndoro* (also heard *ndoor*), "the path between the farms of co-wives." It may also be applied to the path between any two farms. This word, to be translated "border," has an inclusive connotation. The "border" indicates the point at which one wife's rights stop and another wife's start, but the husband has rights in both farms. *Ndoro* expresses the spatial delimitation of rights, in circumstances in which the two persons separated are joined by strong ties in other spheres. *Ndoro,* Tiv say, is a good thing.

(3) The third idea is "the splitting of the *tar*" (*pav tar*). This notion comes closest to the English word "boundary," with its emphasis on the limitation of spatial rights. It is stronger, however, because it emphasizes the isolation between people or groups. *Tar* is always associated with a social group, and the fission of any social group is said by Tiv to be "a bad thing," no matter how justifiable in any given situation. "To divide the *tar*" indicates a breach in social relationships.

Which of these words a southern Tiv will use depends on his idea of how "far" (*cha*) the opposing group is from his group. Two lineages can be "far" in one context and not in another. If an adjoining *tar* is "not far," by definition their lands are separated only by a "border"; they are separated by a "splitting" of the *tar* if they are "far."

There is a time factor inherent in these words. "Border" is a temporary matter; when the crop rotation is over it ceases to exist and is forgotten. "Division of the *tar,*" on the other hand, is a word suggesting a long time duration, even permanency. "Border" lasts for a crop rotation; "division of the tar" lasts until it is redrawn. The border demarcates an area in terms of temporary and equable rights; division of the *tar* demarcates an area in terms of permanent and, for that reason, ultimately inequable rights.

Tiv may call the same boundary "border" and "division of the *tar*" under different conditions. In Iyon, for example, the rather fluid boundary with MbaKunu, the companion segment to the north, is always called "border." Iyon people, by constantly correcting us in this case, pointed up clearly why this boundary must not be called "division of the *tar*": "You and your brother," they told us, "cannot split the *tar:* MbaKunu are our brothers." But the boundary dividing Iyon from Shangev/MbaKyar to their west was sometimes called "division of the *tar,*" sometimes "border." Both lineages have high population densities; each exerts heavy pressure on the other. Part of this boundary once required demarcation by a district officer in order to settle a land dispute. Tiv say he split the *tar,* and each side feels that his decision was unjust.

The boundary of Iyon with the neighboring Uge people is, on the other hand, almost always called "division of the *tar.*" Those people in Iyon who make an effort to speak of the "joining of the *tar*" (*ijuantar*) instead of the division of it, say that they are now at peace with the Uge, so they shouldn't really speak about the "division of the *tar.*" Nevertheless most of them do, especially when citing their grievances.

Among the northerners of Iharev/Ityoshin these words have much the same meaning but different connotative values. Here people use the word "border" almost exclusively; they mark no boundaries, but rather mark the high-water points of the swamps. They also build mounds along paths in this area to indicate "the meeting of the paths" (*ijuangbwenda*)—the point to which people of different compounds are responsible for the upkeep of the path.

All Tiv, particularly southerners, dislike boundaries. Except for pressures created by land shortage or external administration, they would probably not use them at all but to mark off temporary rights to specific farms or to indicate the point of juncture for path clearance. For individual rights to be recognized, boundaries of some sort must be acknowledged, but Tiv consider it important that they be temporary. It is the tendency of boundaries to become permanent (as occurs particularly in areas of land shortage) that Tiv do not like.

A record preserved in a government file from the early 1920's, only a few years after the Tiv had been brought under effective administration, puts the Tiv viewpoint very well. A land dispute which led to fighting was reported between two lineages in northern Kunav (each the size of MbaDuku). A young assistant district officer was sent to settle the matter. His report tells of the reaction to his presence: "Split the *tar?*" the Tiv said to him. "No, don't do that!" The A.D.O. insisted that he had been sent down to make a boundary and that he intended to do so. The elders considered this for a long time, in closed council, and finally emerged with a suggestion for the A.D.O.: "If you really want to make a boundary, why don't you go make it between Gav and Shangev Tiev [some thirty miles away]? They fight about land more than we do; they need a boundary much worse than we do." Try as he might, the A.D.O. could not get enough information to make his boundary.

To Tiv the important fact is that two *utar* adjoin one another. The specific line where they adjoin is of little importance and certainly is not stationary. But when the relative position of a *tar* changes—that is, when it comes to bound someone new, or not to bound a *tar* as it formerly did—Tiv feel not only that their territory has been stolen, but that their most fundamental rights have been challenged. Tiv say that movement is made necessary by increase of population, which is inevitable with passage of time. But they are most emphatic that there should be no change in relative position.

Land Disputes

LAND DISPUTES, like hunger, are acute in areas of land shortage.

In northern Tivland we heard of only one dispute, and that over space in the kitchen gardens for a tobacco plot. This dispute arose between two men who had a good wrangle every few days about something or other. It was settled by relocating a path, the suggestion having come from the wife of still another member of the compound. When we suggested that these men were disputing about land (*kpelan nya*), people agreed, but commented that if they did not dispute about land they would do so about something else.

In the south matters were very different. During the latter half of October and most of November, the "chief" and *mbatarev* (who acted as judges) did almost nothing but settle land cases. Many other such disputes were heard and more or less settled by other arbitrators.

Tiv never dispute whether a piece of land belongs to one man or another. The important right is his right to sufficient land, determined by his place in the lineage system. If agnates dispute a piece of land, the task is to determine a border (*ndoro*), not to decide for one or the other.

Every Tiv compound head has a right to enough land for his needs. He expects his actual geographical location to move from year to year, but he expects to maintain the same set of neighbors. In the land-hungry south, this right to enough land must be defended. The method of defending it is clear-cut. Whenever your most distantly related neighbor begins to pull grass for a new field, you object (*venda*). You first object to him, telling him it is your fallow. He will undoubtedly say that you are wrong, and that it was farmed by his father and grandfather before him (though the whole lineage may have migrated into the area only a few years before). He will also claim that his land goes well beyond the point at which he is pulling grass. You then call in an arbitrator to whom both of you agree (if only tacitly) to listen. If such a person cannot be found (and it is often difficult to find an arbitrator), you ask the chief and judges (*mbatarev*) to decide the boundary of the farm. Almost never (the occasional exception concerns cases of dispute between large-span lineages) is a division of the *tar* created. Neither you nor your adversary wants that. The point is to create an *ndoro* to mark farm rights, not a boundary (*ipavtar*) to demark rights to sufficient land.

When the chief and judges send word that they are coming to settle your dispute, you gather as many of your dependents and kinsmen as possible, to indicate great need, and you all proceed to the field in question and make as much noise as possible—shouting claims and insults, or simply shouting. The judges must be made to think, "Now Andza there—he obviously has a great many dependents; we must give him this farm."

The judges (*mbatarev*) may indicate a border or they may procrastinate. If they give a decision, the parties usually abide by it, though under loud protest. Tiv accept such decisions because the settlement is only temporary; another dispute will come up again in a few years about the same border. Both sides think the decision has been given against them.

If the judges do not give a decision, a compromise is reached just short of the point that either side would be willing to fight for. We recorded two skirmishes growing out of border disputes in MbaDuku in October, 1952, in which the weapons were stones.

Seldom do Tiv take these land squabbles seriously enough to let them interfere with friendships and normal social relations, but this depends greatly on their genealogical distance. If neighboring compound heads are very far apart in the lineage structure, their only relations may be those of hostility.[1]

1. Five representative cases of land disputes were given in detail in *Tiv Farm and Settlement* (Paul Bohannan 1954:53–57); they will not be repeated here.

Tiv occasionally settle border disputes amicably. We discovered one instance in which a MbaDuku/MbaWandia man and his northern neighbor of MbaGbera decided where they would split their fields. The boundary between the large lineages, which both recognized, was on a hillside running in such a direction that laying out fields along it was impractical. The MbaDuku man extended his fields in one direction into MbaGbera, and the MbaGbera man extended his in the other direction into MbaDuku. They felt that both had properly laid-out fields, which would have been impossible had such an arrangement not been made. Such arrangements may be more common than our few instances indicate; in Tivland, as everywhere else, one hears much about disputes, but little about quiet agreements.

Going to the Farm

IF, AS WE HAVE ALREADY NOTED, every Tiv has a right to sufficient farm land to meet his needs, and if every Tiv who needs more land expands against the neighbor most distantly related to him, any increased demand for land causes every *tar*, at every level of segmentation, to expand. Tiv see their country as a "nest of boxes" of expanding *utar*, each containing a series of expanding *utar*, each of which in turn contains a series of expanding *utar*.

Tiv say that population increase is natural, and cite their genealogies (Laura Bohannan 1952) to prove it. Although Tiv are in fact increasing, this is only one of several factors leading to the increased demand for land.

One of the most important additional factors is that Tiv have steadily increased production of beniseed for the "international market." Tiv grew beniseed before the advent of Europeans, but used it only as a condiment. When beniseed became a cash crop, its acreage increased manyfold. Manifestly, there has been an increased demand on the land. The fallow time has thereupon to be increased, which again necessitates more land.

In addition, new transport facilities, as well as safety for travelers and long-distance traders, have enabled Tiv to sell foodstuffs in non-Tiv markets. Every day truckload after truckload of food is exported from Tivland; as a result more land is put into production.

With population growth and increasing demands for land, the agricultural practices of Tiv have led to expansion of many (Tiv say all) *utar*. They speak about this expansion and the migration which is part of it as "going to the farm." This phrase has two aspects, which Tiv do not differentiate. One is a reflex of the fact that a man can build his compound on any piece of land which he claims a right to farm. Therefore, whenever he moves his compound, he quite literally "goes to the farm." The other is a reflex of the fact that Tiv apply to social groups the same idiom they apply to individuals. That is, Iyon (the minimal segment) can "go to the farm" in a way exactly analogous to the way that Ajeje, one of its members, has within the last few years moved out of his half-brother's compound and "gone to the farm" to build his own compound. Thus, since everyone expands at the expense of his most distantly related neighbor, those minimal *utar* on the outside rim of Tivland have a direction of migration which is explicable in terms of expansion of more inclusive lineages and *utar*.

Every lineage segment on the periphery not only feels itself to be an expanding unit, but—since it is a part of a larger expanding unit—to have a definite direction of movement. The moves should be carried out in such a way that there are no changes in relative positions—no new neighbors. Two examples of migratory movements that flouted this norm are enlightening.

MbaKyar lineage of Shangev MbaShaya has been greatly censured for getting out of position in one of its moves. The position of MbaKyar in the lineage system is shown in Table 1; the sketch maps of the migratory movements are shown in Figure 15. Tiv (including MbaKyar) say that MbaKyar should be in a position behind MbaKertiev, the segment which subsumes Iyon. In fact, they were formerly and ideally situated behind MbaPe lineage of MbaKertiev. Some time ago MbaKyar followed parts of MbaKertiev in such a way that it was behind MbaKunu and Iyon rather than behind MbaPe (see Fig. 15[A]). This was not a particularly serious matter, because the spatial relationship of the two equivalent segments in the lineage system—MbaKyar and MbaKertiev—had not changed. MbaKyar bounded MbaKertiev on the east and south and MbaDuku of Kunav on the west. MbaKertiev bounded the Uge on the east, the Udam on the south, MbaDuku on the west, and MbaKyar on the north (Fig. 15[A]). However, about 1922, MbaKyar thrust itself forward in such a way that it now bounds the Udam on the south, MbaKertiev on the east, and MbaDuku on the west. It cut off MbaKertiev from MbaDuku (Fig. 15[B]).

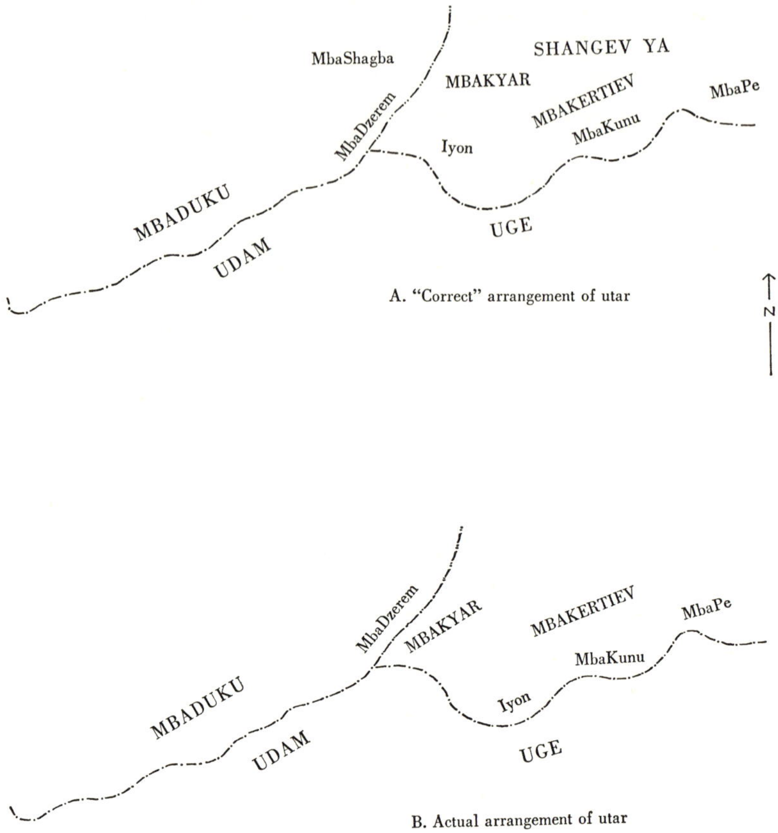

Fig. 15.—Migration of MbaKyar.

The result of this shift has been that all MbaKertiev "knows" (*fa*) that MbaKyar has taken their land. MbaKertiev should bound the Udam and MbaDuku; since it does not, their land has been taken from them. MbaKertiev is a segment of three or four thousand people, all of them very vocal about this matter.

Another case of the same sort occurred within MbaDuku, but only about three acres are involved (as against a possible 1,200 in the above case). MbaYar formerly bounded both MbaGor and MbaWandia of MbaNyam (Fig. 16) and MbaAji and MbaKov of MbaIkaa. Over a period of years, MbaWandia and MbaAji pushed toward one another until they finally met, forcing the MbaYar people out of the area altogether. The effective outcome was that MbaKov no longer bounds MbaYar. MbaKov people cite the fact that they have in the past, and still should, bound MbaYar, as evidence that their land has been taken by MbaWandia and MbaAji. MbaWandia and MbaAji people refuse to talk about it in terms of these lineage depths; they say that MbaIkaa and MbaNyam (or their inclusive lineage, Ikakwer) still bound Mbayar—and since all Ikakwer is "one man" (*or mom*), there is no problem. MbaYar people protest loudly that they have lost land in this area, but when they discuss it they say that MbaItyo took the land.

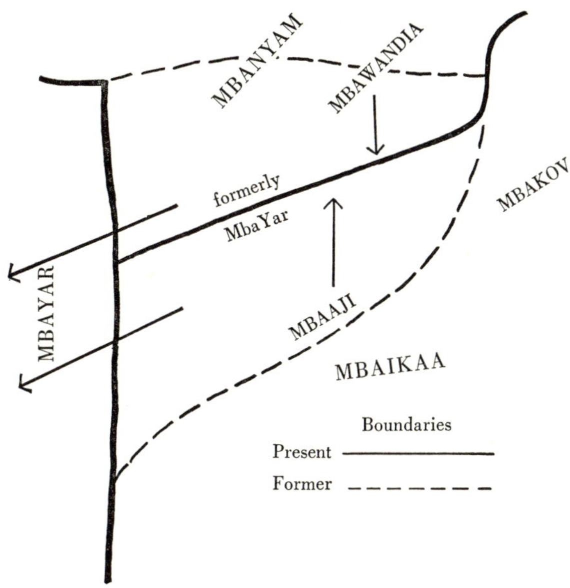

Fig. 16.—Migration of MbaYar.

As a matter of fact, only MbaKov's position has changed. MbaKov formerly bounded MbaYar and no longer does so. Therefore, to MbaKov people, MbaNyam and MbaAji have taken their land. One MbaNyam man said that MbaKov was talking nonsense: if they (MbaNyam) had taken anybody's land, it was MbaYar's, not MbaKov's (thus lining the "reality" up with the morality of the lineage system).

What actually happened in these cases—who was actually on the land in the past—is impossible to determine. It is also irrelevant. Nonetheless, shifts in the juxtaposition of *utar* are greatly deprecated. Generally speaking, the bigger the unit affected by the displacement, the graver the offense. The case of MbaKyar and MbaKertiev is important because all MbaKertiev keep the feeling of injustice alive. In the case of the MbaDuku lineages, the matter is of less importance because only MbaKov (about 700 people) have any interest in it. If, in the Shangev example, Morov (see genealogy in Table 1) had pushed down on Tondov, it would have been an even more serious breach—all Tondov would complain. If Nanev (equivalent to Shangev MbaShaya) had shifted from its position north of Shangev so as to touch Udam, all Shangev would have been embroiled.

In the north, where the pressure of population is not great, there is less compliance with juxtaposition. We know several cases in which a small segment "went to the farm" out of position, and then, after a period of some years, returned to its earlier site. This is not possible in the land-hungry south, where the moment one segment leaves an area, another moves in.

In southern Tivland, land shortage and the firmness of the boundary between the Northern Region and the Eastern Region have had a profound effect on the minimal *tar*. The population of each minimal *tar* in the south is larger, and the density is much greater. Ukusu, the largest minimal *tar* we discovered in the northwest, contained only between 500 and 600 people. Iyon, at the very south, had twice as many and on a smaller area. This change has been brought about, at least in part, by the fact that migration by southern Tiv farther south has been stopped while pressure from inner Tivland has not abated.

Expansion

IYON PROVIDES an almost perfect example in miniature of the pattern of Tiv migration. It has, within a few decades, changed from a single large compound into a minimal *tar* that is ready to split in the course of its next migration (although under present conditions that migration may never occur). Moreover, its position on the very southern edge of Tivland, bordering the Uge people, has led to an eventful history which shows the migration pattern clearly.

Iyon contains 1,050 people on somewhat less than two square miles, or about 550 persons per square mile, a very high density for African subsistence farmers. Before the arrival of the Europeans in 1909, Iyon lived closely packed in a large "village"—the Tiv word means "fence" or "stockade" (*gar*). There were eight compounds—two for each of the four small segments-within-the-hut—within this stockade, some of which are still occupied. The ruins of the rest could still be seen in 1950. The whole surrounded a large market place. Iyon people told us they did not like to live close-packed in a village, but also noted that "if the Europeans leave" they would return to this mode of residence, since it offers their only protection against surprise attacks by the Uge.

As the *Pax Britannica* became secure, the people of Iyon moved out of their stockade, spreading their compounds evenly over their territory, and then overflowing and expanding it. By 1950 each of the four segments-within-the-hut stretched out roughly in wedge shape behind that portion of the stockade in which they formerly lived: this spatial separateness was

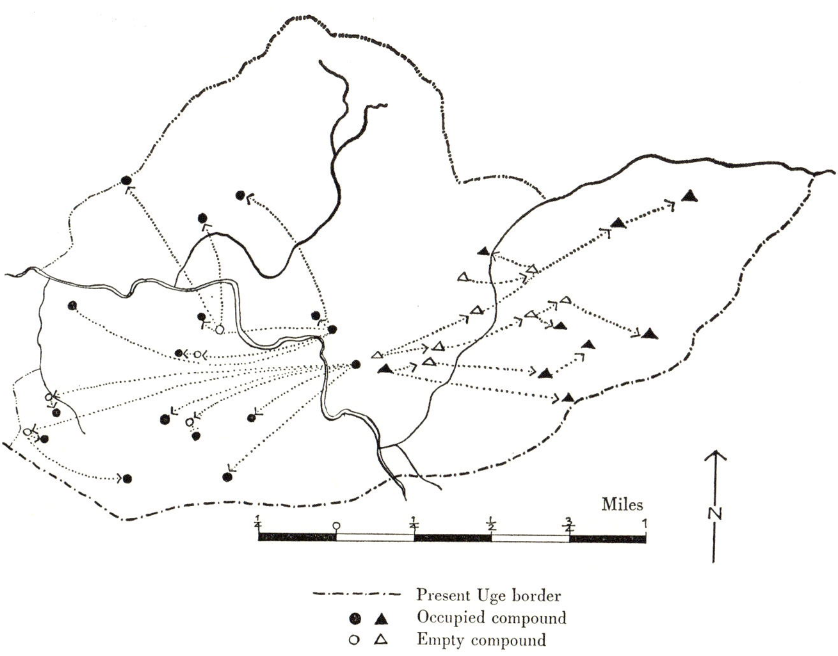

Fig. 17.—Expansion within a minimal tar—Iyon.

not perfect enough for any of them to be *tar,* as their farm land was still somewhat intermixed, even though the residences were no longer so.

This movement from the stockade began about 1925. Figure 17 traces the dispersal of settlement and the accompanying fission of the compounds of two of the segments-within-the-hut. The "boundary" shown was the limit of Iyon farm lands in 1950.

Exactly the same principle of expansion applies to every larger lineage. Thus, for example, the two oldest men in Iyon said that at their birth, Iyon were living as a stockaded village in a position now occupied by Dzev lineage (see Fig. 18). When the elder of these two was a boy of eight or so, Iyon moved south and east, into the territory which is now occupied by MbaKunu, and finally they moved south into the location now occupied by the market place in the center of Iyon.

Tradition tells that each of these moves precipitated a long and bitter

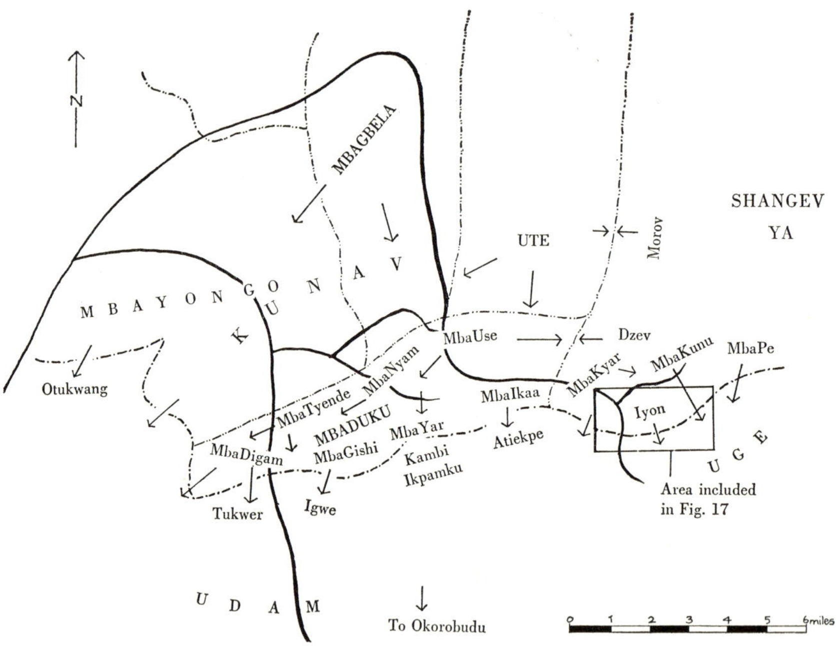

Fig. 18.—Expansion of southern Tivland.

war with the Uge, who were forced to retreat farther south with each forward move of Iyon. It was these wars (endless versions and stories of which can be collected) that forced Iyon to live as a stockaded village. Every time they moved, all the men of Iyon went forward together and built two or three huts of a new compound; then the women came to join them. After that, everyone—man, woman, and child—pitched in to finish building the rest of the huts of all the compounds as quickly as possible. They stayed close together at all times, and never let themselves be divided for more than a few days, for fear of attack by the Uge.

Iyon people say that if the Europeans had not arrived, they would have moved again, perhaps more than once. They did not particularly want to move; in fact, they liked their previous location. But lineages behind them kept pushing in and taking their land; these lineages were in turn being pushed by other lineages. Thus, Iyon had to keep moving on to new farms and continue fighting the Uge. Iyon elders say that in all probability they would have reached points three or four miles farther south if the British administration had not created a boundary over which they could no longer expand.

Iyon itself is a minimal segment expanding its territory outward in all directions; pressure from the expansion of every larger lineage and territory of which Iyon is a part moves both Iyon and every territory subsuming it en masse. Because all lineage territories are expanding in all directions, the minimal territories on the borders of Tivland have a direction of migration explicable in terms of the expansion of more inclusive lineages and territories.

Kunav, a large lineage of some 100,000 people, bounds Shangev. Its southernmost segment, MbaDuku, on which this discussion will center, contains some 10,000 people. Although the various minimal segments of MbaDuku are all attempting to enlarge their territories, the most obvious movement is a directional migration of entire lineages.

This migration was explained in vivid terms by Atesem of MbaDuku. He had about a dozen children between the ages of three and ten, whom he enjoyed having under his feet; when he explained the movement of the lineage segments of MbaDuku, he used a child to represent each segment of MbaDuku as shown in Figure 18 and placed them to demonstrate the actual juxtaposition of the territories on the ground. He picked up the child said to represent MbaNyam and moved it forward, placing it as close as possible to the one said to represent MbaYar. He then picked up the child representing MbaYar and moved it so that it stood as close as possible to the child representing MbaGishi; the child representing MbaGishi was then told to "go over there" (to the southwest), whereupon Atesem turned to us and said, "He is going to Udam country," and the children burst out laughing. Then, by lifting and moving children one at a time in this way, he took the whole group forward. When he had finished, they *were in the same relative positions they had been in when he started.* Atesem insisted that any lineage could start such a move, but he had just happened to start with MbaNyam. MbaNyam might move out, he said, without MbaUse pushing them out, though they probably would not do so; in the same way MbaUse would exert pressure on MbaNyam only after Ute had exerted pressure on MbaUse, and so forth.

We then lifted the child who represented MbaUse and carried him to a position west of MbaDigam. "Can MbaUse move like that?" we asked. Picking up the child and moving him back to his old position, Atesem said such a move was a "bad thing": MbaUse should not get out of place that way.

We can also illustrate the Tiv viewpoint by starting with MbaNyam, a segment of MbaDuku (see Fig. 18). If a man of MbaNyam wants to increase his land holdings he does so, if possible, against the territories of either MbaYongo or MbaGbera, companion segments of MbaDuku and hence his most distantly related neighbors. However—and this point is crucial—MbaGbera or MbaYongo do not see it as expansion of MbaNyam, let alone of that individual. To them it is "MbaDuku encroaching." There are, thus, two points of view: the original man, qua individual farmer, expands his holdings against his genealogically most distant neighbors (MbaGbera, say), but the MbaGbera people affected immediately see the situation in lineage terms, and it is "MbaDuku shoving (*kpolom*) MbaGbera." Now, although MbaGbera are the most distantly related neighbors from the standpoint of the MbaNyam individual, they are not so for MbaDuku as a segment: from that viewpoint the Udam, being non-Tiv, are the most distant of all neighbors. Therefore, MbaGbera people say—and when argued at this level of the lineage system, MbaDuku people agree—that MbaDuku should expand into Udam territory, not into MbaGbera territory.

The effect of this pressure against a foreign people is that each segment along the southern boundary of MbaDuku and MbaYongo has a relationship of traditional enmity with a small Udam group immediately on the other side. MbaDigam of MbaDuku has a traditional enmity with

Tukwer, the small Udam group which bounds it; MbaGishi has a traditional enmity with Igwe, MbaYar with Kambi and Ikpamku, and MbaIkaa with Atiekpe. In the same way, in terms of larger segments, all MbaDuku is the traditional enemy of all Obudu Udam. Similarly, MbaYongo is the traditional enemy of the Otukwang, another Udam people. Each small MbaYongo lineage is the traditional enemy of an Otukwang group, and all MbaYongo are enemies of all Otukwang. MbaDuku are not enemies of Otukwang, but have a friendship pact with them. During wars between MbaDuku and MbaYongo, Otukwang people have been known to assist MbaDuku. However, Otukwang are an Udam people, as are the Obudu; so in still larger terms, Kunav is said to be the enemy of "all Udam" and, in fact, all Kunav have been known to make war on the Udam.

Each Tiv lineage in southern Kunav and in southern Shangev knows its precise destination for the next two or three moves. MbaDuku/MbaIkaa will move into Atiekpe, who—they say—will retreat upon another group behind them, the Okorobudu, and be taken in. But they will thus be crowded, Tiv say; so Okorobudu will move farther south. Then, when MbaIkaa once more need land they will again move in on Atiekpe (who are now where Okorobudu used to be), who will again run to the Okorobudu in their new location, and so forth. This can go on indefinitely. Only two or three moves are known in advance, but it is said that the moves will always be against the same people. Direction, in this context, is described in terms of social, not geographical, direction.

Lineages that do not bound foreigners directly know just which lineages they "follow" (*chir*), and, though they are likely to be impatient and push or shove (*kpolom*), they will assist those in front to take land from the foreigners.

When the boundary between Northern and Eastern Regions was drawn in the early 1920's, the situation of migrating Tiv was radically changed. In the colonial period, a territorial rather than a genealogical citizenship was imposed. Tiv had to obtain permission, by decree, from the Udam to settle in their country. Unless they had Udam kinsmen, this was usually not given. Even those who lived with their mothers' agnates in Udam felt at a moral and legal disadvantage. Tiv said that they did not get justice in Udam courts, that their taxes were misused, and that they were restricted by people who are their inferiors, culturally.

Because of pressure from the British administration, MbaDuku elders tried to dissuade people from going to Udam and also tried to defend the boundaries between themselves and the other Tiv lineages "at their back." But we have also heard elders try to persuade MbaDuku people to move into Udam country, leaving more room for expansion. The dilemma was created by colonial fiat and by Tiv inability "to become Udam."

At the time of our study, pressure from inside lineages on those outside had not abated. MbaGbera still pushed on MbaDuku, and in turn was pushed by other lineages behind. Population density continued to increase along the southern boundaries. The only answer Tiv had discovered was to reduce the size of their fields and the length of fallow periods.

In most areas, future moves are not described in such narrow directional terms. People in the east and in the northwest—in Ukum and in Iharev/Ityoshin—insist that a small lineage has a right to move in any direction. They agree that a lineage should not get out of position, but their reason is that if one does get out of line, it will be cut off from its companion lineages and therefore be open to all sorts of dangers.

Fission and fusion of lineage territories often accompany expansion movements. MbaKor, a large central lineage, are moving into the area left by Nongov/Kambe as the latter move north across the Benue. Segments of each of the two large lineages of MbaKor fused (perhaps temporarily) into a single large territory. Fission is perhaps more common.

Summary

In their "land policy," Tiv have carried several simple propositions to logical conclusions. The community is the basic reference point. Every member of the community has security of farm tenure and rights to sufficient farm land to meet his needs. When farm land is not sufficient, one expands against one's most distantly related neighbor. The result is a type of "territoriality" reminiscent of some birds rather than of Western "land tenure"; this produces a complex of boundary disputes and ultimately an expansion outward in all directions of all the component segments in the system.

This simple set of conditions was somewhat blocked by the arrival of the colonial powers, who thought in terms of titles, boundaries, and containment, even as they set out under Lugard to safeguard the African land rights they did not thoroughly understand. They established boundaries that ultimately made the Tiv system unworkable and changed the type of migration to a disjunctive one that could have destroyed the social structure associated with the old system (Paul Bohannan 1955).

Thus land, although it is always a factor of production, is also a spatial dimension of society. Among the Tiv, the community and its spatial milieu are of primary importance and give the fundamental values within which production takes place. Land and the market principle are as divorced as it is possible for two aspects of a culture to be.

8

Resources, Tools, and "Capital"

Aside from the land and labor that have been described in the foregoing chapters, Tiv resources and "capital" are not extensive. This chapter gives first a brief description of resources other than land and labor, then an examination of the agricultural tool kit of the Tiv, and finally a short investigation of the idea of "capital."

Water, Mineral, and Bush Rights

When asked about rights to trees, water holes, fishing sites, or mineral deposits, Tiv say that such things belong to no one and are for the use of all. Their practice, however, does not follow their doctrine.

They claim that water does not belong to anyone. However, each compound has a pool in a stream (*wagh*) from which its women normally take water. In those parts of northern and central Tivland where only a few watercourses do not dry up in the dry season, many compounds share a single pool; usually it is said to be that of one compound, the others having asked and been given permission to take water from it. Such permission is said never to be refused.

Anyone may improve any spring that nobody else claims, and thereafter it is his. Before anyone else can regularly draw water from it, his permission must be requested, and he sometimes refuses. Springs are occasionally protected by *akombo,* especially in the dry season. The *akombo* is not intended to stop people from taking water from the spring, but only to stop anyone's stepping in or bathing in the spring when seepage is slow and several hours may be required for it to settle and clear.

In Ityoshin, every compound has one, sometimes two, fish dams, usually near the pool where water is drawn by the women of that compound. These dams and traps are constructed by men for their own use and the use of those to whom they give permission. Tiv say they may construct a fish dam anywhere they like; actually they do so only in the minimal *tar.*

Women sometimes fish with baskets or nets; they may or may not use poisonous plants to stun the fish. These women also say that they can fish anywhere they like. We have never seen them go outside the minimal *tar,* though all insisted that they could if they wished.

The only minerals Tiv use are pottery clay, clay used for house plaster, and whitewash. They say that they can collect these items any place. Women in MbaAliko and Ukusu in Ityoshin made regular trips of fifteen miles to a source of tough, white clay called "porridge" (*ruam*), returning home with the clay to be worked. They made no payment and asked permission of no one.

Pottery clays and plastering clay are fairly common throughout southern Tivland and women need go only short distances to obtain them. No payment is ever made. We noted, however, that only women of MbaGishi (of MbaPwa/MbaDuku) sold whitewash in the local markets. Tiv insisted that anyone *could* take whitewash from their site, but that it was too far to go for that special purpose, so MbaGishi brought it to market. MbaGishi women agreed with this. The fact remains, however, that whether or not they could do so, Tiv from other parts of MbaDuku did not go into MbaGishi to collect whitewash.

Bush rights can be subdivided into (1) hunting rights, (2) gathering rights, and (3) rights to trees.

Men say that they can hunt anywhere they please, and they often do hunt far from home. In southern Tivland people say that hunters should give a foreleg of any antelope they kill to their chief (*tor*) or to the official called *ortaregh*. They seldom do so. This practice in the south has nothing to do with land: the token should be given because the chief is "father" of the hunter. It is given to one's own chief, not that of the *tar* on which the game was killed.

Gathering rights are more restricted, but only in some circumstances. Women gather mushrooms and wild plants (for use in preparing sauces and soups) from cultivated land only when it is their own. Food plants which grow wild are always left at weeding. Women say that they would not gather such plants from the field of another woman without asking her permission but that this permission would readily be granted if the owner didn't want the plants for herself, or if there were plenty. No such exclusiveness is held in regard to fallow land or bush, however; women can gather sauce-plants or mushrooms on any piece of fallow. We have seen women coming home from the farm, from market, or from a social call, picking plants off any fallow they pass. This is not limited to the minimal *tar* or to any other *tar*. Certain types of caterpillars and termites are also gathered wherever they are found.

Trees are the most valuable bush products. The important "economic trees" are locust bean (*anune, parkis filicoidea*), *gbaiye* (*prosopis africana*), and, in the south, *ive*. In some areas, particularly the southeast, the raffia palm is important; oil and wine palms are found in some places in southern Tivland, but are rare. So far as we know, Tiv do not tap palm wine from the trees, though they drink it gladly when it can be bought from peoples farther south. In those areas where they exist, the same usages seem to apply to oil palms as to locust bean and *gbaiye*.

Should any woman have a locust bean or a *gbaiye* tree growing in her fields (and many have several), she and her children or attached wives harvest the fruit. One old woman in MbaDuku, who was a stickler for propriety, had her eldest son put an *akombo* on the locust bean trees on her farm to ensure that no one would steal the fruit. Everyone laughed behind her back and said that she always did everything too formally and precisely.

Tiv in the north, where there are plenty of locust bean and *gbaiye* trees, say that once a woman has established rights to one of these trees it is hers or her children's. Most of these trees are owned in Ityoshin, but southerners say that when the land on which the trees are growing is not in cultivation, fruit may be gathered by any person who lives in the compound whose members last farmed there.

Trees growing near disputed boundaries are themselves sources of dispute. When we were in MbaDuku there was an argument between two compounds in MbaWandia over rights to harvest locust beans. In settling the matter, the elders did not attempt to decide which compound had the overriding right to the land or trees, but rather what distribution should be made of the fruit. They eventually got the people of the two compounds to agree to an almost even division.

Wars have been precipitated by arguments over locust bean trees. In 1934, a war broke out between Shangev MbaShaya and Kunav (Akiga 1939:415–16). Land on the boundary between these large lineages was in dispute; when Kunav men came to pick locust beans (said by Shangev to be unripe) from trees growing on the disputed land, it was taken by Shangev people as a sign that Kunav were trying to claim the land and a war resulted.

In the course of another argument (settled peaceably) between MbaKyar and Iyon over locust beans picked by MbaKyar people from trees on disputed land, one of the elders of Iyon gave the following rules as rights in these trees: (1) a man shall not pluck the locust beans that grow on your farms; (2) a man shall not pluck the locust beans that grow on your fallow land; and (3) if your kinsman sees locust beans on your fallow land, he will come (to you) and say, "My brother, give me those locust beans," and you will give them to him. This particular case was interpreted as an indication that MbaKyar, in harvesting these locust beans, was initiating a step toward taking over Iyon land in its southward movement. In the settlement (in which the locust beans were handed over to Iyon), Iyon said that if the boundary with the Eastern Region had not existed, they would not have objected to MbaKyar's taking the locust beans; they would have moved farther south to make room for themselves and for MbaKyar and the other lineages coming behind them.

Ive (a fruit prized for its mucilaginous pits) and the few palms which grow along streams belong to those compounds which can plant rice or other paddy (*pagh*) crops beneath them or whose land adjoins the stream bank. We did not encounter any disputes or arguments about such trees, and none of our Tiv informants could recall any. Rights in these trees are not protected with the same avidity as are those in locust bean and *gbaiye* trees.

Tiv in the areas we know do not plant any of these trees, though they may plant lime, orange, or pawpaw trees in their kitchen gardens, in which case the tree belongs to the man who planted it.

Gbaiye provides a fruit used for seasoning. It is also a valuable, hard,

Fig. 21.—Aga.

Capital

THE WORD "CAPITAL" sits obtrusively in most studies of subsistence economy because it must be used metaphorically. "Capital" is the base of capitalism, but not of "economy," when that word is used broadly and comparatively. The word "capital" began to be used in the early seventeenth century to mean "accumulated wealth employed reproductively." A little earlier it meant only the original trading stock of any commercial enterprise. The history of the word in Western economies must be remembered: its present-day use grew up not merely with political economy and the science of economics, but with the very culture that spawned such a science.

Obviously, if capital means "accumulated wealth employed reproductively," there will be some economies in which the word can be assigned only minimal meaning. Tiv have such an economy: they do not, as a goal, accumulate wealth to employ it reproductively. Their wealth can be except that the handle is usually a little longer in proportion to the blade. When a hoe has worn down to about half its length it is taken back to the smith, who adds two or three strips of metal to the end, restoring it to its original length.

The Tiv say that they formerly used wooden hoes (Akiga 1939:65). We have never seen one of these wooden hoes, nor have most contemporary Tiv.

The digging stick (Fig. 20) is almost always made by the local blacksmith. It has a heavy metal point which is hafted by nails or pegs into a smooth wooden handle. When new, the blades may be as much as 18 inches long; they are resharpened and used until only about an inch of metal remains. The over-all length of the digging stick is 3 feet, or a little less. Some women use digging sticks of the same general proportions, but without the metal point.

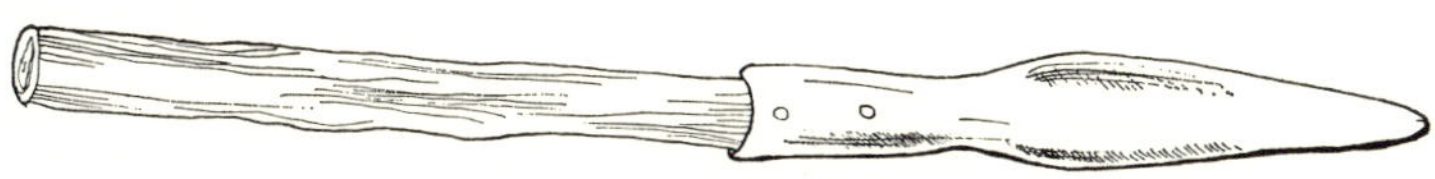

Fig. 20.—Digging stick.

Only women use digging sticks in agriculture for planting and harvesting. Men use them in house construction for digging post holes and in hollowing wooden mortars and drums.

The *aga* (Fig. 21) is made from an approximately straight branch of wood, some 6 or 8 feet long, with a piece of rope attached to each end of the branch, forming a loop. It is used by grasping the loop of rope in both hands, placing the foot (usually the left) on the wood, then moving slowly forward, using foot pressure on the wood to press long grass, millet, or guinea corn stalks to the ground.

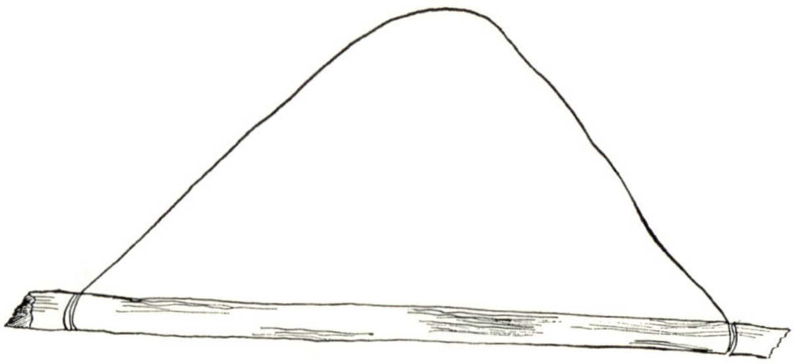

Fig. 21.—Aga.

Capital

THE WORD "CAPITAL" sits obtrusively in most studies of subsistence economy because it must be used metaphorically. "Capital" is the base of capitalism, but not of "economy," when that word is used broadly and comparatively. The word "capital" began to be used in the early seventeenth century to mean "accumulated wealth employed reproductively." A little earlier it meant only the original trading stock of any commercial enterprise. The history of the word in Western economies must be remembered: its present-day use grew up not merely with political economy and the science of economics, but with the very culture that spawned such a science.

Obviously, if capital means "accumulated wealth employed reproductively," there will be some economies in which the word can be assigned only minimal meaning. Tiv have such an economy: they do not, as a goal, accumulate wealth to employ it reproductively. Their wealth can be the horse) is *ilegh;* its plural, *ilev,* means "livestock." The word can, however, refer to sheep, goats, and fowls to distinguish them from cattle (*bua*); when that is the case, the inclusive word is *izendenya,* "things that walk on the earth."

The animals of most economic importance to Tiv are without doubt their chickens (*ikegh*). Cocks are usually owned by men; hens may be owned by either men or women, though predominantly by women. Chickens are seldom fed. Tiv who have seen European chickens admire them, but will not care for them, even when they know what to do. Several attempts were made by the Agricultural Department to "improve the strain" of Tiv chickens by distributing Plymouth Rock, Leghorn, or Rhode Island Red cockerels to "deserving farmers." Only an occasional color throwback resulted.

Chickens are the usual gift to a distinguished visitor. The bigger the chicken, the more prestige (*shagba*) is acquired by both the giver and the receiver. Tiv sometimes use eggs in sauces, but many people claim that they do not like them, and eggs are a common avoidance for some entire lineages and for women of other lineages during pregnancy. At the time Akiga wrote, chickens called "tens" or "twenties," according to their sizes, were the usual form of payment for ritual services. By 1949–53, most payment to ritual experts was made in money.

Tiv seldom merely butcher chickens. Rather, they sacrifice them; however, one can always find an excuse for sacrificing a chicken. Chickens are so much the usual sacrifice that the word "chicken" is used by Tiv witches (*mbatsav*) as the "double talk" word for a human sacrifice. The value of a sacrifice is increased by increasing the number of chickens until the approximate value of a goat is reached. Except that chicken feathers are sometimes used in making fetishes (when the feathers are a reminder of the chickens sacrificed to activate the fetish), no by-products are used.

Tiv keep black-and-white Muscovy ducks (*ikeghdam*) for butchering only. Ducks are not sacrificed. The price of a full-grown duck in southern Tivland in 1950 was about 7*s*. We have never seen Tiv eat duck; when we asked about it, they said they considered it too expensive.

Dovecotes are not infrequent in Tivland. Tiv say they like to have pigeons around, and they sometimes eat them, although they prefer chicken. Doves were said to be on sale for 3*d*. each, but we were never able to buy any for that price, and only once for any price.

The domesticated mammal of greatest importance is the goat (*ivo*).

The vocabulary referring to goats is large and reflects their relative significance. Goats are kept in the compound and are allowed to wander freely in search of food; they seldom go far. In some Iharev compounds, the goats outnumber the people, but in southern Tivland there are few enough goats that it is not necessary to fence crops in the kitchen gardens behind the huts.

Goats are the most common sacrifice when a chicken is of insufficient value. They are often butchered to celebrate occasions such as marriages.

Since goats are in constant demand as sacrifices, most Tiv do not keep their own in evidence. To avoid the just demands of one's agnates, one "releases" (*tuhwa*) one's goats either to a non-agnatic kinsman some distance away or to a friend, who then releases his goats to you. The return is usually one kid in three, ordinarily a female. The network of debt created by this practice is almost as dense as that created by exchange marriage, but is of little or no economic moment. If a "released" goat dies, the keeper should go immediately to the owner and explain its demise, preferably with a witness and with the hide. If the death of the animal occurred under mysterious circumstances, a replacement may be demanded. Such occurrences sometimes lead to court cases.

Tiv keep some sheep (*iyongo*). They are sacrificed, so far as we know, only in rites of the fetish known as *akombo iwa.* Sheep are butchered at market, but usually not until they are old or ill. Like goats, sheep are allowed the run of the compound.

Pigs are the only animals about which Tiv are ambivalent. They keep them in the compound as scavengers but say that they make as much dirt as they clean up and that they are stupid. Pigs root out and eat yams, sweet potatoes, and peanuts, and will go much farther afield than either goats or sheep. Pigs may be confined by a pole stockade or by being tied up. However, they are never adequately fed when they are tied, so eventually they are released. Tiv make shelters and wallows with small thatched coverings for pigs, and sometimes even haul water from the streams in the dry season "so that the pigs can have a bath."

Tiv like pork but say they should not eat it. They claim they will not eat any animal which eats anything they would not themselves eat—except pigs—but "pork is sweet with one heart." Pigs are never, so far as we know, sacrificed, though they are butchered on festive occasions. Every man owes each set of his affines a pig (but will probably never pay it). A pig is the conventional fine which an age-set levies on its members. Pigs are more common in the north and west of Tivland (Iharev and Masev) than in the other parts.

Most of Tivland is infested with trypanosomiasis. Therefore, the only cattle that thrive are the small dwarf variety that Tiv call *bua* (humped cattle from the north are called *agundu*). In southern Tivland there are usually no more than three or four *bua* in each minimal *tar*. In the north there are many more. The largest compound in Ukusu, in Iharev/Ityoshin, had 18 dwarf cattle, and some compounds north of the Benue were said in 1952 to have 60 or 70 head.

Tiv do not use milk, and indeed consider its use as food disgusting. Cattle are not worked. Tiv say that a cow is a thing full of prestige; therefore everyone wants one. Cattle are never sacrificed on fetishes, but they are butchered on important occasions. Every man wants to kill a cow for each of his wives, but few ever manage to do so. If they do, the meat is divided among the husband's age-set, the wife's agnates, and the elders of the husband's lineage. Cows are also killed by "big men" on great occasions. In the reception hut of Adamgbe, who was the government chief (*tyo-or*) of MbaYongo when we began work in southern Tivland, there were 56 cattle skulls. Adamgbe claimed that he had killed them all on the day that he was made "spokesman" of Kunav South in the 1920's. Skulls of cattle are found in many reception huts; sometimes they are hung up, sometimes used as stools.

In the south cattle are usually staked just beyond the kitchen gardens and at night are brought into the compound, where they are tied to a tree. People are particularly careful with their cattle during the time seed yams are on top of mounds or stacked in a field, for the seed yams devoured by a cow must then be replaced by its owner. In the north, where cattle are more common, one occasionally finds a log corral (*igar bua*) in the middle of the homestead. Corn husks, the tassel of sorghum, or the spike of millet from which the grain has been removed—all are thrown over the fence to the cattle. Since it is dirty already, it is sometimes used as a latrine at night. Tiv pen cattle to keep them out of yam fields and for no other reason.

In those areas of Iharev in which water is scarce, cattle are allowed to wander during the dry season. Usually a scout determines their general location every week or two, but even this may be neglected. The owners say they will come home when the dry season is over and water can be found close to the compound.

Horses are even more prestigious than cattle, but since they are not butchered or eaten, they are of even less importance in subsistence economy. Since there is more honor in having bought a horse than in actually owning one the turnover is rapid. Horses usually live only a few weeks in Tivland, but it is on the main trade route along which horses are moved from Gobir and other northern areas to the Ibo, who butcher and eat them. Occasionally one will be sold to Tiv for a sum of from £15 to £20 (1952). The purchaser, all of the members of his compound, and members of closely related compounds dress themselves in their best, decorate the horse, and lead it around to various compounds, accompanied by a drum band. Songs are made up about what a wonderful horse it is and how much prestige the purchaser has acquired. Collections are often taken up; sometimes, we were told, one gets enough to pay for the horse. Then, after a couple of weeks—or indeed, after only two or three days—the horse is resold to another Tiv in as distant an area as possible, and the whole procedure begins again.

Eventually the horse dies. "The man on whom the horse dies," to put it in the Tiv way, reaps full and lasting prestige. The horse is buried on the main path nearest his compound, the head having first been removed. A ring of trees is planted around the grave and the jawbone of the horse is hung on the trees. The skull is put into the reception hut, where it is used as a stool or a chair by the compound head. To have a "horse's grave" is a sign of a prestigious man.

Hunting and Trapping

THE TIV WORD *nyam* means "animal" (and also "meat"), but it is usually applied only to wild animals (*nyam ken toho*). All Tiv agree that there are only four (we should say five) "great animals": leopard (*anyam*), bush cow (*yar*), elephant and hippopotamus (both called *nor;* the latter, "water *nor,*" if it is necessary to distinguish), and the ape (*ika* or *ikya*—baboon and gorilla). We asked on several occasions about lions (*begha*). Tiv say that they exist in Jukun country but that they have long since disappeared in Tiv country. There is not enough game in Tivland to support lions.

All the rest are "small animals." Some of the animals hunted in southern Tivland are cob (*ikpam*), antelope (*ivambe*), a small antelope (*iho*), duiker (*ikuran*), and reed buck (*kwantera*). Tiv also include such rodents as the hare (*alom*), the large bush rat (*viha*), and a large ground squirrel (*hinga*). Others are the serval cat (*abiem*), a large cat called a "chicken leopard" (*anyamikegh*), the civet cat (*ijev*), another wild cat called *shimbe,* whose skin is used to make tobacco or tinder pouches, and the *congo* (translated as "fox" in the dictionaries).

Tiv seldom hunt the "great animals" but have a considerable lore about doing so. Tiv are more afraid of leopard than of any other animal because it lurks in trees, usually unseen, and kills people. Iyon elders claimed that about 1940 a leopard appeared in the area and stayed in a tree in the swamp for several weeks. All were terrified. When the leopard finally went away, the tree was cut down and burned.

Elephants have disappeared from Tiv country within the last fifty years or so. One man of forty-five or so told us that his grandfather had hunted elephants but his father had not. Those Hausa resident in Obudu in 1949 claimed that their forefathers had come to southern Tivland a few years before the Europeans arrived, to hunt elephants which then still existed on the hills of southern Tivland and in Obudu District.

Tiv today say that elephants are "bad things." They trample and spoil crops; they even come into the compound and trample down houses. When hunting elephants, several men climb a tree large enough that one elephant cannot uproot it, and then other men drive the elephant toward them. The elephant is shot with poisoned arrows from the treetop. Killing an elephant was ground for great celebration because there was so much meat.

Elephants were sometimes caught in pit traps. This subject came up only when we were shown an old spear (*iwanger*), said to have been the type used in such traps. A pit was dug to a depth of some fifteen feet and poisoned spears planted upright in it. It was then covered with saplings, leaves, and dust. An elephant or bush cow that fell into this trap was killed, if the poison was still fairly fresh. Any large animal that was upsetting the neighborhood might be driven into it. The trap also served as defense in time of war.

If an elephant was caught in the pit, the tusks and the entire head went to the man who organized the making of the pit. If the elephant was shot with arrows, the head and tusks went to the man whose arrow was the first to hit it, and a foreleg to the man who organized the hunt (as is the case in all community hunting).

Baboons live in southern Tivland, but are relatively rare. They are more numerous in the area around the Benue and on top of most of the hills with which Tivland is studded. We were told throughout Tivland that baboons are nasty-tempered animals who will rape women alone on the farm and who delight in emasculating men as they lean over doing farm work. Baboons are cooked and eaten, as are all other monkeys—a habit which the neighboring Idoma deplore and use to "prove" that

Tiv are cannibals. Baboon skulls are kept in the reception hut, along with the skulls of other game and of cattle.

Few bush cows are left in Tivland today. In the south they come down from the hills every five or six years. Tiv reminded us that Mr. Osmund Weld, one of the early district commissioners in Obudu, had been killed by a bush cow, in the station, in 1910; the D.C. in Obudu in 1950 told us that the last bush cow to come into the station had done so about 1944. Tiv opinion differs about whether or not a bush cow is dangerous. One man said that a bush cow would not hurt one unless it was molested. At another compound nearby, when we asked what would happen if one met a bush cow on the path, we were told calmly, "He would kill you." In Iharev bush cows are fairly common, and warnings about their location are systematically sent from one compound to the next.

Tiv traditionally hunted small game in a communal drive (*ibem*). Such drives were made illegal about 1930 because so large a number of manslaughter cases resulted from them. Tiv recalled in 1951 that in the old days several hundred men might take part in a drive, but that today they consisted of "three men and a dog." In the past the men formed a large ring, perhaps several miles in diameter; on signals from the horn (*kor*) they began beating all game toward the center, shooting what tried to escape past them. No one was supposed to begin to shoot until another signal from the horn was given. Occasionally fire was used in the drive.

The "big man" who initiated the invitations would provide beer—always drunk in the bush, never in the compound. In the division of the meat, every man got the animals he killed, but he had to give a foreleg of each animal to the man who called the hunt. Tiv say that only a big and important man would call a large hunt, but that calling one did not in itself add to his prestige.

Today, with game nearly gone in southern Tivland, hunting has become an avocation of an individual, usually pursued in Udam or Uge territory. In northern Tivland, where game is still comparatively plentiful, most hunting is done by individuals with smooth-bore guns. During our stay in Ukusu (in 1952), the most successful hunter was Apia, who bagged an antelope every week or so. After smoking it, he sold the meat or gave it to his agnates, sending some to his fathers-in-law. Apia was always teased if he went hunting and came back empty-handed, especially if the sound of gunshot had been heard. He carried, besides his gun, a bow and a few poisoned arrows stuck point upward into an armlet (quivers were rare in Iharev).

Game is divided in accordance with a few strict principles: the man whose arrow or bullet first hits the animal gets the head, the hide, and the hind legs. He gives the heart to his mother. The man who next hits it gets a foreleg. Southerners say that if a third man also hits it, he should be given a foreleg as well. The rest goes to the man who hit first, to be given out to the other hunters. Disputes between hunters are rare. As one of our best informants in MbaDuku put it: "Look, the two of us go to the bush hunting. I say that it was I who hit the animal, you say that it was you, and we begin to quarrel; so we take the animal and go to the chief (*tyo-or*) and ask him. He says to me, 'Was it you who shot this animal?' but before I can answer, you say it was not. If we can't agree, the chief takes the whole animal—the whole animal completely—and we go home empty-handed. My mother says to me, 'Kpash, what did you do with the heart of that animal that you do not give it to me?' and she scolds and my wives scold me. If we had not argued and fought about it we could have divided it evenly—so that is what people usually do. We eat it quietly (*te te*). Even if we do not have an argument, we may eat it quietly so that the chief doesn't know."

Hunters keep the skulls of animals they have killed. They put them inside the reception hut or string them through the eyeholes and place them beside hunting *akombo* on the path just outside the compound.

Tiv complained in 1952 that hunting was no longer taught by fathers to sons. Since it, like war, had been outlawed by the administration (in fact, only the communal hunt was outlawed), they regretted that knowledge of hunting was disappearing. Hunting is the symbol of men's work, but economically it is of little importance except for a few individuals in northern Tivland.

Nowadays, most hunting is for rodents. Grown men do not admit hunting rats and mice, but they do not avoid catching those that "want to be caught." They hunt hares, larger bush rats (especially those called "cutting grass" by Englishmen), and the large ground squirrel. They also hunt porcupines and anteaters in areas where these animals are still to be found. The usual method is to locate the animal runs and lie in wait for them. Sometimes arrows are used, sometimes traps.

Both land and water turtles are caught and eaten. The Tiv metaphor for plenty is that food is so abundant you can use turtles as braces for grinding-stones. Turtles today are rare and expensive in Tivland. They are a necessary sacrifice for the fetish called *akombodam*, and are so scarce that it is usually necessary to borrow a turtle shell (which is always kept), stuff a chick into it, raise its value to that of a turtle—

some 3*s.* 6*d.*—with money, then sacrifice the chick with the loud statement, "This is a turtle." Turtle sauce is said to be among the richest and finest sauces that can be made.

Finally, Tiv eat snakes when they kill them but claim that they never hunt them.

Fishing

FISHING TECHNIQUES VARY with the fisherman. Boys use European fishhooks, line which may be either imported or locally spun from local cotton, poles cut from bamboo groves, and floats whittled from light, dry wood. Women fish either by poisoning the water and picking out the stunned fish or with baskets and nets in a form of communal enterprise called the *suwe*. Men build fish dams and traps, and in some areas near the Benue and Katsina Ala rivers they have taken over harpoon fishing from the Wurgum and other riverine peoples.

Fish dams are relatively rare in the south, but in Iharev almost every compound has one or more. These dams are substantial constructions as much as 24 feet wide and seven feet high. During the early rains, while the water is rising and the fish are coming upstream, traps are set into the dam with their mouths downstream; sometimes mud fish as much as eight feet long can be caught in such traps. Later in the season, as the water begins to fall, a different type of trap is set into the dam, mouth upstream. The dams in Iharev were tended twice a day.

Boys fish during November when the dry season is setting in and the water is beginning to recede. The catch is usually big—we have watched one boy with one hook and earthworms get fifteen small fish in about three hours.

Groups of boys, usually from one compound, go fishing together. They sometimes cook and eat the fish, but most often they take the catch home to their mothers. Children often hunt crabs along the banks of small streams; sometimes as many as two dozen are caught, taken home, and boiled for sauce. If only one or two are caught the boys usually roast and eat them.

Women's fishing is done later in the dry season because it demands fairly still pools—December and January in Iharev, February in the south. The catch is usually limited to a few fish four or five inches long and a few crabs.

Community fishing (*suwe*) is common throughout Tivland in the dry season. Large nets are used to wall off a stretch of stream and as many as 800 people—men, women, and children—dip out fish, sometimes up to two feet in length, with nets. These fishing parties are organized by "big men" as communal hunts formerly were; the organizer gets a portion of each person's catch, which he shares with chiefs and *mbatarev*.

Collecting

COLLECTING OF WILD PRODUCE is a small but significant aspect of subsistence. Women collect termites and a particular caterpillar called *ikosun.* In August and September, these caterpillars are found on locust bean and prosopis trees. They are bent double and the insides squeezed out, washed, dried, fried in red palm oil, and eaten as a snack that Tiv compare to peanuts.

Men gather wild honey, which is eaten as a sweet. They claim that they do not smoke out the bees, but merely take the honey. Everyone eats swarming termites. Tiv also eat the mineral-rich earth of termite mounds under some conditions (Vermeer 1966).

An exhaustive list of the sauce plants which Tiv women collect would be prodigious. Individual women collect plants, almost casually, including several sorts of mushrooms. There are, however, harvests of several wild products which do figure in the annual cycle of subsistence. The most important of these is the harvest of the locust beans and *gbaiye* beans in the late dry season (February and March). Men often help gather locust beans by climbing the trees and knocking down the fruit. Women and children take them home to dry for a few days, then pound them in mortars to crack them. The pods are filled with a spongy yellow substance between the beans, which must be washed off, usually by putting the beans into a colander or a basket, taking it to the stream, and washing repeatedly. This yellow substance badly pollutes the streams, which are extremely low at this time of year. People sometimes put up protective *akombo* on their springs to keep people from washing locust beans in them.

Men gather flint and several types of scoria to make grinding stones or fire-striking stones. These can be picked up anywhere. Basket weavers and carvers gather their own materials in the bush.

Tiv use their environment extensively, and their lives and diets are much poorer when they live in towns where they cannot hunt, fish, and gather. Yet the basic subsistence of Tiv comes from their agriculture, and it is there that their major emotional concern lies as well.

10

Domestic Economy

TIV ECONOMY IS, as we have seen, primarily institutionalized in the form that has been called "householding." Although no household actually creates all the products it uses, Tiv are firmly of the opinion that it could do so, but that reciprocity among kinsmen and marketing among non-kinsmen are easier. In this chapter we deal with the domestic economy: that part of the economic process that centers around domestic allocation and consumption. In the following chapters, we take up reciprocity and marketing.

Control of Agricultural Products

FROM JULY TO DECEMBER, Tiv eat mainly yam porridge; during December through about May they finish the yams and depend more and more on guinea-corn porridge. From the last of May until the new harvest of yams in July, millet is the staple. In the north, in areas of plenty, there may be enough yams and guinea corn that little millet need be eaten as porridge. Maize is an important crop from the standpoint of nutrition, but Tiv do not consider it a staple because it is seldom eaten as porridge.

These staple crops are "feces-forming foods." The sauces into which they are dipped, as well as all fruit, peanuts, palm oil, fresh vegetables —everything except the staple foods and meat—"form urine." Tiv say that a man must have at least one meal of feces-forming food a day in order to remain healthy.

Tiv often say without being asked that yams and guinea corn belong to women but that millet belongs to men. In Ityoshin people added that women are entitled to some millet because of the work they have done on it. Millet there is stored in shacks (*tumbe*) on the farms, is never guarded, and apparently is never stolen. In central and southern Tivland, where food is not so plentiful, millet is stored in a man's reception hut.[1] It is a hunger-breaker, and is controlled by men, to be doled out to dependents according to their needs until new yams are ready to eat. Thus, during the tense time of the consumption year, the food supply is controlled by men. In the occasional year in which the millet harvest is very good— when there is no chance whatever of running short—some men divide the millet among their wives for storage and use, but most issue millet to their wives every few days.

Millet is the basis for the best beer, and beer is made by women.[2] A woman may sell millet which her husband has given her (so long as she can feed him and her children adequately), or she may brew from it beer for sale at market. Women also buy millet for brewing, usually from other women. Women thus control small quantities of millet, but seldom large quantities and seldom the millet used primarily for food. If a man needs a supply of beer for entertaining or paying an obligation, he gives his wife millet; in return she gives him what beer he needs, though she usually has some left which she either drinks, gives away, or sells.

Tiv in the south drink more beer during the hungry season (mid-May to mid-July) than at any other time. They have been roundly condemned

1. The photograph facing p. 62 of Akiga 1939 shows a store of millet inside a reception hut, apparently in central Tivland. In amount it resembles the millet stores of Ityoshin; no Kunav farmer ever had such a fine supply.

2. Brewing technology is described in detail in Bohannan and Bohannan 1958:271 ff.

for this, but their reasoning seems sound: if a man's resources are running low, he can give his wife only a few handfuls of millet to last a day or two. If she makes porridge, it will scarcely fill one person; if she uses it for brewing beer, it gives ten people the illusion of being full.

Yams belong predominantly to women. Ityoshin women can sell theirs if they want to. A man can ask his wife to give him yams to sell and she will do so "because it was his back that hurt" (from hoeing the yam mounds). In fact, few yams are sold because there are few markets in this part of Tivland.

Since yams do not belong primarily to the husband, he can only ask his wife not to sell them. If yams therefore run short, the man can beat or publicly ridicule her if she has sold yams. In the south, where such occasions are frequent, men often grow yam patches of their own. Yams are in the effective control of senior wives and mothers-in-law, who can take physical possession away from their junior attached wives and daughters-in-law. Such control is seldom exercised. Young wives, before they have children, sometimes sell much of their crop. One senior wife in MbaDuku said it made her sick at heart to see young women selling yams to buy perfume, powder, and cloth. "They will not sell food," she prophesied, "when they have seen their children hungry and have only water and earth to give them." When the profligate wife's food is gone she asks the husband for more; he then has to give her food or money which should be reserved for the use of the entire compound. This gives rise to great resentment—not against the husband, who is fulfilling his recognized obligation to provide members of his compound with food—but against the irresponsible wife who has "spoiled the compound" (*vihi ya*).

A husband can request yams from his wives to give as presents. Men make such requests from the wife who is best equipped to give them: she who has the most yams in proportion to the number of her dependents. Most men tell the recipient of such presents which wife supplied the yams. It is good form to thank her, though the gift is from the husband, not from the wife.

A woman occasionally refuses to give her husband yams. We know of two cases in which this has happened. In one the man forcibly took two yams from his wife and gave her a cuffing. He said that if his request had been unreasonable he would not have dared such a thing or she would have left him. In the other case, the head of a large compound (eighteen adult males) needed a large supply of yams to feed a group of people who were coming to his compound for a ritual and feast. He asked his own wives to give him a portion of the yams that would be needed. Each complied ungrudgingly. He then asked the other men in the compound (exclusive of his own sons, whose wives' shares came from their mothers-in-law) to ask their wives for yams. The food was to be pooled and cooked publicly by all the women. One woman refused. The compound head told the husband that she could not refuse; the husband took the food and punished the recalcitrant wife. All the women of the compound agreed that she was wrong—on such an occasion she should have given her husband the yams required because the hospitality and the rite were for the benefit of the entire compound.

Seed yams present a more serious problem than seed of other staples. Seed yams are expensive and are, Tiv say, always short.

The first planting season after her marriage a woman gets seed yams from her mother-in-law or her senior wife. If she has no mother-in-law or senior wife, her husband supplies the seed yams; he may give them to her out of those he controls or give her the money to buy them. It is considered a fine gesture for a senior wife to brew beer for all the women of several neighboring compounds, each of whom gives the new wife seed yams. This is a friendly act of well-wishing; the senior woman herself gets a great deal of prestige from it, and the husband and compound head are pleased because their women are cooperating. Gifts received in this way from co-wives are never sufficient; the remainder must be purchased.

A woman has a right to all the seed yams her farm produces. If she has more than enough to plant her next farm, she can sell the extra ones. If it does not produce enough, her husband supplies the difference: it is part of his duty of "hoeing his wife a farm." Thus, once a man has supplied his wife with seed yams, he ceases to have a primary right to them or to the produce which derives from them, unless she leaves him. Responsibility for seed yams varies widely from one part of the country to another. Three patterns are given below: they are typical, but there are others.

(1) In the south when a woman dies, the crops and seed yams from her farm are inherited by her daughters; lesser amounts are given to her full sisters who attend her funeral. If she has adult daughters, neither her widower, her sons, her attached wives, nor her daughters-in-law have a right to them.

(2) In Iharev/Ityoshin seed yams are inherited by a woman's sons'

wives, not by her daughters. If a daughter is badly in need, she may ask her brother for some of their mother's seed yams; such a request would not be refused. But the woman, as daughter, has no right to them. If a woman leaves her husband, she forfeits her rights to seed yams. In the rare event that she is "given the road" (divorced by her husband), she may take at least some of the seed yams with her.

Women in Iharev say that their goal is to "accumulate" (*kem*) a larger yam field. Here, where land shortage is no problem, the only limits on the size of farms arise from shortages of seed yams and labor; most people, indeed, made more mounds than they had seed yams. "Accumulating" a larger farm is difficult because one always has obligations to give seed yams to a daughter-in-law or to some other new woman in the compound, and the more seed yams a woman has, the more she must give.

(3) In central Tivland (Jengba/Ipav), a woman's seed yams are inherited by her married sons. If she had previously given seed yams to one of her sons' wives, that son has no claim to any seed yams because he had already received his "inheritance" (*dyako*). In this area a woman during her lifetime normally gives seed yams to wives of all sons except her youngest, who ultimately gets the largest share.

In both central and northern Tivland, a man whose mother dies before he has married is seriously handicapped; when he does marry, he has no seed yams. Several people are obliged on kinship grounds to give him seed yams, but they never give enough. One woman in Iharev/Ukusu had only eight rows of yams because her husband was orphaned as a child. His father's brothers had appropriated the seed yams of his father's wives, and now no one helped him "accumulate" a farm.

A man controls any seed yams produced from the yam fields which he has retained for himself in anticipation of taking another wife; he simply gives this farm to the new wife, often in mid-season. In this case, of course, he forfeits primary rights to the seed yams.

Guinea corn, like yams, is controlled by women. Although it plays an important part in Tiv subsistence, it is not so highly prized. In southern Tivland, little guinea corn is sold, though a little is traded by women (a small calabash-full is worth 2*d.* or 3*d.*). They keep separate the grain grown for food, and buy guinea corn or millet for resale or for beer intended for sale. There are some areas farther north (one center is Adikpo market) where guinea corn is an important cash crop.

Tobacco is said to be a crop of either men or women, but it is almost always the men who plant, transplant, tend, pick, dry, braid, roll, sell, and smoke it. Women assist in the picking and rolling. Almost all Tiv women smoke; almost none has her own tobacco. Women either beg a pipeful or take a few pulls from someone else's pipe; we stopped giving Tiv women tobacco because it was so often taken from them by their husbands.

Side crops (except cassava) belong entirely to women. What money a woman needs to buy salt, palm oil, an occasional onion, or a pennyworth of soap is earned by trade in side crops or in food prepared from them. Such items include chili peppers, fresh and dried vegetable and soup leaves, sauces of *gbaiye* and locust bean, seasonings such as ginger root, fruit (bananas, pawpaws, and mangoes are the most common, occasionally also oranges and pineapples), and some vegetables (peanuts, boiled yams, boiled corn on the cob). The most important prepared foods sold are *akpwekpa* (a mixture of ground peanuts or green maize or flour of various grains, palm oil, and pepper, wrapped in a large leaf and steamed) or fritters made of cassava batter fried in palm oil. Women differentiate this trade in side crops, or products made from them, from speculative buying of yams and other foods. This "sauce trade" (*yuwe yamen*) is a highly complex economic arrangement. Tiv men claim they do not understand it.

Consumption of cassava in southern Tivland has increased with land shortage. There is also a ready cash market for cassava because Hausa and Ibo women buy large quantities of it (as well as other products) from southern Tiv. Tiv say they have increased, and are continuing to increase, cassava acreage to meet this demand.

Although cassava is usually controlled and sold by women (a woman owns all that is grown on her yam farm, of course), there is a growing tendency in this area for men to grow it for sale. It is beneath the dignity of a man to sell cassava at market, but he can give it to his wife in lieu of money or food of another sort, and she can usually convert it into money or the desired foodstuff at market. He may, on the other hand, give a wife a large loading tray (*gbange*) full of cassava, saying, "I want 1*s*. 3*d*. for this." The wife takes it to market, sells it for what she can get (always more than the price her husband asks, or she would not accept it in the first place), and keeps the difference.

All Tiv with whom we have discussed the matter associate cassava with land shortage. Near Makurdi and Gboko, where farm land is at a premium, cassava acreage has increased. There is a little, but very little, cassava grown in Ityoshin.

Preservation and Storage of Staples

We have already noted that Tiv build granaries for food storage, that they sometimes store yams in the ground or in the fields in areas of plenty where theft is unlikely, and that they often store millet on platforms in their reception huts.

Yams and sweet potatoes are sometimes sliced and sun-dried. The dry slices can be stored in baskets or kept on open trays on the platform above the fireplace in either the sleeping hut or the reception hut. The slices (called *akpo*) are ground into flour and boiled into porridge. Tiv say that such porridge is a welcome relief from grain porridge, but is not as good as porridge made from fresh yams.

Sorghum and guinea corn are stored unhusked on outdoor platforms four or five feet square and seven or eight feet high, one of which is usually built for each group of linked wives. The grain heads are tied into bundles and stacked on the platform. Since harvest is early in the dry season, there is little problem of protecting the grain from moisture. When Tiv neglect to build platforms, or when wood is too scarce, guinea corn is stored in large baskets, as much as four or five feet in diameter, set into forks of trees in the compound. Toward the end of the dry season, any remaining guinea corn is husked and the grain stored in pots, covered with a calabash, and sealed with mud. This form of storage is widely used for vegetables, tuber foods, and cloth. Most Tiv women have a shelf in their sleeping huts on which from two to two dozen pots and calabashes of dried food are stored (the usual number is five or six).

Several varieties of beans are stored, the most important being *ahuma* (*sphenostylis stenocarpa*), which are dried without being cooked, and the type of bean known as *adenger* in Kparev and *filifi* in Iharev, which is first boiled in the pod and then dried. Beans are used primarily in sauces; occasionally they are ground into flour to stretch out and enrich porridge. Various sorts of gourd and watermelon seeds are dried and stored in pots or calabashes; large quantities of green pawpaws are sliced and dried.

Locust beans and prosopis beans, once cleaned, are boiled, dried, and put into a calabash lined with *tindikula* leaves; a layer of *kpaghnindi* leaves is added and the whole is sealed with earth.

Condiments

Tiv know of three kinds of salt and a sort of potash substitute for salt. Most salt used by Tiv is imported. A bag of salt, bought when one is affluent, is considered a good investment. Since salt takes up moisture from the air, the usual practice is to make it into cakes and fire it, in which form it is more resistant to moisture. The salt is moistened, molded in small milk cans (which are usually bought empty from Hausa, who use the milk), and turned out onto green leaves. A pile of these molded salt cakes is then surrounded by dry grass or old thatch and fired. The cakes come out of the first firing yellow in color; by the end of the third (they are likely to break unless they are fired three times), they have a purple-brown crust that preserves the salt from moisture.

Before imported salt was available there were two sources. Salt was produced in Ogoja Division and put into containers of woven reed, about three-quarters of an inch in diameter and several inches long. In southern Tivland these containers were still seen in 1952; though more expensive and dirtier than imported salt, it was bought for ritual purposes. Northern and eastern Tiv got salt from the salt marshes and springs in what is now Jukun country. We were told that these salt springs are still worked, but we have never seen salt from this area.

Potash (*mtsem*) was made and used when salt was impossible to get—one occasionally still sees it made today, for Tiv say that it has a different flavor than salt and is good for a change. Yam or sweet potato peelings, pawpaw leaves, *saghegh,* and many other leaves and locust bean pods are burned. The ashes are put into a pottery colander; then water is allowed to soak through and drip into another pot. This solution is then boiled and evaporated, and the remaining salts are used as condiments. Hausa traders also sell potash in the markets. Tiv use Hausa potash only in making snuff, not in cooking.

Tiv grow several sorts of pepper and use it liberally in their cooking. *Mkem* is the name of the small red chili pepper and also the general word for all pepper. Peppers are sun-dried and stored in pots or calabashes. Cherry peppers (*bukem*) are also common and are also dried and stored. *Ikehegh* is a pod, the seeds of which are hot and sometimes used as pepper, but it is more commonly used in ritual.

Southern Tiv make some of their own palm oil, although most of it is imported from farther south. Palm fruit yields two sorts of oil. When the fruit is crushed and boiled, the red oil rises to the top and can be skimmed. The kernels, when cracked and boiled again, yield "white palm oil." Few Tiv extract the "white" oil; they rather crack the kernels and munch them as snacks. Palm kernels never get as far north as central and northern Tivland, and palm oil there is very expensive. It is replaced by peanut oil or sesame oil.

Cooking

ALL DOMESTIC COOKERY is done by women. Most women's sleeping huts have a cooking fireplace, but cooking is generally done outside or in the reception hut. One cooks standing, bending over from the waist, knees stiff. Tiv women can grasp a pot of boiling water under the neck rim and lift it directly off the fire without burning themselves; they have, in fact, no other way to get it off. Few Tiv men can do this.

Women cook meat that was bought in the market; they also cook game and fish. Men cook the meat of sacrifices if it is to be eaten as a ritual meal and may cook the animals brought as "payment" (*tia*) for the performance of the ritual. Men say women cannot cook meat at all and that most women do not take enough care to make food taste good.

Chickens are sometimes killed for food in the richer sections of northern Tivland, but even there most chicken is sacrificed. Chickens are usually not plucked; the feathers are burnt off over a fire and the chicken is then disjointed. The feet, legs, and wings are cut off; the body is cut into four pieces, each of which may be further subdivided almost innumerable times. The feet are then skinned, and stuffed through the opened beak, extending out through the neck. The giblets are then spitted on the claws. The pieces are boiled in water with salt or potash and two or three tablespoons of palm oil or peanut oil. Palm oil turns chicken a bright yellow, mottled with black where the feathers were singed off. Good cooks sometimes add onions, tomatoes (which Tiv say they got from Europeans and enjoy very much), and ground beniseed (in place of the palm oil).

Young goat or lamb is not skinned, but is singed to get rid of the hair. It is then boiled in the same manner as chicken.

Most fish is dried or smoked before being cooked, even when it is caught locally and eaten soon. Fish is boiled with palm oil or put into a vegetable sauce, like any other meat.

A sauce "should" have some substance, either vegetable or animal, to make it sticky (Tiv food is inclined to be mucilaginous), as well as some sort of oil and plenty of pepper. A small piece of meat can be added to almost any sauce to improve the flavor. One woman said, "If you make it sticky and don't leave out the salt, people will eat almost anything." [3] Okra is a common and appreciated sauce base; many other greens are grown on the farm (sorrel is probably the most common) or collected in the bush.

The focus of attention in cooking, however, is on the staple porridge called *ruam*. Yam porridge is made by first boiling the yams, then pounding them in a mortar until they are stiff enough to mold. The porridge may be flavored or adulterated (southerners say the former, northerners the latter) with sweet potatoes or cassava. Grain porridge is made by stirring flour into boiling water. Tiv admit that it is difficult to avoid lumps in grain porridge and praise women who achieve a good texture. Grain porridge may be flavored or adulterated with flour made from dried yams, dried cassava, or dried beans. Guinea-corn porridge is preferred to that made from other grains—millet or corn.

Tiv are generous with cooked food. They are all but exhibitionists about the amount of food they harvest, but often secretive about the amount of food they save and store. The head of a large compound (and, rarely, some other man) generally has a granary in which he is said to store food for the use of the entire compound, but into which no one else dares look because he also keeps his most potent magical apparatus here. This apparatus may take a variety of forms (and in many cases there may be nothing there), but the sanction against a woman's looking at it is usually barrenness, while if another man—especially a son—were to look upon the apparatus during the lifetime of his father, he would likely be hounded by bad luck, sterility, and perhaps death.

There are no similar sanctions against looking into a woman's granary, but it would seem to be just as rare an occurrence. One old woman confided that a granary was the safest place to hide money. Children are warned, after helping their mothers transfer yams into the granary, that "we Tiv never talk about how much food is in the granary." Most women have secret caches of food, kept mainly for their children in case of shortage. It is secret because in a severe shortage what food is held openly must be shared equally by all within the compound.

3. Recipes, as well as detailed information on cooking, are given in Bohannan and Bohannan 1958:257 ff.

Domestic Economy

Tiv are considered by their neighbors, themselves, and most European observers to be good housekeepers, to have large, neat, and airy houses, and to be well off in subsistence.[4]

4. Details of house furnishings, housekeeping operations, and the other mundane considerations of domestic economy are presented in Bohannan and Bohannan 1958.

11

Allocation: Reciprocity

In spite of the fact that each Tiv household could, with some adjustment, become self-sufficient, the rigors of self-sufficiency are in fact mitigated by several modes of allocation of produce. Three sets of institutions must be examined in order to understand the allocation of produce among Tiv: gift-giving, inheritance, and the peripheral market system. This chapter deals with gift-giving and inheritance; the following four chapters consider the peripheral market.

Gift-Giving

Gift-giving among Tiv is a duty of certain kinsmen and of special friends called *hur-or*. Although there is undoubtedly an equilibrium to be found in the system, it is in no sense a simple reciprocity between givers and receivers. Tiv all insist that they do not reciprocate gifts punctiliously unless they are from strangers or from co-wives.

Gift-giving, seen from the social analyst's point of view, is a movement of goods among people who already stand in some sort of social, but not economic, relationship to one another. The gift either ratifies the relationship or is one of the jural obligations connected with it. On the other hand, exchange establishes a relationship among people only loosely related or not related at all. The relationship between the parties to the exchange is primarily in terms of the item changing hands. The relationship between parties in a gift-giving situation is not primarily in those terms. Tiv agree with the almost universal dictum that people do not "sell" to kinsmen. Gift-giving relationships and exchange relationships are felt to be antithetical. Gift-giving is a validation of an existing social relationship, whether *quid pro quo* equivalence is suggested or not; exchange is an act which creates a relationship in terms of the things exchanged, and equivalence of some sort is implied.

In southern Tivland, food moves across the country during the lean season among kinsmen who "send hunger" to one another. Yam harvest can begin a week or two earlier in southern Tivland than in central areas, and baskets of yams are, on request, sent north in early August, just as grain had been sent south earlier. This movement of food is not organized; it takes place on a kinship or friendship basis and is morally important to Tiv. The actual amounts of food are impossible to estimate, but are probably small.

Tiv insist that these gifts are not reciprocated, even though a person is more likely to "send hunger" to someone who has formerly sent food to him than to someone who has not. The idea among Europeans that gifts should be reciprocal has led to misunderstanding of gift-giving among African peoples. On important occasions in the life of a man of prestige (*shagba-or*), Tiv present him with gifts. Some of these occasions are of a political nature. Tiv do not consider these gifts indemnification for future favors. European observers, assuming the *quid pro quo* which Africans do not, interpret such gifts as "corruption." A gift of this sort is called *iwua*. There are several other words for describing gifts by the various situations of giving. All of them either validate or symbolize existing social relationships.

Inheritance

Inheritance among Tiv does not form a single cultural pattern. We have, in fact, had to discuss inheritance in connection with compound fission and succession to status in Chapter 6 and inheritance of seed yams in Chapter 10. As we have seen, Tiv do not have a word for "inheritance" but only one for property which can be transferred from one person to another, *dyako*.

Inheritance of personal property—clothing, prestige items, and rights in women—is most often carried out at a meeting of the moot of a small lineage about a year after a death, but it may be done on the illness of one of the persons involved in the transactions. Inheritance of *akombo* is largely metaphorical; insofar as it is not, it is done in the context of the moot. Inheritance of subsistence is fairly simple in principle, but to set forth precise rules is to overconcretize. In principle, a woman's crops are taken by her husband for the benefit of her children. "Children" includes her married daughters, but their portion is usually a token. Children who are still "in her hut" take the most, the sons who have married often taking only a token amount.

The rights of a woman, and, by extension, of her children, do not lapse when her husband dies. There is, therefore, no actual movement of crops or seed on the death of a man except in the case in which he has kept a small yam farm for himself, when the seed yams are inherited by children in sets of full siblings.

A man's livestock is inherited by his sons and brothers: usually the stock is inherited by the sons, and gifts (*iwua*) from it given to the brothers of the deceased. As in other Tiv affairs, the criterion is to maintain approximate equity, not to follow set rules.

A man's cloth and weapons are inherited by his sons or, failing sons, by his brothers. Again, the inheritance is divided equably according to age, seniority, and need. "Rules" in the matter are always *ad hoc*.

In the 1930's and 1940's, the colonial government searched for "heirs" in the sense of successors to political office. Their assumption was that the right to rule was actually associated with a material *akombo* symbol, and hence heritable. They took the Tiv word "receiver" (*orngohor*) and made a sort of "heir apparent" notion of it, creating a new office in its name.[1] An association of such an idea with indigenous Tiv ideas seems to us doubtful.

The most important "property" to be redistributed on the death of a man is his rights in women, both wives and wards. Inheritance of widows

1. One often sees this word written *angahar* in official reports. This word means "feet" in Tiv and is an erroneous rendering of *orngohor*.

is usually settled informally, by action on the part of the women themselves. Either a woman returns to her guardians and parents (in which case the heir must get the bridewealth returned and use it to find himself a new wife), or she selects one of the possible heirs (sons, brothers, father's brothers' sons) and establishes a new household. A woman has a right to "follow her heart"—to cohabit with any or all of the men who might inherit rights in her and then to select one at a public moot held about a year after the death of her husband.

The inheritance of marriage wards is usually considered in a context of illness and witchcraft, whenever one of the wards or one of the potential heirs of a ward is taken ill. Then, the fact that the wards are "on the ground"—i.e., in dispute—becomes an issue, because the role of guardian is necessary in all moots and rituals relating to the welfare of wards. Inheritance of rights in wards is a matter that requires a meeting or moot of the ward-sharing lineage group (the "*ingol* group").

Both of these institutions—gifts and inheritance—are morally important for Tiv, but account for little actual allocation of goods. The major allocative institution, in spite of the relatively small amount of goods involved, is the market place, which harbors a free market.

12

The Traditional Market Place

MARKET PLACES are among the most obvious and easily observed of Tiv institutions. The market principle is of less importance in Tiv economy and is, in fact, analytically elusive.

Southern and central Tivland are dotted with market places that range in size from small gatherings by the pathside or in the central yard of a compound to those in which several hundred people gather in an open space some fifty yards square, to the large gatherings which at their peak hours contain several thousand people in a space as large as two New York City blocks.

As one listens to Tiv discuss market places, one hears such sentences as these: "The market is the day." "A market cannot be sold, for a market is the people." "A market is trading." "My market was no good yesterday." "A market is when people meet." These descriptions clearly refer to different functions that market places perform. To Tiv the market is not merely, and sometimes not even primarily, a place to buy and sell. It is just as importantly a social institution where one dances, celebrates, and meets friends. It is a political institution of major importance because in a society that is dominated by the lineage principle it is a means of overriding that principle.

Tiv markets can be called peripheral (Bohannan and Dalton 1962). Important beyond their economic importance, only a small amount of produce that any compound produces or consumes ever enters the market. Tiv markets are peripheral because they show two important characteristics: they are multifunctional and the factors of production do not enter the market.

History and Organization of Ticha Market

ABOUT 1939, a man named Kwakar, of Nanev lineage, was teaching in MbaDuku at a Catholic school which no longer exists. Kwakar claimed that all of the markets held on one day in the five-day market "week" were too far away and that the people of the neighborhood should have a market of their own on that day. Furthermore, the students at the school, he said, should have a market where they could sell the things they made at school and the produce of the school gardens. He established a market to meet on this day in MbaGor on land that had recently been the farms of Butu, who cooperated with him.

At first the market was no more than a clearing under two large locust bean trees. It was kept free of grass by the schoolboys, who also swept it the day before the market was to be held.

This market began as a "beer drink," but it "grew," as Tiv put it, and before long was well enough entrenched that Kwakar began to levy tribute on the people who came there to sell. He claimed the right to make a levy because it was "his market." This action was frowned upon by the people, and particularly by the government chief. Kwakar was not a man of much authority and most people could evade his levies, so they continued to attend his market. Then Kwakar began to charge admission. The chief brought the matter before the District Officer and was given permission to move the market from its original location to another site under a great *ndur* tree just outside his own compound.

Soon after the market was moved Kwakar was transferred to another school some distance away, and not long after that he grew tired of teaching and became a policeman in the Native Authority. The market originally was called the "teacher's market." After its move it continued to be called *Kasua Ticha.*

The only lineage in MbaTyough of MbaDuku (see Table 1 for the genealogy) which had a right to a market was MbaAji. When Kwakar's market was moved from its original site, the chief offered this market to MbaAji. MbaAji elders refused it because it was not on the right day; the defunct MbaAji market, called Tsar market, had been what we call a "consecrated market." Because Ticha market met on another day, it could not be considered the same market. MbaAji said they did not want two markets.

Therefore, the chief took Ticha market himself. The oldest man of each of the two component lineages of MbaNyam began to act as its overseers: these men were Kuka of MbaWandia and Nor of MbaGor (Table 1). These two men had a small shelter erected in the market place and spent most of the market day in it, settling disputes, sharing out and drinking whatever beer was given to them, and acting as a clearinghouse of information. The chief seldom attended the market. When he did so, he sat in one of the market shelters built for trading stalls, not with the two market overseers.

The two elders were called "those who look after the market." They kept an eye on everything that went on. If any group wanted to dance in the market, they had first to make arrangements with "those who look after the market." Permission was seldom (probably never) refused, but sometimes dancing parties were made to wait until later in the day than they would have liked, so that marketing would not be disrupted. And sometimes the dancers were disarmed.

As an "honorarium" for their services in looking after the market, the two elders got the liver of every animal slaughtered for sale there. Both denied that this was the reason they did the job. The market, they said, had to be looked after by responsible people, and therefore it was their place to do it. Their answers were the same as those of civic leaders everywhere.

Ticha market was policed by the Native Authority policemen who were stationed with the chief, and sometimes by his Native Authority messengers as well. There was a "stranger" in MbaDuku named Iyoitiev ("field snake") who also helped with the policing. He was said to be the chief's direct representative. Iyoitiev, a man of fifty or so and unmarried, was what Westerners would call neurotic, and he was sometimes resented by Tiv, although he did indeed help in maintaining order.

Women sit in the same positions every market day. The seating arrangement of the women is an approximate duplication in miniature of the locations of their compounds. Should a fight break out in the market, a woman would not have to traverse the market place in order to reach the path leading home. Tiv say that all Nigerian markets have this sort of seating plan.

In the case of men, the commodities being sold dictate the seating arrangements. Cotton-sellers, sellers of dried fish, all tobacco-sellers sit together in distinct groups. All the butchers are on one side of the market place; all the foreigners' stalls are on another.

A large number of Udam and Ibo women come from nearby Obudu to Ticha market to buy yams. They always collect in the southeastern corner of the market place, the point nearest their path of exit. These women said that within the group they maintained their relative positions: the Ibo women together, the Bafum women to their right, and the Udam women behind them. The main market business of these women was selling bulk salt or buying yams and cassava, which they either carried home or sent on the bicycles of young men. Cassava was peeled in the market place before it was carried away; this made it a little lighter and was also the first operation in making *gari* flour, a type of tapioca which is an Ibo staple.

Ticha was never a "big market," even though two thousand people or more sometimes came to it. A "big market" is determined not by its size, but by whether or not it has a protecting fetish. Therefore, although Ticha became, from an economic point of view, one of the most important secondary market centers in south central Tivland, Tiv refused to call it a "big market." Its economic importance resulted from two facts: (1) rice became an important cash crop soon after it was introduced, and (2) Ticha was in a geographically advantageous position as a bulking center for rice.

Describing a Tiv market is difficult. Here is an attempt, quoted directly from our field notes, to describe Ticha:

> Each market day, Ticha market goes through a definite cycle. First there are the early morning crowds (until about 8:30 or 9:00) which consist mainly of the butchers who get there early to do their slaughtering and get the meat cut up and spread out on leaves for display so that they can be squatting down behind the meat to give full attention to sales by the time the main buying crowd arrives. The rice sellers also come early; they want to get rid of their loads as soon as possible so that they can have cash in their pockets and spend the rest of the day in the beer stalls. The butchers also cook the offal of their animals (except the livers, which go to "those who look after the market"), which they sell in penny lots through the day. At the market people also eat *kpekpa* (steamed peanut or corn cakes, with pepper and palm oil), and sometimes some fruit or a coconut, although they

usually take coconuts home with them. There are plantains in the market (which Tiv eat raw) and you can occasionally get fried cassava fritters and bits of baked yams.

The morning phase passes into the phase in which the cotton sellers, the fish sellers, the kola sellers, and the tobacco sellers do their major trade; this is also the time at which women do their biggest selling of yams, palm oil, sauces, locust beans, peanuts, millet, and *kpekpa,* and the time when the butchers sell their meat. The Hausa also do their biggest run of business at this time, and the sewing machines are all busy.

Then, some time between 11:00 and 12:00, the beer has arrived. It has been dribbling in all morning, but it is always there in plenty after 11:00. I have seen some women who live nearby make two trips to market: one about 9:00 with sauces and oil, another about 11:30 or 12:00 with beer. The serious business of marketing then gradually fades away, until about 2:30 there is very little left except the beer stalls. The Hausa pack up and go home about this time, and the sewing machine business is over.

The market is usually over by 4:00. A few solitary drunks may still be around, but most drinkers have gone to the compound of someone nearby to continue drinking, and the market place is deserted, littered with debris, and left for the buzzards and stray dogs, who are picking and licking around the spots where the meat was killed.

Ticha market was "killed" in 1950 when MbaAji's market, Tsar, was reopened. However, MbaDuku people soon afterwards tried to get the chief to revive it because they said that having a large market was all very well, but they also wanted a small market, where there were fewer people, where they could drink beer in more comfort, and where they could "trade ha'penny ha'penny" instead of trading seriously. The chief and the *mbatarev* refused to reopen Ticha market. People who wanted to go to market on that day had to walk the four to seven miles to Iyon market, or even farther to some other market.

The chief alone had the right to make a levy on (*sav*) Ticha market. However, he did so only four or five times a year: levy was made at Ticha on 22 November 1949, and another on 31 January 1950 (there had not been one between these two dates). One of the market elders put it this way: "On the day of levying the market every woman offers us something, but for the most part we turn it down because we are the big men around here."

A somewhat more detailed, and perhaps less biased, picture comes from our notes:

31 January 1950: As I walked into the market place about 10:45 this morning, I was met by Anyo [a "son"—physical grandson—of Kyagba (the

chief)] with a load of yams on his head, who shouted, "We are levying the market!" Kyagba was sitting in front of the market shelter at the corner. The two policemen, the scribe, the two elders who were the market masters, and Iyoitiev, Kyagba's personal representative, went around the market, taking a bit of salt, rice, or a yam from everybody who was selling. The items were then carried back to where Kyagba was sitting surrounded by his various sons and grandsons.

There were 75 yams collected on this day: 10 were given by Kyagba to each of the two market elders, 5 to me, and the rest Kyagba took home and distributed among his wives. In addition, there were sweet potatoes and cassava, salt, palm oil, rice, the tripe of every animal killed that day: a representative sample of what was being sold in the market. On two occasions women came up to Kyagba to complain that the men making the levy had taken too much. To one of them Kyagba listened and returned 3 yams; the other he sent packing. One of Kyagba's young wives, sitting on the arm of his chair, told the children who helped carry the produce to go get her some soap—it was the only thing they had not got, she said, and she needed it badly.

When it comes to assessing the size of the crowd and the amount of trade, severe methodological restrictions are felt. Although we tried it on several occasions, counting the people who come to market is difficult. On 31 January 1950, four assistants helped us count the people who came to market. They were "stationed" at the various entrances to the market place. On that day we counted from 7:00 A.M. to about 10:30 A.M. (see Table 14). We stopped too soon, thinking that rain had "spoiled the market." It grew much larger about noon.

TABLE 14
ATTENDANCE AT TICHA MARKET, 31 JANUARY 1950
(7:00 A.M. TO 10:30 A.M.)

Path No.	Women Carrying: Rice	Yam	Cassava	Nothing	Loads	Total	Men Carrying: Rice	Nothing	Loads	Total	Sum Total
1	54	38	13	62	29	196	77	92	24	193	389
2	6	10	4	28	45	93	11	104	25	140	233
3	5	31	9	22	55	122	14	43	9	66	188
4	4	19	3	33	30	89	30	28	4	62	151
Total	69	98	29	145	159	500	132	267	62	461	961

There was a difficulty involved in the fact that almost everyone carried a variety of miscellaneous goods as well as the staple by which he was classified. Some women carrying yams also had some rice, and almost all had a small collection of miscellaneous items in addition to the staple. Thus, when to put a person into "yams" and when into "miscellaneous" was necessarily an *ad hoc* proposition. It must be remembered that small quantities of the staples entered the market in addition to those mentioned in the tables; further, most marketers brought a variety of goods.

The next market day, 5 February 1950, we took another count. This one was more complete (see Table 15).

TABLE 15
ATTENDANCE AT TICHA MARKET, 5 FEBRUARY 1950

Path No.	Women Carrying: Rice	Yam	Cassava	Misc.	Nothing	Total	Men Carrying: Rice	Misc.	Nothing	Total	Sum Total
1 & 4	106	91	35	30	39	301	106	55	97	258	559
2	14	16	83	91	116	320	10	119	118	247	567
3	16	110	89	121	47	383	25	24	110	159	542
Total	136	217	207	242	202	1,004	141	198	325	664	1,668

There are several things to be learned from these figures: it would seem that rain keeps more women than men away from market; it would also seem that on a good day almost twice as many women as men attend a market like Ticha (this is definitely not the case with the larger markets, such as Tsar, although we do not have detailed data to document this fact). Perhaps the most important thing the data show, however, is that rice moves from north to south. Men and women coming from the north loaded with rice far outnumber those coming from the south.

When it comes to estimating quantities of goods, one is on even less secure ground. Our estimates are very rough. On 2 December 1949, we counted the rice bags in the market at 8:45 A.M. We knew, on the basis of counts we had taken at the preceding three market meetings, that almost all rice had arrived at market by that time. Serious selling, however, was just beginning: the women had sold their small caches of rice worth 2*s*. or 3*s*., but a man does not like to be the first to sell his rice, because the price might go up if he holds out. There were 472 bags averaging roughly 40 pounds each. We weighed about 20 bags and then classified the rest into three sizes as we counted. There were also 81 bags of about 60 pounds each—a heavy load for a strong woman. Most of these bags were in the hands of Ibo and Udam women who sat in the southeast corner of the market in a group. They usually bought rice by the calabash- or basin-full, then bagged and transported it either to the motor road a mile and a half away, or to Obudu, about four miles by bush path. This figure of 81 included 7 bags of rice belonging to one Ibo middleman, but it excluded 9 bags of from 100 to 150 pounds, each of which had already been bought by Beba Bie, one of three full brothers who bought a ton or two of rice every market day. Beba and his brothers took it home, where their wives worked very hard husking it in mortars (they were paid for this); the brothers then resold the husked rice several months later at a much higher price.

Although rice was, in season, far and away the most important item sold at Ticha market, there were many others (see Table 16).

TABLE 16
NUMBER OF PEOPLE SELLING AT TICHA MARKET, 8 OCTOBER TO 12 DECEMBER 1949

	Number of People Selling on						
Product	8 Oct.	17 Nov.	22 Nov.	27 Nov.	2 Dec.	7 Dec.	12 Dec.
Cake salt		7	2	10	5	2	4
Bulk salt		5	9	7	7		5
Cotton	16	8	4	3	5		
Tobacco			5	10			
Kolas				3			
Butchers:							
goats	7	5	4	3	8	3	5
pigs	1	2	0	0	0	0	0
cows	1*	1*	1*	1*	1*	1	1
Smoked fish		5		11	12	21	9

* This was meat left over from the slaughtering done at Asawa market the day before.

On 8 October, of the 9 butchers, 6 were local MbaDuku people; the other 3 all lived within seven miles. On that day, of the 16 cotton sellers, 13 were local MbaDuku people, 1 lived about five miles away, and we did not interview the other 2. The cotton came from distant areas, however: Nongov in the Benue-Katsina Ala confluence, upstream as far as Ibi, and from Idoma country to the west.

Far fewer women than men go to market empty-handed. It is not considered proper for women to go to market without business there, even if that business is only the sale of a few greens or peanuts, or a handful of millet or pepper. Following is a description of what three representative women had to sell at Ticha on 22 November 1949. The first had 2 big calabashes of Bambara groundnuts, 9 sweet potatoes, and an old beer bottle about a quarter full of red palm oil. The second had 5 small piles of sweet potatoes, containing from 6 to 10 potatoes each;

each pile was to sell for ½*d*. She had a pound of rice in a small calabash, a few gourd seeds that she was shelling, and some pats of sauce made from fermented seeds of prosopis. She also had a small bit of meat which she had bought, 3 pennies, 3 ginger roots, 3 yams, and 2 small empty bottles. The third woman had 2*d*. worth of okra in ½*d*. stacks, peppers, half a medium-sized calabash of millet, a pot of oil, 2 plantains, and a spool of hand-spun cotton thread. As many as 500 other women were similarly loaded. Many of these women brought 3 or 4 yams to sell. At this time of year yams ran about 3 farthings per pound in this market, though they were not, of course, weighed—by anyone but us. They were bought by other Tiv women and particularly by Ibo women from Obudu and Ibo men with bicycles, who took them away and resold them at the next Obudu market.

There were usually two or three Hausa medicine traders from Obudu who attended Ticha market with their wares. On 2 December 1949, there were four. Here is an inventory of the goods that one of them displayed: about 40 onions, perfume, a sort of bean used as medicine, 7 knives with wooden handles, potash used in making snuff, 3 sizes of nails, 3 grades of pipe, earrings, small fishhooks, 3 long bars of Nigerian-made laundry soap, 4 brass finger rings, string, *ilyoho* (a dried berry strung on a stick and said to be used for soup), antimony for cosmetic use, magenta crystals (sold by the pennyworth as dye), washing blue, and ground camwood. This man sold his perfume at 2*d*. for a .32 caliber shell-casing full. It was then poured into whatever bottle the customer had brought with him.

Tiv petty traders have a different sort of produce: safety pins, Lifebuoy soap, exercise books, sugar lumps, string cut into yard lengths, matches, moth balls, shoe polish, mirrors, scissors, hinges and hasps, pencils, mentholatum, Passion Flower perfume, and face powder.

It is notoriously difficult for Europeans to discover correct prices in an African market. The following prices for 12 December 1949 are probably fairly accurate, because it was known that we did not buy these particular commodities. We weighed 14 lots of yams for sale—they came out to about 15 pounds for a shilling. Rice is more difficult: on 12 December, one of the most successful rice middlemen told us that he had paid £1 3*s*. for some 200 pounds of rice; the week before he had got over 300 for £1 5*s*. (we believe that this statement may be wrong). The same week, another middleman told us he had paid £1 15*s*. for approximately 150 pounds. When we asked about the difference in price, we were told that some people are better at marketing than others.

Price documentation in an African market is probably the most difficult information to acquire and evaluate. Prices on all items are usually more or less the same throughout the market place, though no one fixes prices. There is no authority or traditional figure who says what the price is, and there never was. There is no practice of selling throughout the market at the price first accepted after the market's opening, as has been found in Dahomey and in other places. The price is reached purely in the operations of selling and buying. However, in the absence of precise measurements or of any mode of communicating the going prices (such as those to be found in many Ghanaian and Yoruba markets where going prices are posted), there is sufficient variation that "market luck" is still a factor. For example, about a fourth of large beer bottle full of palm oil usually cost 3*d.* The price was well enough known that when a woman one day bought that amount for 2*d.*, she paraded her bargain and was congratulated all round. Yet the women who sold it to her claimed that the quantity in the bottle was less than the traditional amount sold for 3*d.* The price of yams, since they are not weighed, cannot be anything but approximate, for they vary in weight from one-half pound to as much as 15 pounds. Thus, the investigator can weigh the yams that actually changed hands and, if his sample is sufficiently large, arrive at a definite price. Tiv obviously do not do so; rather, they bargain for each collection of yams. They do it, however, in such a way that the "price" is quite uniform.

A tremendous amount of food was exported from Tivland, some of it from Ticha market. On 17 November 1949, we ourselves checked path 2 (see Fig. 22) in an attempt to estimate the produce that was being exported in headloads and bicycle loads to Obudu. Table 17 presents a summary of what was found. Some food probably left the market after noon, although by that time most of it had been carried away and the market place given over to beer drinking.

Although the data presented are accurate for numbers of loads, it is almost impossible to calculate the quantity of goods which left the market. A few of the bicycle loads, for example, were no larger than good-sized headloads, although most of them ran from 100 to 150 pounds. We would estimate the headloads at between 50 and 60 pounds, although some were smaller. We tried to divide the yam headloads into large and small, but the figures are necessarily rough.

Food export is largely in the hands of Udam and Ibo women and young men. Until they are seventeen or eighteen years old, the young men carry headloads; by that age most of them have acquired bicycles.

TABLE 17

Export of Produce from Ticha Market on the Path Leading South, 17 November 1949

	Before 9:00	9:30	10:00	10:30	11:00	11:30	12:00	Total
Headloads:								
Yams (large)	1	1	9	12	9	5	7	44
Yams (small)	—	—	4	9	7	1	3	24
Cassava	—	6	4	8	2	8	5	33
Rice	4	6	18	24	13	12	4	81
Sweet potatoes	—	—	—	—	—	2	—	2
Peanuts	—	—	—	—	—	—	2	2
								186
Bicycle loads:								
Yams	10	2	8	2	3	—	1	26
Rice	2	—	1	—	—	—	—	3
Cassava	—	—	1	—	1	—	—	2
								31

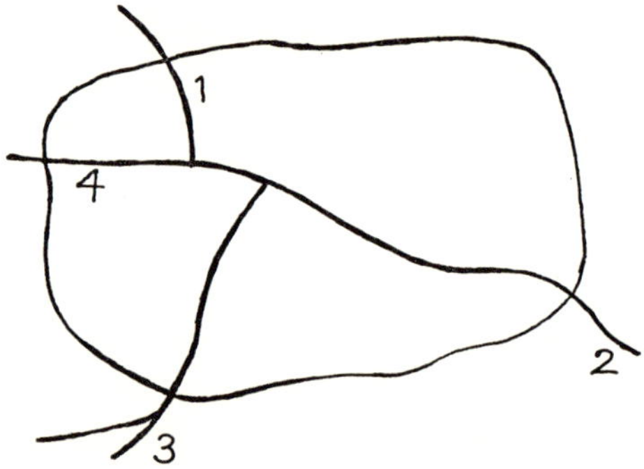

Fig. 22.—Ticha market place, showing incoming paths.

Some Udam women make two or even three trips a day; some young men with bicycles make four or five. In January of 1950, a group of

Ibo in Obudu acquired an old lorry chassis, stripped it down to the frame and wheels, and pushed it to a position on the road near Ticha market. After Ticha was "killed" and trading took place at Tsar market located on the road, this vehicle, pushed by eight or a dozen young Ibo, made two or three trips every market day.

Again, it is difficult to say how much of this food was for resale at Obudu market, eventually for export to Iboland, and how much of it was to be consumed by the Ibo and Udam in Obudu. Both Ibo and Udam say that most of it is resold; in the case of the Udam, at least, an attitude of pride in food production, and shame at buying food for consumption, must be taken into consideration when evaluating these statements.

Organizationally, Ticha market was simple. Tiv considered the fact that it was economically important as beside the main point: it was a "beer drink" because it had never been consecrated and had only a minor political history.

History and Organization of Iyon Market

Some six or seven miles to the east of Ticha market is Iyon market, held on the same day; it meets in the central open space of one of the biggest compounds of Iyon lineage. Iyon market was begun many generations ago, by a man named Ugur, when Iyon was a small "segment-within-the-hut" of Shangev Ya lineage (which was said at that time to have formed a single community of about the size that Iyon forms now [see Chap. 7, Table 4, and Figs. 7 and 9]). A text dictated by the master of Iyon market explains why and how it was started:

> Long ago, it was very far [in time], we of Iyon sat alone. The Uge were not close to us. MbaDuku was not close to us. We were all in the bush. If we went to Uge, they killed us; if we went to MbaDuku, they cut off our heads. But hunger seized us; hunger seized MbaDuku. So we all gathered in the farms. MbaDuku brought salt [which they were able to buy from MbaYongo, who in turn got it from the Chukwan Udam, who got it from people still farther west in Ogoja Province]. The Uge brought palm oil and yams. We sold cloth, and the Uge bought it [Uge did not, and still do not, weave]. We bought salt and palm oil. We sold sweet potatoes, and MbaDuku bought them. Thus, we met in the farms of Ugur.

In this way Iyon market started as a meeting place for hostile groups to exchange their produce. Iyon people today say that they met in the farms instead of in a compound because the damage would have been less had a fight broken out. People sat by lineage segment or by tribe, with their backs to their own territory.

After Ugur's death the market passed into the hands of Darshom, a man of MbaChaku lineage of Iyon (see Table 1). No one in Iyon will give either Darshom or Ugur an exact position in the present-day genealogies. They all say that Darshom was more recent than Ugur, "because in Darshom's time we already bounded MbaDuku; we already bordered the Uge." That is, there was no unfarmed bush between them. But MbaKyar lineage had not at that time come down between Iyon and MbaDuku (see Chap. 7).

The market grew, and after a while it became altogether too big for a person such as Darshom. So "he took it and gave it to Ityodugh." Asked to explain that statement, one Iyon elder said, "If my small son goes into the bush and meets something too big for him, he comes and calls me. He gives it to me, for he can't cope with it. It was thus with the market."

In the past even more than today it was necessary that a market, if it were to survive, be in the hands of a strong, capable man of influence who could keep the peace, ensure safe conduct to travelers, and protect those who came to trade. The market master's story continues:

> Ityodugh was a man of great influence. They knew him in MbaHura [a lineage of Shangev some five miles to the northeast]. If MbaHura [people] came here, they went to Ityodugh. They paid him tribute so that the people of MbaKertiev [a lineage which included Iyon] would not seize them and sell them as slaves. When the market became too much for Darshom, he gave it to Ityodugh. All Shangev was summoned, all MbaDuku and all Uge. They all came.

It was during Ityodugh's time that a market *akombo* was erected. Transfer of a market from one man to another is a purely domestic matter, but the consecration of a market involves all those lineages or tribes whose members habitually attend it.

The consecration of a market puts it on a new footing—indeed, the term for the establishment of a market fetish (*ver*, or *tim, akombo kasua*) is sometimes abbreviated to "establish a market" (*tim kasua*), a phrase never used for an unconsecrated market. In its purely secular, commercial aspect, the market may be "summoned" or "invited," or the people may be said to "meet to buy [and sell]." One Tiv told us, "There are several places for buying and selling in MbaGbera, but Jape and Aichwa are the only markets there." He meant that only these two markets had been consecrated.

The avowed purpose of the market *akombo* is to assure the maintenance of peace; breaches of the peace are thenceforth subject not only to the market courts but to automatic punishment by the *akombo*. The form of a market *akombo* is closely related to this purpose.

Market *akombo* generally consist of two elements: (1) *swem*, or one of the other *akombo* of stature associated with justice and capable of instruction; and (2) the *ishohough* tree, which is an emblem of one aspect of the great *akombo swende*. *Swende*, in all its aspects, is associated with bloodshed. This particular aspect of *swende* kills with sores.

Swem is a great *akombo* known to all Tiv, but its most popular use is in Jechira. Markets like Tsar, Angur, and other Kunav markets had *swem* in their market magic. Iyon, farther east, did not have *swem* in its market fetish. Rather, it had an *ishohough* tree for *swende*, and for its second element a silk-cotton tree (*vambe*), the emblem of the *ver akombo* which most commonly manifests itself in the various symptoms associated with the deficiency diseases that are common among vegetable-staple peoples in Africa.

In early 1950 the *akombo* in Iyon market place was represented by the dead hulk of a silk-cotton tree. Though wood often fell from it, no one ever used it for firewood—this in a country where firewood was at a premium. When we asked about the *ishohough* tree, we were told that it had died and disappeared. No one was concerned about the disappearance of the physical tree—the *akombo* was still there and was still operative, even though the tree which was its emblem had died. Should it ever become necessary, it could easily be replaced.

At the time these fetishes are set up, they are "instructed." Tiv say that although new market *akombo* are no longer set up, their establishment was done in precisely the same way as that of the *akombo* set up and instructed for the bridge built by Iyon across their stretch of the Aya River (Bohannan and Bohannan 1958:325–39). On that occasion the fetish was instructed to kill anyone who defecated on the bridge, who willfully damaged the bridge, or who shed blood on the bridge. Such instructions are delivered publicly in a loud voice by the elder who performs the *akombo* ceremony.

As in all *akombo* ceremonies, sacrifice is necessary to make the emblems potent. In the case of a market *akombo*, the sacrifice is said to consist of human beings, members of the lineage associated with the market who are "killed by night"; that is, killed mystically by witchcraft. Lives are said to have been given for all market fetishes. We believe that the sacrifices were of the same nature as the "sacrifice" made on Iyon bridge: just after the bridge was finished, a drunken youth insisted on bathing in the river, which was at flood; he drowned. The entire countryside claimed that he had been bewitched (*tambe*) and killed by the elders of the lineage in their roles of "witches" (*mbatsav*) to consecrate the bridge.

The most important function of a market *akombo* is to mark an association among several lineages for purposes of meeting peacefully for trade. However, there are other associations which should be considered briefly before we turn to the "pact" aspect of the market *akombo.* The market *akombo* punishes breaches of the peace in the market place. It works primarily in the market place itself or in the territories of the lineages or tribes who participated in establishing it. If the market place is moved, the market *akombo* may or may not be moved physically to the new site. Iyon people said that when they moved, a slip from the old *ishohough* tree was planted, and the old tree was destroyed as soon as the slip was flourishing (the latter part of this statement is almost surely not to be taken literally).

Tiv compare their market *akombo* to "a piece of paper from the European given to show whose market it is." The stump of the silk-cotton tree—and, more importantly, the right to erect the *akombo*—is the charter of Iyon market. In the case of Tsar market, to be considered below, the presence of the *akombo* and the holding of papers from the administration were considered of about equal importance in determining rights to that market.

The market *akombo* is much more strictly associated with a day than with a place; the day that the *akombo* is erected, and every fifth day thereafter, is that fetish's day. Once a market has been consecrated, it is irrevocably linked with that day. As the *tyo-or* of MbaDuku once said of Tsar market, "The day of Tsar market is in the order of the universe." The day of an unconsecrated market like Ticha can be changed. In fact, it was once mooted that Ticha be revived on another day. The day of a consecrated market like Iyon or Tsar cannot be changed.

In the course of describing the histories of markets, Tiv often judge whether two markets are "the same market" by the association of the day and the *akombo:* of the surviving one of two consecrated markets, always held on the same day, they say uncompromisingly that "there were always two markets; one died." If the defunct market was consecrated and the surviving one was not, the caretakers of the surviving market are said to have "stolen" the consecrated market. In a few cases the *akombo* itself is said actually to be "stolen"; that is, a slip of the tree is transplanted.

By far the most important function of the market *akombo*, however, is that of a peace pact among the various tribes and lineages who subscribed to it and attended the ritual of its establishment. Market "pacts" are something like peace treaties (*ikul*) (Bohannan and Bohannan 1953: 30–36), with two major differences: the latter are always bilateral, whereas the market pact may involve many parties; the market pacts concern peaceful and honest trading relationships and not a series of prescribed acts and prohibitions which give the *ikul* treaties aspects of pseudo-kinship.

The elders of the lineage in whose territory the market is held invite the elders of those lineages whose people attend the market to participate in establishing the market *akombo*. It is today not possible to collect any sort of "standard" version of what is involved in the ritual. We were told, for instance, that an elephant was necessary; Hausa or Kanuri elephant hunters were said to have played an important part in the Tsar *akombo* and to have been present in the founding of the Iyon *akombo*. We were also told that, like *ikul* treaties, the market *akombo* requires daytime sacrifice of slaves (in addition to the "nighttime" sacrifice of lineage members) by each participating lineage or tribe. It would seem, however, that the slave may, in some cases at least, actually be replaced by a dog (to Tiv and to most Udam a dog is the ritual equivalent of a human being if other wealth is added to "make it taller"). Iyon people told us that the blood of these sacrifices was poured over the emblems of the fetish, particularly the *ishohough* tree. Obviously we do not know whether these were the actual sacrifices when these rites were carried out. We suspect that they may not have been because an idiom of the number and value of lives is the normal one in which Tiv discuss any hierarchy of *akombo*.

All members of those groups whose elders participated in establishing the market *akombo* were bound to proper behavior in that market place. Misbehavior such as brawling or stealing between members of two such groups in a consecrated market was, in the early 1950's, both more heavily fined than the same offense against members of groups not involved in that market pact and also treated as a moral offense.

The consecration of Iyon market *akombo* was attended by representatives of the various lineages of Shangev, by MbaDuku, and by the Uge. We have heard a recalcitrant MbaDuku thief in Iyon market confess, give his name, and pay his fine when it was pointed out to him that the woman from whom he had stolen was Uge and not Udam. He said in his defense that he had thought her an Udam. (They did not participate in this market *akombo* and hence were neither bound nor protected by it.)

Tiv themselves speak of some of their consecrated markets as much in political terms of interlineage relationships as they do in terms of trading. Such markets as Iyon are considered appropriate places for intergroup negotiations conducted between the participating lineages through their elders, especially the heirs of the persons who were instrumental in establishing the original *akombo*. When the government chief of Shangev Ya, a man of Morov lineage, wished to negotiate with the Uge, he came to Iyon market and dealt with them through the market organization.

To understand the internal organization of Iyon market with its market chief and market judges, it is necessary to complete the history of the market. Ityodugh, to whom the market was "given," was both capable and well liked. One of the cornerstones of his power was that he controlled the market and ran it extremely well. He even had to fight several wars over it.

The war which we call the "first" war between MbaKyar and MbaKertiev (see Table 1) did not begin over markets at all. A MbaKunu man whose mother was from Iyon (both lineages are MbaKertiev) was killed in this war; and no one was killed in retaliation because all Tondov lineage, which included both belligerents, became embroiled in a war with the Uge. At about this time, MbaKyar was still "behind" Iyon, and had a small unconsecrated market which met on the same day. This market had originally met in the compound of Agan. On Agan's death Kwange became the most influential man in MbaKyar. Kwange claimed that he had been "given" Agan's market, and he moved it to his own compound. Soon after, MbaKyar, in a regular migration movement, "got out of position" and came in between Iyon and MbaDuku. This meant that the two markets, held on the same day, were in direct competition, and that Kwange and Ityodugh were also in direct competition for the position of power in this part of the country.

The Uge-Tondov war broke out before this rivalry between Kwange and Ityodugh was settled. During the war Ityodugh stopped the trading at Iyon market, which was dangerous because Iyon bounded the Uge directly. He told his people to go to Kwange's market. It was later claimed by Kwange's people that Ityodugh had "given" Iyon market to Kwange. Kwange, now at the communications hub, "refused to return the market to Iyon." That is, when Ityodugh announced the reopening of trade in Iyon market place on the same day, Kwange used all his influence to keep people from attending.

The "second" war between MbaKyar and MbaKertiev arose from this dispute; it is today discussed entirely in terms of market rivalry. If Kwange could "keep" the market (that is, if he could arrange matters so that everyone came to his market on this day), he would gain in political influence over Ityodugh. All the negotiations between the Uge, Kunav, and Shangev would take place in his sphere of influence rather than in Ityodugh's. Iyon, again joined by the rest of MbaKertiev, set out to make war on the pretext of avenging the killing that had occurred in the "first" war. Iyon elders today say that they killed Butu of MbaKyar in revenge for the original killing and that they succeeded in getting their market back.

By the time the earliest members of the southern Nigerian administration arrived in nearby Obudu about 1909, Ityodugh was dead, but the market was being held in his compound. Kwange (described in reports of these early administrators as "chief of the Shangevs" and "a very old man") was without a market. Kwange's compound was a couple of miles closer to the new station than Iyon, and it was directly on the path from Obudu to Katsina Ala. By their own reports, some of the early administrators encouraged Kwange to start a market, because the provisioning of troops and government staff at Obudu was always somewhat difficult. Today Iyon people speak of the ensuing struggle as another—a "third"—war "fought in the Europeans' courts." The disputes between Iyon and MbaKyar centered on two points: MbaKyar's move into land that Iyon considered its farms, and the right to the market. Many years after the area was transferred to the Northern Region, a district officer was called to settle the dispute. He left the market with Iyon on the grounds that it was in their possession already, and he left MbaKyar in their position on the ground—again, because they were there already. Both sides are still indignant.

Ityodugh's successor in Iyon was a man named Ayakwagh. He was equally forceful but not equally liked. Since his death, Iyon lineage runs itself "in committee," so to speak, and takes precautions to avoid the emergence of another such leader. Much of the concern of the elders centers in the market; the complexities of the present organization arise out of this conscious effort to achieve a balance of power. The members of Ityodugh's compound (Ugbajir's in 1952) today have the duty of keeping the market shelters in repair and of sweeping the market place (actually the central yard of their compound) on the evening before market day. In return they get a share in the market levy which is these days taken by the market chief and the market judges.

All Iyon consider it their responsibility to keep the roads to the market open. Iyon market is on the east bank of the Aya River. In the late rainy season the Aya becomes unfordable. The major task is to keep the bridge in repair. When the bridge collapsed during the dry season of 1949–50, all Iyon worked to rebuild it. The Uge, as their part of the venture, furnished certain rare vines used as ropes for hauling the huge planks into place; these vines grow in the forest surrounding the hilltops of Ugeland but grow nowhere in Tivland. To Iyon's disgust, MbaDuku not only made no contribution in labor, but they refused to pay toll to cross the newly completed bridge. MbaDuku's excuse was that they kept open all the paths and small bridges leading to Iyon. They also added that Iyon was involved in Tsar market just as much as they were involved in Iyon market, and since Iyon did not help with building the market place of Tsar, they did not help with building the bridge of Iyon. Traveling traders, especially Bafum and Bamenda, were asked to pay toll for using the bridge and actually did so if the water was high; it was seldom requested if the water was not high.

In the early 1950's the market *akombo* belonged to Ugbajir, the only living son of Ityodugh; he was, in 1952, a grouchy old man. Iyon elders said quite openly that they did not wish either Ugbajir or Ipevnor (who was appointed *ortaregh* of MbaKertiev by the government) to control the market. They therefore "invited" Ayatyavga to become "market chief" (*tor kasua*).

Ayatyavga is a "child of a woman" (*wankwase*) of Iyon. That is, he is related to Iyon through a female link. Therefore, Ayatyavga was a kinsman rather than a lineage brother. Selecting him as market chief was no accident; Iyon elders have told us that Ayatyavga was an efficient market chief but could never be more because of his genealogical position. His job was to keep order in the market, to settle disputes that were not important enough to be brought before the "market judges," and to make announcements. He had a tremendous voice, with which he called for quiet, made announcements for the elders, and ended with

"Go ahead with the marketing!" after which the noise would begin again.

In keeping order Ayatyavga was assisted by two husky men in their early thirties—one from MbaKunu and the other from MbaPe, the companion lineages of Iyon in MbaKertiev. They were called "those who seize people who are making trouble." They were sometimes joined by one or two junior elders of Iyon if more men were required; several young men sometimes "helped" the market chief and police by running errands for them, carrying in the produce when a market levy was in process, and chasing down defendants in market cases. Ayatyavga took his duty of keeping order especially seriously when there were dances or market levies. As the witness par excellence he had to be present at all important negotiations among the groups bound by the market *akombo*.

Whereas in Ticha market the "men looking after the market" performed the job of the Iyon "market chief," and of a court as well, Iyon had a special group of elders who were the "court" (*jir*) of their market. Iyon's market judges were any of the elders of MbaKertiev who cared to sit in the market court shelter. In fact, however, four of the most senior and influential elders of Iyon were present almost every market day and did the bulk of the work; MbaKunu and MbaPe elders attended more sporadically and took less part when they did attend. In addition, at least one Uge elder was to be found in the court shelter. He sometimes judged exclusively if a dispute came up between Uge; he took part in Tiv cases as well. He occasionally acted as interpreter, though most Iyon elders know sufficient Uge, and all Uge coming to this market have a smattering of Tiv.

Each party to a dispute paid a halfpenny court fee. The money was pocketed by the market chief who, at the end of the market, divided the total among the elders who had acted as judges that day. No women or children, unless concerned in a case, were allowed in the shelter. Cases were similar to the market cases discussed in *Justice and Judgment* (Paul Bohannan 1957:152–57), although at Iyon there was a wider range than in most. For example, spoiled food—especially meat and fish—could be confiscated and buried by the market chief and judges; although this may well happen in other markets, we witnessed it only in Iyon.

An *akombo* is kept hanging in the shelter to be used for oaths if no agreement is reached by the parties to the case. This *akombo* is not the market *akombo*. The market *akombo* is generalized, inevitable, and limited to the participating members of the market pact. Its justice is ultimate and cannot be invoked by individuals; its concerns are limited to those injunctions laid down to it at the time of its establishment. Any *akombo* used for swearing in a market court reflects the favorite fetish in the area concerned. Iyon market (frequented by Tiv, Uge, Udam, Bafum and Bamenda travelers, and by Ibo and Hausa who are settled in Udam country) used as a fetish for taking oaths a composite of one Tiv *akombo* (the *akombo iwa*) and one Uge fetish (*ikam*) which is shared by the Udam: that is, it will attack Udam as well as Uge who swear on it falsely. The *iwa* is symbolized by a crab's claw ("because if you pierce this *akombo* it hurts like a crab's claw biting at your flesh"); it can make a perjurer impotent and eventually strike him dead by a thunderbolt. The *ikam* is represented by a bundle of grass; it brings leprosy to perjurers.[1] Tiv refer to the *ikam* as "the *swem* of the Uge."

Tiv and Uge both say that no special fetish is needed for the Hausa, Ibo, Bafum, and Bamenda, as those who do not live in Udam country must cross enough Tiv, Uge, or Udam territory on leaving the market to give one fetish or the other time to work on them. As Uge and Udam especially can leave Tiv territory within a few minutes' walk, they could escape the action of the Tiv *akombo* acting alone.

When cases arise between Tiv and non-Tiv, every effort is made to ensure good interpretation and to find an elder from the foreign group to act as advisor and, especially, to concur in the verdict. In cases of theft, restitution must be made on the spot and a fine paid to the market court. If the thief is unable to pay, a call is sent out through the market giving his name and lineage and announcing that unless the fine is produced he will be sent to the Native Authority court. In every case we know, a lineage mate of the thief, or a kinsman or age-mate of the thief's father, paid the fine and gave the thief a lecture far surpassing in severity, insult, and ridicule anything said by the judges.

Besides the market court in its shelter there is another shelter for the *ortaregh*, the government official. His presence is necessary to the smooth running of Iyon market. Many arguments occur at market that are not directly concerned with the market business and hence are not directly the affairs of the market chief or the judges.

The judges of Iyon market refused to hear cases which were not in the market domain, saying, "We are just the market elders. We cannot

1. Among Tiv, leprosy is said to be caused by witchcraft and never by fetishes. Tiv say of the *ikam*, "It is an Uge fetish; it is the Uge and Udam only which it seizes. With Tiv it is different; with Uge it is different."

hear cases of the *tar*"—that is, cases of political importance, or those which the elders think should come before the Native Authority courts. Relatives of those owing bridewealth or other debts are often seized in the market place by creditors who happen to be there—who, indeed, may go there for this specific purpose. Such men are brought not before the market court but before the chief and other men of influence who gather around him.

The business taken before the government chief, however, does not include negotiations between MbaKertiev people and the Uge. These matters are handled by the elders. It is clearly a disadvantage for the chief that he is made to sit separately and await notification by the market elders of events between these groups.

In the old days strangers sought protection with men whose power and influence depended on their ability to grant safe passage; markets were one of the main institutions through which they achieved and maintained their positions. During the colonial era the situation was different. A sphere of business of the sort handled by "government courts" and meetings of the tribal council was set off by the elders as belonging to the "chief." The elders retained for themselves anything which they considered outside that sphere. Thus, the Iyon chief was cut off from the market, except insofar as he used its organization as a means of communicating between the people and the British administration. Since the *ortaregh*, a fairly young man, was not part of the organization of elders, and since he was not one of the possible heirs of the *akombo*-holders, his entry into the group was resisted.

At all times, questions of relations between Shangev Ya as a whole and the Uge were discussed through market channels. Minor Tiv-Uge disputes, such as theft, whether occurring in Tivland or in Uge country, were settled out-of-hand by the elders at market with the avowed purpose of keeping disputes out of the courts. Court action across this boundary (Uge are in the Eastern Region of Nigeria) was expensive and time-consuming. The agreement that Tiv could cut wood from Uge land was reached through the market organization. Negotiations to allow land-hungry Tiv of MbaKertiev to rent land from the Uge were initiated through market channels, but colonial officials who were approached to make it "legal" squelched the arrangement.

In time of war between two groups, a market such as Iyon can play the role of neutral embassy. During the Kunav-Udam wars in January and May of 1950, the Uge kept Shangev Ya informed of Udam decisions, policy, and negotiations with their administrative officers and among each other, with the intention that this information should be handed on to the combatant Kunav Tiv. Similarly, Tiv of Kunav gave information to Shangev elders to pass on to the Uge, eventually for the information of the Udam, whom they were fighting. Although the entire story of the market wars between Kunav and Udam is a part of the history of Tsar market (Chap. 13), the following incident provides a specific example of Iyon market used as a neutral embassy.

Just after the first Kunav-Udam war in January, 1950, the Udam placed an embargo on all trade into Tivland and asked the Uge to close their market to Tiv. The Uge elders appeared at Iyon market to discuss the affair with Tiv elders. They first explained the internal debates after which the Udam decision had been reached, detailing the names and imputed motives of those who had been for or against. Then they said that for internal political reasons they felt they had to comply with the Udam demand, at least partially. They also said they were afraid that fighting would start if Tiv came to their market, for the Udam had promised to send young men to Uge markets to chase Tiv away.

Therefore, the Uge elders suggested that the Tiv agree to stay away from Uge markets for two market days (the Uge markets are held every tenth day instead of every fifth day as in Tivland). Meanwhile, the Uge would attend Iyon market, bringing enough palm oil and camwood to offset both the non-attendance at Uge markets and the Udam embargo on direct trade with Kunav.

The Tiv elders approved, with some enthusiasm, and announced the situation to the government chief, who also approved. The market chief then called for silence in the market and announced the decision. The news was widely spread and well obeyed. Just in case someone should not have heard of it, Tiv and Uge elders sat along the main paths from Tiv to Uge market on that morning. They told us later that only one Tiv got as far as the boundary.

The organization of Iyon market is reflected in the list of people who refer to it as "my market." Ayatyavga, the market chief, says, "Kasua Iyon is my market," because he has been appointed market chief by the elders of Iyon. Ipevnor, the government chief, says, "Kasua Iyon is my market," because he has been selected as *ortaregh* of the lineage which includes Iyon and because he attends it every week and hears disputes. Ngough, the oldest of Iyon elders, says, "Kasua Iyon is my market," because he is the oldest and because he occupies the central position when market cases are heard. Ugbajir, the surviving son of Ityodugh, says, "Kasua Iyon is my market," because he is the heir of Ityodugh and as such is the only man who can claim the market *akombo*. Each group or person—market chief, government chief, elders, and *akombo*-holder—has an interest in Iyon market. Its organization is a balance of these interests.

We have yet to look into the movement of goods through Iyon market. The arrangement within the market place is a combination of arrangements by goods and by lineage location. On 7 March 1950, when we were still new enough in Iyon to be escorted formally around the market, the various groups were pointed out to us: "Here are the Uge, and there the Ibo and police from Obudu and the Hausa over at the end. Those selling tobacco are here and those with fish there in back and those with beer under the big mango tree." As usual, most of the women sat in geographical position.

The number of sellers of particular commodities appearing week after week is enlightening (see Table 18).[2] It is noteworthy, too, that there is almost never any fresh meat to be bought in Iyon market, that there is no livestock offered for sale there, and that there was no rice in evidence.

What rice was to Ticha market, guinea corn and cassava were to Iyon market, though on a smaller scale. Most of the guinea corn comes from MbaHura and the area to the northeast; most of it is bought by the for-

TABLE 18
NUMBER OF PEOPLE SELLING AT IYON MARKET PLACE

Product	7 Mar. 50	6 Apr. 50	26 Apr. 50	6 May 50	15 July 50
Cake salt	5	1	3	3	3
Bulk salt	19	**	15	25	10
Tailors with sewing machines	6	5	5	6	6
Cotton	12	15	8	8	7
Dried fish	11	15	12	19	3
Tobacco	**	**	**	12	15
Cloth	**	**	**	**	4
Raffia	**	**	**	**	34
Coconuts	**	6	8	15	6
Charcoal	**	8	**	6	**

** Not noted.

2. The most complete record we made of items for sale in Iyon market is for 11 May 1950 (published in Bohannan and Bohannan 1958:342–45).

eigners in Obudu, a little by the Uge. Cassava is grown locally and almost all of it is bought by Ibo to make *gari* flour. Iyon market is also an important center for palm products, both because Iyon itself has a few palm trees and because the Uge bring their palm produce there. The lineages north and west of Iyon are not self-sufficient in palm produce, and some of their supply is bought at Iyon.

Iyon market passes through a series of stages on each market day, just as Ticha does. The guinea corn and cassava trade starts early—about eight o'clock—and is usually finished by ten. The ordinary produce trade goes on, reaching its height about eleven o'clock or so. A little after noon, trade gives way to beer drinking; the market breaks up around four o'clock, though the last drunks may still be around at dark.

Prices at Iyon market are better documented than those at most mar-

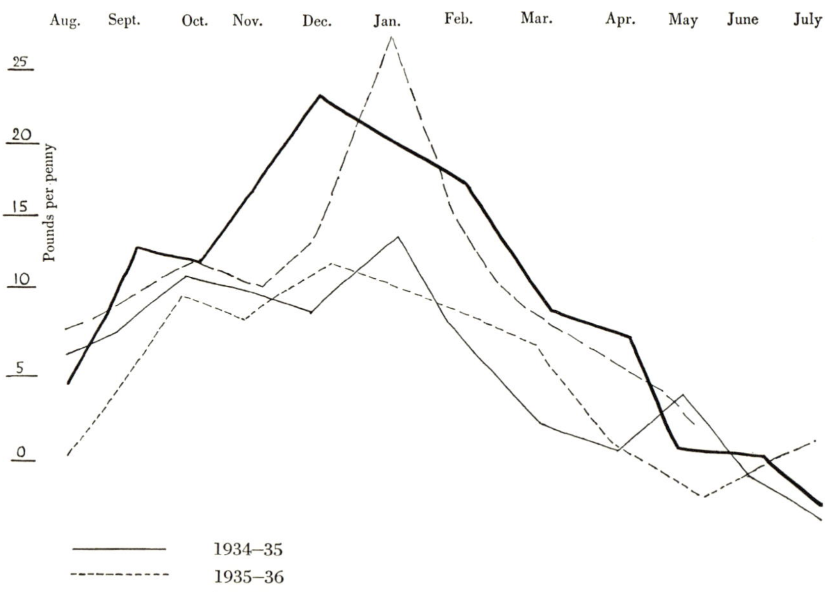

Fig. 23.—Price of yams in Iyon market, 1934–38.

kets, because records were kept for several years by assistants in the Agricultural Department; a small demonstration farm had been maintained near Iyon for some years, even though it had disappeared several years before our field work, and we could see no evidence of its having had any effect on farming practices. The fluctuation in the price of yams between 1934 and 1938 shows both a seasonal change and a change from one year to the next (Fig. 23).

These figures assume a special importance when it is remembered that yam harvest begins some time in August and is finished in December. In every year graphed, one got many more pounds of yam per penny in December and January than one did in June, July, and August, the hunger season. Moreover, in some years the price of yams is higher than it is in others; 1937–38 is remembered as a very bad year in which there were many crop failures and considerable famine.

As in Ticha market, all prices at Iyon are the result of supply and demand and the operation of a free market; no person or group can fix a price. In fact, how can a price on commodities such as yams and grain be fixed unless there are fairly precise and widely recognized methods of measurement? Tiv buy yams by the lot, and although they may buy small grains by the calabash-full there is no standardization of calabash sizes or of any other measurement.

13

The Dynamics of the Market Place

IYON AND TICHA markets both represent traditional types of market places; both were fairly stable until the sudden reopening of Tsar market, which was one of the dozen or so major markets in Tivland. This series of events (which must be recorded in some detail before a generalized description and analysis of the qualities of Tiv market places and the principles on which they operate are given in the next chapter) give us an unparalleled opportunity to analyze the dynamics of Tiv markets.

Tsar market was "killed" about 1912. When we first went to southern Tivland we heard nothing of it. However, following disputes between the MbaDuku Tiv and the Udam to the south of them—disputes that were not economic in nature—preparations were made by Tiv to reopen Tsar market rather than to continue attending Asawa market in Obudu. The dispute over Tsar market preoccupied most of MbaDuku and the leading men of many other Kunav lineages throughout the spring of 1950. Consideration of Tsar market allows the ethnographer not only to see the market system in action but also to study the growth in the market sector of the economy that took place during the colonial period.

History and Organization of Tsar Market

In November of 1949, the Hausa colony in Obudu was divided into two camps by indecision in face of a fiat from the administration of Ogoja Province to move the location of their market place. Part of the Hausa moved five miles, into southern Tiv Division, thereby removing themselves from the jurisdiction of the administration of the Eastern Region. The other part of the Hausa community, somewhat over half, stayed on in Obudu. The market picture, which had centered on "Hausawa market" (called "Asawa" by Tiv) in Obudu, was upset; war between Tiv and Udam broke out on two occasions. Asawa market, also called Obudu market, was one of the largest in the area; although it was located about two miles outside Tiv Division in Udam country, Tiv constituted its major contributors.

In early November of 1949, representatives from the Hausa colony in Obudu came to the *mbatarev* and elders of MbaDuku saying that they were unable to stay in Udam country because they were being made to move to new homes and to a new market site, both of which were unsuitable and expensive. Consequently they had decided to "bring MbaDuku's market back." The Tiv elders agreed that they would indeed welcome the return of the market, would give the Hausa land on which to live (see Chap. 6), would build compounds for them and construct stalls in a new market place to be located on the motor road which traverses MbaDuku on its way to Obudu.

Within a few days the name of Tsar market was on everyone's lips. Tiv were congratulating one another that they were getting Tsar market back. During the next few weeks, while the Hausa "town" and the market place were being built, we gathered many stories about Tsar market, its origin, and how it happened to be "in the hands of the Hausa in Udam country."

The gist of the story is that until about 1912, Tsar market in MbaDuku had been the largest market in this part of the world. However, in the early days of the administration in Obudu, administrative officers brought Hausa to Obudu to set up a market in Obudu itself, so as to provide rations for the garrison and, later, for the battery of clerks and assistants that administration requires. The head of the new market was a Hausa known to Tiv and to Udam as a great hunter. At the invitation of the colonial government he came from Takum to establish a Hausa colony and a market in Obudu.

The "market" in Obudu did not at first have a "day." Rather, it met every day. With the "opening up of the country" there was a great exploitation of formerly untapped areas. The extension of communications greatly widened the trading horizons of European companies and their African retailers. This expansion of the European-centered market undoubtedly had repercussions in the indigenous market systems; the presence of a large market in Obudu, linked directly to sources of trade goods, undoubtedly affected Tsar market adversely.

About 1912, a fight broke out in Tsar market place. We gathered several versions of this story. In one version a fight between a Tiv woman and an Udam woman over the price of pepper turned into a free-for-all. In another, a Tiv man and one of the wives of the Hausa headman disputed a marketing transaction. The reconstructed details of the fight, as they were told in 1950, were almost endless. All versions agreed that in order to settle the matter the administrative officers of Obudu sent the Hausa headman, Adamu, and one Audu (then government interpreter and in 1950 head of the Hausa community) to quell the riot and take steps toward establishing peace. As a result, Tiv were forbidden by the administration to hold Tsar market. Everyone who had previously gone to Tsar market now went to Obudu. Thus, not only did the center of trade move to the new market place, but with the closing of Tsar, Obudu market took on all the rest of the functions of Tsar: it came to have a "day" on which the market was held (though some marketing continued on other days as well), and it became the center at which Tiv and the various Udam tribes, as well as the Hausa community, settled their disputes, for Tsar had been a consecrated market. In the 1950's the *ishohough* trees of its *akombo* still stood in the former Tsar market place.

On Adamu's death, the Hausa community in Obudu had agreed on the appointment of one Abraham as his successor. Then, a few years after Audu had retired from his position as interpreter, he was selected as Abraham's successor, for Abraham had grown senile and blind. Audu had, in 1950, a handwritten letter from one of the district officers of Obudu saying that he had been elected by a majority of the Hausa householders.

All went well until some time shortly after World War II. Then, in the organizational refurbishing which Nigeria underwent with the influx of ex-soldiers as administrative officers and with the emergence of nationalism, the old Hausa settlement (*abakpa*) in Obudu was condemned, and both a new settlement and a new market place were planned. The

Hausa in Obudu had for many years paid a rental of 10*s.* a year per house plot to the Treasury of the Native Authority of Obudu District; they had, however, controlled the market place and levied tribute on sales in the market, themselves taking the lion's share of the levied goods. Now they were asked to pay the same rental for new plots, and also to rebuild their compounds (each contained several huts and some fairly substantial entrance buildings) at their own expense. They were, further, losing control of the market tribute which under the new scheme would go to the Treasury: an annual rental of 10*s.* per market stall would be charged, and no one would be allowed to do anything except petty trade without a stall.

The Hausa objected to the 10*s.* a year—Obudu administrative officers were quite right in thinking that the Hausa would rather have paid more, in the form of a few pennies each market day. The Hausa objected even more to the fact that they were, in their own eyes, losing control of the market. It was, they said, "Asawa" (Hausawa) market, and it was theirs; government officers had asked their fathers and some of them to leave their homes many years ago to establish this market. They had done so. Therefore, it was their market. The government attitude, on the other hand, was that this was Udam land, and consequently everything on it was Udam; Hausa were therefore guests and strangers.

It was against this background of misunderstanding and the overriding of their wishes that the Hausa decided to leave Obudu. Unfortunately, they could not agree on where they would go. One group of Hausa had intermarried with, and in other ways come to consider themselves close to, the Udam. These people said they wanted to move their colony either to Usan market in Uge country or to Chukwan market several miles west of Obudu. Another group of Hausa had close ties with the Tiv rather than with the Udam. They wanted to go north to Tivland.

The condition of permanent enmity with intermittent cease-fire which had existed between Tiv and Udam for generations made it impossible for the Hausa community, with their different alliances, to agree. Finally, those with Udam connections made an open break from those with Tiv connections and appointed a new leader for themselves. It was at this time that Audu, the headman, and Akwashi (son of Adamu, the original founder) made their overtures to MbaDuku.

Work was very soon under way on the new colony. It proceeded with some dissension within MbaDuku, as we have seen, and considerable agitation on the Udam side, among both the tribesmen and the administration, until late in January when an event completely outside the market issue gave a new turn to the situation.

The Market War

On 27 January 1950, a fight occurred between two Tiv women and an Udam man and woman in a stream which, along one portion of its course, marks the division between Tiv and Udam. The Tiv women were fishing and had muddied the water. The Udam man and woman objected, for they drew their drinking water from this stream. The ensuing argument led to blows, and one of the two Tiv women received three relatively serious matchet wounds on her left arm. She and her companion raised the alarm by ululation and hurried back to our compound for first aid, streaming blood and shouting that the Udam had "killed" them.

Although some 600 armed Tiv warriors were called out, and presumably Udam were also up in arms, no serious fighting occurred that day. Two nights later, however, a group of Tiv who had been on a trading expedition to Ubi market, some twenty miles or so south of the border in Boki-Udam country, were returning through Buabi-Udam area, just south of Obudu; they were set upon, beaten up, and their loads stolen. They were taken to Obudu dispensary for medical treatment. Furthermore, a Tiv woman and her adolescent son, who had been in Udam, disappeared, and, we believe, were never heard of again. Udam said that there were also atrocities committed by the Tiv.

There were two results of this affray: the affair was considered serious enough that the two administrations became involved and formalized the settlement along lines which they considered seemly; more importantly, Tsar market was revived precipitously, before the market place or even the traders were ready for it to begin.

On the first day of selling in the new and uncompleted Tsar market place, there were 18 men selling Tiv cloth, 10 selling tobacco, 5 with dried fish, and many Hausa stalls. The lorries, which were an important part of Obudu market, all stopped, either on their way to or from Obudu or both, for passengers and to buy rice and other produce. Obudu market was said to have been very much smaller than usual—indeed, since none of the Kunav Tiv were there it was certainly 2,000 or 3,000 people smaller. The fact that Tsar market and Obudu market were held on the same day thus became a *fait accompli*.

The attempt of the two administrations—Northern Region and Eastern Region—to settle the affray had a profound effect on the outcome of the market issue. It is our belief that Tiv and Udam could have settled the issues at stake in the fight, in the Buabi attack, and the other violent actions of the time. However, it is characteristic of Europeans that when they settle something they want to settle it completely and for good and all, or at least to give themselves the impression that they have done so. They therefore linked the two issues. The Tiv and Udam could not or would not reach a settlement about the market. Therefore, no settlement was reached on any issue.

For the next few months, instead of a disturbed market situation, there was an openly acknowledged rift and almost complete lack of direct communication between Tiv and Udam. Communication went either through the administration or through Iyon and Chukwan markets. The lorry drivers from trading centers such as Onitsha and Bansara soon accommodated themselves to attending two markets instead of one, and the "export trade," so to speak, suffered little. However, the waning of Obudu market had far-reaching repercussions in other markets. The Uge, east of Obudu, and the Otukwang people, with Chukwan market to the west, came to the Tiv and said that they were not involved in the dispute which centered on Obudu. Just as Uge said that they would continue to attend Iyon market to bring kolas and some palm produce, so Chukwan elders came in a group to MbaDuku, with assurances that if MbaDuku continued to bring cloth to Chukwan market and to buy palm products, everything possible would be done to insure safety and good marketing relations.

Eventually, the various tribes in Obudu District were brought together in their anti-Tiv feeling. By pressure from such groups as the Obudu District Council and, Tiv said (though we do not know with what justification), the Obudu police, Chukwan market was closed to the Tiv in all but name. It was at this time that the Uge asked Tiv not to come to Usan. They promised—and they kept their promise—to bring the equivalent amount of produce to Iyon market or to Tsar market where Tiv could get it. By February 25, Tsar market was fully established and its attendance was stable.

In an attempt to settle the dispute, the Tiv evolved a plan according to which the Udam would revive one of their old markets, called Fember, which had met on the day preceding Obudu/Tsar. If Fember market were reopened, they said, and Obudu market "killed," trade could proceed with benefit to both sides. During the same period, the Udam evolved a plan for re-establishing the *status quo ante*. They suggested that the Tiv should hold their "new" market, Tsar, on Ticha day—when they had "always had a market": Ticha market would become Tsar market, the only change being in the market place.

At the turn of February/March, the administration again tried to bring Tiv and Udam together by calling a meeting. The Tiv put forth their plan for opening the new Obudu market place, built by the "Governor," on Fember day, leaving Tsar day to the Tiv. The Udam put their counter plan: Tsar market should be held on Ticha day and Tsar/Obudu day left to the Udam. MbaDuku refused the Udam plan because, they said, the market *akombo* was tied to the day, and Tsar *akombo* would be no good for a market held any other day; without the *akombo* it would not be Tsar market but only a beer drink. The Udam proposed that the Tiv make a new *akombo* for this day. That idea was rejected as impossible: it was, they said, no longer possible to get elephants, and an elephant's leg was a necessary ingredient of the *akombo* magic. Furthermore, to set up a new *akombo* would be an act defiling the original one. The Udam, for equally cogent reasons, refused the Fember solution. Again, the meetings achieved nothing.

A subsequent meeting was held early in March on the boundary. Before this meeting MbaDuku "killed" Ticha market by declaring it closed. No one came to the market place; everyone was told to go to Iyon market on that day. "Killing" Ticha market was a bridge-burning for MbaDuku. They refused unconditionally now to have any market at all save Tsar. "It is our day," they said single-mindedly. "We gave Tsar market to the Hausa. If your brother gives you a cloth to hold and someone tries to take it from you, is that not a bad thing?"

The Udam were as adamant about the day as were the Tiv. The contest, thus, came to hinge on acquiring the trade advantage on that day. The Udam threatened to cut off all forest produce, not only via Obudu but through Chukwan and Uge markets as well. The Tiv replied that they had some of their own lineages (Turan and Ikurav Ya) within the forest zone, and that they could get forest produce from there and from as far away as the Egedde (in Idoma Division of Benue Province), should it be necessary. The Tiv added that all Ibo lorries had to cross southern Tivland to get to Obudu, and they would stop them and fill them with yams so that they need not proceed to Obudu, either to sell their goods or to buy export produce.

The Udam thereupon declared the boundary closed to Tiv. The administrative officers present informed them that they could not legally close the boundary, as Nigeria was an entirety and offered its citizens freedom of movement within its confines. The two administrations, after much discussion, told the Udam and the Tiv that there was nothing in the law to keep them from having their markets on the same day but that they were fools to do so.

During March and the first half of April tensions continued to grow. There was no weakening in the alignment of either side. With every market day, Tsar market grew and flourished; as its organization and policing were settled, it began to function smoothly. Obudu market grew smaller but its Ibo and Hausa core kept a part of its established export trade going.

Chukwan, with their market, and the Uge with their Usan market, were facing a dilemma. Each tried to solve it by improving its own position at the expense of both disputing parties. Chukwan wanted Tiv produce to sell. They wanted to become the largest Udam market. Therefore, they tried to retain their Obudu trade and at the same time to make overtures to the Tiv who would (they reasoned) continue their accustomed trade and also bring to Chukwan the increased trade which was no longer going through Obudu market.

Uge had something of the same idea, but more of a dilemma; culturally and traditionally they were closely associated with Tiv. However, they were now politically affiliated with Udam. They therefore had obligations to both sides, and from time to time they considered themselves the enemies of each. Immediately after the breakdown of negotiations at the boundary meeting, in which Uge government chiefs had sided with the Udam, the Uge elders sent assurances to Iyon market that they were not concerned in the market disputes of Obudu and Kunav.

The Uge continued to bring palm produce and kolas to Iyon market, but they did not allow Tiv to attend Usan market. Uge elders said at a meeting in Iyon market that they had been persecuted and fined by "committees" of Udam for allowing Tiv in their market.

Similar tactics were used by Obudu against Chukwan market: Mba-Yongo people (who lived near Chukwan and had matrilateral kinsmen or affines among the Otukwang) told us that the Otukwang were paying £2 to the police each market day that Tiv attended, but they considered it worth it. Hence, not only was Tsar market a threat to Obudu, but both Chukwan and Uge were trying to keep Tiv trade without alienating Obudu, so that their own markets would be increased in size and prestige.

MbaDuku and its hinterland did not, in fact, go short of palm produce during these months. They got some from the southeastern part of Tivland, but most of it came from the accustomed sources. However, instead of going through Obudu market, it now went through Iyon and Chukwan markets.

This situation continued until May 21, when the second Tiv-Udam affray broke out: a war (*tyav*) that lasted five days. On the two sides, about 40 people were killed. About 1,100 huts were burned down and crops were ruined. However, the affray settled no economic questions; indeed, aside from the destruction of property which made for a poverty-stricken season or two, it had no economic results. The economic changes had already taken place. The war was, for the Tiv, merely an attempt to maintain the new *status quo*. They considered that they had won the war since they were able to keep their market.

The reestablishment of Tsar market

The reopening of Tsar market cast a new light on the political situation within MbaDuku. The chief summed it up one day by saying that within MbaTyough, MbaNyam "ate the chiefship" and MbaIkaa "ate the market" (see genealogy in Table 1). He added that "Nyumamgbagh was different," meaning in this context that Nyumamgbagh had its own market, Tortyar market, which had become Betse market (see p. 200).

Tsar had consecrated his market with a proper ritual. A Hausa hunter, at what would seem to be about the turn of the century, had used Tsar's compound as his base of operations. He rewarded Tsar with the elephant's hind quarter which was necessary for the erection of a market fetish. The ritual of the consecration was attended by representatives of not only all the lineages within MbaDuku, but of all the other lineages of Kunav equivalent to MbaDuku. There were also representatives from Shangev Ya and the Uge. The chief claimed that even the Utange MbaShaya were present at this ritual; but when we asked the Utange, on Easter of 1950, none of them knew anything about it.[1]

With Tsar's death, some years after the "removal" of his market to Obudu, the *akombo* descended to his successor as compound head. Actually, since the market was no longer physically present, no one paid much

1. The Utange MbaShaya live in a remote hilly region of Obudu Division; they say they are not Tiv. Utange MbaShintiev live in central Tivland and consider themselves Tiv.

attention to the *akombo*. At the time the market was revived, the head of what had been Tsar's compound (though the descendants of most of its members were now living separately) was a man of about fifty named Wanshosho. Because he was Tsar's grandson, he claimed the *akombo*.

During early 1950, Wanshosho emerged from obscurity and became, at first in name only, the "market chief" of Tsar. The chief of MbaDuku had an interest not merely in the return of the market, but also in associating all MbaDuku—at least all Ikakwer—with it and hence claiming it as his market. It could be called his market in the sense that it belonged to the lineage of which he was the most important and influential man (if he, a member of MbaNyam, could keep reference to it associated with Ikakwer lineage or higher).

Since Wanshosho was inexperienced and the chief wanted to keep his hand in, Tsar market was at first run by committee. The committee consisted of Wanshosho, as *akombo*-holder, Dar (the eldest man of MbaAji), and Udu (the eldest man of MbaKov, both lineages forming MbaIkaa). To them the chief added Nor and Kuka, who were not only the oldest men of the component lineages of MbaNyam (which with MbaIkaa made Ikakwer), but who also had been overseers of Ticha market for several years and therefore had experience in market management. The rest of the committee was made up of representatives of the new Hausa community.

As in all Tiv markets, the seating arrangement followed a geographical arrangement. The Shangev people sat at the extreme east of the market place with the Uge seated amongst them; Shangev "sponsored" the Uge at Tsar market, much as the Uge "sponsored" the Tiv at Usan market, making themselves at least nominally responsible for their safety. Chukwan people sat with MbaYongo and other western Tiv lineages on the western edge of the market in a similar arrangement. The northern part of the market place was taken up by people from those lineages to the north: Ute, MbaGbera, and the various Kunav lineages "behind" them. We asked where Obudu would sit if they were to come to this market and were told that they would sit separately in the southern part of the market. As always, the seating arrangement was rationalized on the possibility of a fight in the market place.

By the beginning of May the committee had, in fact if not in name, dwindled. Wanshosho was "market chief" and Dar was the only Tiv elder who took an active part in assisting him. The two Hausa representatives, though they had no part in the market court or in the policing of the market, did allocate places and stalls and took care of the details pertaining specifically to marketing business.

Sitting entirely separate from the market "administration," the chief attended almost every market day. If the *mbatarev* attended market they sat nearby. They spent the day settling disputes, hearing news, and drinking palm wine. Any matter which was involved enough to demand a court case was postponed for a formal meeting of the Grade-D court, but a great deal of minor business was taken care of in the market.

As Tsar market grew larger, the problem of thieves became serious. Tiv gave two reasons: in a large market there are inevitably a great many strangers and strangers are difficult to control; moreover, the number of policemen was inadequate. Both of the Native Authority policemen attached to the chief spent market day working in the market place.

One policeman was also present the day before, because Tsar was becoming so large a market that trade in its market place was not limited to one day in five. Some people were beginning to camp for two or three days at a time in the market place. Every effort was made to discourage them; but the police force had its regular work to do and thus was not sufficient to stop them altogether. Several men from Ikakwer lineage were appointed to help the police and did in fact ease the situation, but without uniforms they experienced difficulty in carrying out their job. Before long, sales began to be made on the day before market. Tiv were anxious to sell their produce on that day so they would have money in their pockets on market day. Moreover, Hausa women from the new settlement (none of these was in purdah) sold food on all days. A few Tiv women also engaged in this food trade to make money.

On 13 July 1950, the day before a Tsar market meeting, there were some 30 Ibo women from Obudu buying cassava from about 100 or so Tiv women. There were two Ibo lorries which, between them, carried off about three tons of husked rice. One elder from Iyon who wanted to buy an animal trap to catch the civet cat that had been carrying off his chickens discovered to his amazement and disgust that a suitable iron spring trap, made by an Ibo blacksmith, would cost him 14*s.* on Tsar market day. He eventually got it for 8*s.* on a non-market day. Everyone agreed that prices were higher on market days than on non-market days.

The Hausa settlement had "become a town." Tsar was no longer "merely a market." In spite of the prosperity of their market and settlement, the Hausa were concerned that there were no resident Ibo. The Tiv had, when they accepted the Hausa, agreed to let Bafum settle, but they stipulated that no Ibo were to live there. The Hausa had accepted these terms. Now they remembered that no large Nigerian market could (at that time) be fully effective without an Ibo contingent: Ibo had access to certain types of trade goods—especially European cloth and hardware—which were available only at much greater expense from other peoples. Ibo were also good carpenters and blacksmiths; their tables, lock-boxes, and hoes were much cheaper for comparable quality than those of most other carpenters and blacksmiths. Finally, there was the vast kinship-like network of Ibo between the trading outposts (such as Obudu) and the large market centers (such as Onitsha). If a market is to be fully successful, the Hausa said, it must have Ibo with kin or business connections in these large trading centers. The Tiv, however, stood adamant: no Ibo.

Tiv distrust of Ibo in their markets is more than the mere "traditional enmity" of the two tribes. Hausa traders are, to Tiv, a "good thing"; Ibo traders are a "bad thing." The main difference is that Hausa export almost nothing from Tivland save coin, and they import items that for the most part can be considered luxuries. They settle for long periods of time as permanent residents of the various *abakpa* settlements in Tivland and throughout the Benue Valley. The Ibo, on the other hand, are mainly transient; they come to buy as well as to sell. Ibo have established a long-distance, "wholesale" trade, so to speak, in the very items which Tiv market. The effect of these traders is that the area of distribution of food and produce is vastly increased, and the local market becomes subject to a supply-and-demand mechanism covering a very much wider geographical area than formerly. Tiv see merely that when there are Ibo in the market the price of food rises, often beyond what Tiv can pay; also, out of "greed," as Tiv say, they sell more than they ought, creating a strain on supplies and inevitably lengthening the hunger season. Tiv market their produce; Ibo trade in foodstuffs. Tiv see the results, without understanding them fully, and say that "Ibo spoil a market." Since they cannot (or at least do not) take any sort of economic action against Ibo, they sometimes resort to political action against them.

The Tiv finally did allow Ibo to come to market. The Obudu Ibo took advantage of the situation in which they found themselves by dividing their goods and sending one of their "young brothers" to Obudu market and another to Tsar market. They were thus selling in both places. On 11 March 1950, there were 19 Ibo blacksmiths' "boys" at Tsar, selling hoes, knives, skewers, and Tiv-style spear points, as well as 26 men (mostly Ibo, though a few may have been Bafum) selling Manchester or Indian cloth, and two men selling secondhand European clothes. Most of them had come from Obudu.

The Ibo also initiated new trade. Since Obudu and Tsar markets were held on the same day, since the Tiv and Udam did not attend one another's markets, and since the types of goods were different in each, the Ibo became the middlemen who bought in one and sold in the other. Bicycles soon proved inadequate. The lorries from Abakaliki and points south were pressed into this service by their Ibo merchant "contacts" in Obudu. A lorry sometimes made two trips from one market to the other before loading for the trip south. The Ibo, like the good merchants they are, turned the Tiv-Udam dispute to their own advantage. The stripped-down chassis that formerly came to Ticha was now loaded with food and other goods from Tsar for the use of the Ibo colony in Obudu; it brought back palm products and Ibo-manufactured items.

That Tsar was becoming a major market was evident in the type of goods offered for sale. As many as 35 men sold dried fish; as many as 200 animals (mainly goats) changed hands in one day; there were as many as 18 sewing machines and 60 or 70 Tiv selling locally woven cloth.

The rice trade from Ticha was now carried on at Tsar, where Tiv buyers found direct Ibo competition. A typical Ibo buyer (and there might be as many as 12 or 15 each market day) bought three to eight bags of rice, each weighing some 150 pounds. These were loaded onto the lorries and sent to Bansara, Abakaliki, and points south.

The grain trade expanded, and so did one of the major industries of Shangev boys: weaving raffia bags. They had formerly sold bags at Asawa market; now they stepped up their production for sale at Tsar. During the season—from July until about February—there were as many as 400 bags sold every Tsar market day, each holding something over a hundred-weight of grain when full. Bundles of raffia were also abundant.

By the middle of March, 1950, attendance at Tsar market had reached some 2,800 people by our partial counts and estimates. By the end of June it was 4,000. On 5 May 1950, we counted 12 lorries loaded (indeed, overloaded) completely at Tsar (and there may well have been others). There were two horses for sale at the market, though neither was sold. The owner of the horses was the Ibo district clerk for Obudu, who sold his horses by private arrangement after he discovered that even a Tiv who wanted a horse would not buy it at market, for horses, a prestige item, traditionally do not enter the market.

Although the war in May gave Tsar market a minor setback, it had relatively little effect over a period of three or four months. However, administrative officers did not allow the situation to develop fully on its own. Toward the end of 1950 and early in 1951,[2] government officers from both Northern and Eastern provinces decided that, while they could not now ban a market the size of Tsar, they should remove the Hausa from it. They agreed that the competition and hard feelings between the two Hausa communities would keep alive the tensions between the Tiv and Udam. Therefore, the administration of Benue Province formally welcomed the Hausa to Tivland but said that they were too close to the border and must move about nine miles north to VandeIkya and build a new settlement. The news came as a great blow to them. A few returned to the Obudu group, but most went to VandeIkya and again spent the money necessary to build new compounds.

With the departure of the Hausa, the "town" in MbaDuku returned to farm land. Tsar market, however, did not suffer. The Hausa came down from VandeIkya on market days; there was almost no change in the wares available or the prices paid. With the exception of meat, comparatively little of the non-market day trade had been based on the Hausa.

Marketing—1952

When our household returned to MbaDuku in July of 1952, we took special care *not* to arrive on Tsar market day, as we were apprehensive of the size and drunkenness of the market and of the thieves who were now said to be a serious menace there. We arrived the day before market. When the truck carrying us and our camp equipment drove by Tsar market place, we thought we had mistaken the day because it looked as though a full-scale market was in progress. Oba, one of the

2. In the spring of 1951 the Udam and the Tiv held a peace ceremony on the boundary, with their administrative officers looking on.

policemen, came out of the crowd to greet and welcome us. He said that people now came regularly to start trading the day before market, that they would all stay the night, that everyone in the neighborhood had gone into what we could only call the "hotel business," and that many other people slept in the market shelters.

That afternoon, after our truck had unloaded, we returned to Tsar market place by bicycle. Tobacco and beer were being sold in large quantities; the grain and cassava trade was booming. There were no cloth merchants, Ibo stalls, or blacksmiths, no fish or cotton, and no livestock. In short, the Tiv were selling and foreigners buying. Only tobacco and beer were there for Tiv consumption. The rest was being bought by Ibo traders and Tiv middlemen.

The next day was by far the largest Tsar market we had ever seen. There had been several innovations during the preceding 21 months: there was now gunpowder to be bought in plenty; there were two "restaurants" run by Tiv, where one could go in, sit down on a bench at a rude table, and order Tiv food: yam porridge and sauces. There were two Fulani cattle slaughtered and all the meat was sold; there were four pigs and about 30 sheep and goats butchered. The Hausa who now lived at VandeIkya still considered themselves the heart of Tsar market as well as of VandeIkya market (which had been Kasua Agbo before they arrived there).

Prices had also rocketed. Tiv cloth that would have cost 3*s.* or 4*s.* in 1950 cost double that in 1952. Chickens were priced from 4*s.* to 6*s.* (in 1949–50 we had spent 2*s.* 6*d.* or 3*s.* on chickens for our table).

The amount of livestock was phenomenal; there were over forty pigs and several hundred sheep and goats sold on that day. The number of Ibo cloth-sellers had almost doubled. Oba, the policeman, said that even at night people had to be chased out of the market place or they would go right on trading. Actually, however, by sundown the only ones left were the drinkers. About one hundred Tiv spent the night in the market place because, "our loads are too heavy to carry when we have been drinking beer."

Wanshosho was now the "market chief" in fact as well as in name, and no one was "helping" him. He spent most of his time in the market place, hearing market disputes and trying to make the youngsters of his small lineage, MbaAji, sweep the market place. They seldom actually did so. The cassava rinds lay six inches deep where the Ibo women peeled the roots before taking them back to Obudu. Peanut shells lined the walks between stalls.

The chief, in summing up the changes, said that thieves were much worse and prices much higher. He blamed the Ibo for the high prices; they were, he said, "ruining our market" because they came in droves and would pay such high prices that Tiv could no longer afford to buy chickens and grain and yams in their own markets.

Iyon people said that their market had been greatly improved by the new situation. Iyon market had not lost the increase in attendance that it attracted on the demise of Ticha market.

Tiv in 1952 said that marketing was brisk and that their country was much more prosperous than it had been before the change in the market situation brought about by the events of 1949–50. It seemed to us, however, that their view was a short-range one. There was, indeed, much more market activity going on in southern Tivland after the reopening of Tsar market. There was correspondingly less immediately to the south, in Obudu. The total volume may have increased, but not as much as the growth of Tsar made Tiv think. From the standpoint of trade routes, there had been little change; the point of exchange had merely moved a few miles north. That was all. The same goods flowed in the same directions. Tiv probably did increase their production of cassava, and perhaps their production of yams. Certainly the production of rice had increased a great deal in these two years; it was no longer primarily a cash crop for women, but had been taken over by the men. By and large, however, only the purely local picture had changed. So far as the Ibo lorry owners were concerned, the flow of goods had not been interrupted, and their trade routes were the same as always. However, more lorries came in 1952 than in 1950.

By 1952, Ticha market was all but forgotten. The fact that we remembered it and asked about it made us old-timers. Tsar market and its re-establishment can be seen as a miniature of Tiv economy, including both production and allocation, and of the impact the increasing marketization of the whole of Nigerian economy had on the subsistence economy of the Tiv. We shall examine these points in the next chapters.

14

Structure and Characteristics of Tiv Market Places

SEVERAL OF THE CHARACTERISTICS noted in the histories of the three markets given in the preceding two chapters deserve further analysis. They are of two sorts: the economic and the non-economic aspects of the market places. On the one hand, Tiv market places are institutions within which goods and money are exchanged on a free market. There is absolutely no institutionalized price fixing; the price of subsistence goods fluctuates with the season, the quality of the crop, and the geographical area from which demand is drawn. On the other hand, there are political, religious, and recreational aspects of the market places. Although the market place is justified in terms of allocation of goods, it is nevertheless widely used for non-economic purposes.

This chapter presents a summary of the non-economic functions of Tiv market places; the next chapter will examine how market places fit together into rings and "solar systems," after which we will be in a position to summarize the whole of Tiv economy.

In the two preceding chapters facts have been marshaled under four main topics: social structure; the goods that pass through the market; the set of repetitive events that characterize each market; and, finally, the values concerning what a market should be and how people should behave in it.

Every Tiv market shows a formal organization, although it differs from one market to the next. In some a committee of elders operates the market and settles disputes that arise within it. In others there is a "market chief." Whole lineage communities vary in this same way: some have a committee of responsible civic leaders, while others follow a man with charismatic power. It seems likely that both communities and market places swing from one of these forms to another.

A market official may be an elder in the lineage with which the market is associated; he may be specially appointed by the elders of a lineage; he may be the founder of a market; he may have inherited the *akombo* associated with it; or he may "receive" or "steal" his market if it has run some other market out of business.

In return for the fulfillment of their duties, the market administrators have a right to a share of market tribute. The position affords a certain amount of influence. There is, in fact, a traditional association between markets and men of influence. In the pre-European days their primary tasks were assuring the physical safety of those who came to trade and settling disputes among traders.

Political Aspects of the Market

THE MARKET ORGANIZATION performs many tasks besides running the market. It fulfills political functions, most importantly in the peace pacts that center about the market *akombo*. Since such an *akombo* is equally binding on all of the several participating groups or lineages, the market place is politically second in importance only to the lineage moot. Indeed, market places provide a political avenue in situations in which the lineage principle presents an impasse. The lineages or tribes who are parties to a market *akombo* need not be equivalent segments. In the case of Iyon *akombo*, the contracting parties were Iyon, its two equivalent segments in MbaKertiev, the remaining segments of Shangev on the level of Mba-Kertiev, MbaDuku, which is a lineage of Kunav distantly related to Shangev, and the Uge, who were a separate tribe. These groups of neighboring lineages and tribes can meet and settle their problems in Iyon market place; otherwise it would be necessary for a dispute between Iyon people and MbaDuku people to go up to the lineage level of Shangev and Kunav or even to Jechira and Kwande "septs" (to use the administrative term). Theoretically, at least, in a dispute between an Uge and a man of Iyon, the entire tribe of both the Tiv and the Uge would be concerned, were it not for Iyon market or something like it. Obviously such a top-heavy arrangement would be quite impractical in settlement of minor disputes. The market organization provides a channel by means of which non-equivalent political groups can interact more or less on the basis of equality.

Just as a moot or *jir* is an institution for controlling activities within a lineage (Paul Bohannan 1957), a market is the most important and widespread of several indigenous institutions for controlling activities between lineages. The moot can function only as the moot of the lineage which includes both parties. In the market place, on the other hand, each lineage retains its separateness and each participant group meets as itself, not as a segment of a single lineage which also includes all the others.

A market place may become a "neutral embassy" between warring groups, even if they are not subject to the market *akombo.* We saw that Iyon market acted in this capacity in the dispute between Kunav, of which one segment is a member of the market pact, and the Udam, who are not members at all.

The market, then, is one of the institutions which, in the days before the *Pax et Lex Britannica,* permitted Tiv to circulate; it was an institution with a meeting place and personnel capable of arbitrating disputes, and a mechanism for handling relations between groups, whether they were Tiv lineages or foreign groups.

Because consecrated markets have this political aspect and are at the same time the hub of communications, their control is the object of much political intrigue. Tiv say that no wars were ever fought over unconsecrated markets. We interpret this to mean that markets important enough to be fought over were sooner or later consecrated either in fact or in accepted myth. Every history of a consecrated market is rife with fights and intrigues and with stories of complications surrounding the inheritance of the market *akombo.* If the heir is not a strong character or a man of influence, then an important elder, or, in the colonial period, a government chief, assumes control of the market. This is not inheritance, but rather the attraction evident between a man of influence and a market. If the people approve, they say that the man of influence was given the market. If they do not, they say that he stole it.

Although markets are fought over, the fights concern the pivotal points on the routes over which goods flow, not the goods themselves. These fights and disputes are usually set forth in a religious idiom of *akombo*, rather than in an economic idiom of material goods. In the Tsar dispute capture of the trade was important, but the dispute was overtly about correct settlement of the fetish question. Nobody wanted to change the over-all flow of goods, but each sought by religious means the advantage of having his market place at the hub. Everyone used religious pressure, as it were, to underwrite his position at the center. As we have seen, the flow of goods continued throughout the Tsar dispute, even though the channels were changed. An earlier settlement, economically not unlike that finally made, could have been arrived at had the Tiv agreed to hold their market the day after Obudu, or the Udam the day before.

Social Values of the Market

THE VALUE SYSTEM associated with Tiv markets is set forth in terms of ethical precepts. There are right and wrong reasons for going to market, right ways to run markets and to behave in markets. There are also right ways to carry out one's marketing exchange, and there are ideas about the marketer's luck and how marketing should be fitted into other aspects of life.

The characteristic of "neutrality" is behind some of the most important values people hold about markets. For women the neutrality of the market place is most important in the sphere of kinship. Tiv households are virilocal. When women go to visit kinsmen, most husbands are uncomfortably aware of how tenuous their marriages are. At market, however, a woman may meet her kinsmen, visit with them, and get all the news without creating role conflict in herself or arousing her husband's apprehension.

Men are drawn to markets on the basis of quite different relationships, largely non-kinship. Markets are excellent places to confront debtors, to arrange witnesses, and to "settle out of court." The jurisdictions of market masters always include settlement of disputes which have arisen in the course of the market and may be limited to that. However, they may be broad enough to include any small matter which both principals want judged. If the market masters themselves do not settle these minor non-market disputes, there is usually a group or a person present who will:

the elders at Iyon market, the chief of MbaDuku himself (occasionally at Ticha, and more or less regularly at Tsar market after its revival), and the "market chief" of Tsar after he had consolidated his position.

Elders are attracted to market because it provides an opportunity for them to meet as a body without a specific agenda. Meetings of elders of the lineage as a moot or *jir* are never called without a specific cause. These same men may meet informally as beer-drinking groups in the markets. Thus, the market provides the guiding powers of the community an opportunity to meet informally and regularly.

The market is an ideal place for people to hold celebrations or to show off musical and dancing accomplishments. The celebrations and dances are, in their turn, irresistible to young people and attract still greater crowds to the market place. Markets are marked, more often than not, by celebrations of one sort or another. When a wedding occurs, particularly a man's first wedding, dances must be held in the compound of the groom. These dances continue for several days; routines are worked out, and dance leaders appointed. Such a group of dancers often proceeds to a market place to display its accomplishments. Most market masters will not allow dancing to begin until the peak of the trading has passed. Then, at his signal, the elaborately dressed dance group forms on the edge of the market. With drums, gongs, and perhaps some wind instruments, they dance into and around the market place. A crowd follows them, and admirers run in and out among the dancers, pressing pennies onto the sweating foreheads of those they most admire. After the exhibition has continued for a while, the dancers are joined by others. Dancing and drinking may continue for several hours.

Work parties may celebrate the completion of their task by publicly proclaiming it in dance at the market (see Chap. 5). Sometimes groups of dancers practice together for weeks for no purpose save touring markets to perform. These people do not consider themselves professionals. The few pennies they receive as gifts for their dancing are not considered as fees and are certainly not sufficient to keep them in food during the tour. It is fun to dance, and it is fun to go to markets.

Obviously the Tiv *kasua* is an institution similar to the weekly fairs that marked Europe from the Middle Ages. The fair does, indeed, contain a market. But it contains much else of social, political, and religious significance.

The morality of the *kasua* is a simple one. These markets are in themselves valued institutions. Tiv have a penchant for classifying everything into good things and bad things, and markets are good things. However, going to market without the good reasons of trading, seeing kinsmen, settling cases, or dancing is frowned upon. In fact, some men refused to give us information on markets, saying they did not know because (it eventually emerged) they did not want us to think that they "ran around to markets" (*dzende kasua*) when they should have been at home doing their farm work and attending to other important business. If a man goes to market every day without good reason of the sort mentioned, he thereby neglects his other duties, particularly his farming. This sort of running about to markets is said by all—even those who do it—to be a bad thing. Elders say that youngsters have been spending too much time at markets since the British pacification of the countryside has increased the number of market places.

Two points about the event sequence of markets should be noted. First, Tiv markets are held every fifth day or, in a few cases, every seventh day; they therefore provide a rhythm for daily life, and provide one of the pivotal points in a calendar of Tiv activities. This recurrence, and some activities associated with it, will be discussed in a later chapter. Second, there is at each meeting of the market a typical, repeating sequence of events. It usually begins with the sale of cash goods, grain, or some other staple. This is followed by the more general trading of sauces and subsistence, which is followed in turn by the buying of imported items, which is followed by beer drinking. Finishing the beer usually closes the market. The event sequence of any specific market may be different from that of some other, but all of them follow a more or less general pattern.

Obviously, Tiv markets are highly institutionalized. The next question becomes: Is there an institutionalized *system* of markets and market places?

15

The Network of Market Places

TIV MARKETS, as we have seen in the foregoing chapters, have qualities associated with fairs, cosmographical aspects through the association with days, religious aspects through the presence of fetishes and rites of consecration, jural aspects through the market court, and political aspects because markets are often the locus of political activity.

The next step is to determine the extent and type of connection among the individual market places. We shall examine the same aspects of integration and interconnection of markets that we examined for each individual market place: first the event sequence, then the idea system, and finally the system of material culture and goods, including trade routes and the routes which goods take in traversing the countryside.

Market Neighborhoods and Market Cycles

TIV, IN EXPLAINING MARKETS to a foreigner, tell him that "The market is the day." If a stranger asks, "What are the markets hereabouts?"

the answer is almost always, "Today is X," giving the name of a market. The informant will usually go through a list of four additional markets and end by repeating the name of the first: "Today is X." Alternative market names are seldom volunteered in this context, and their mention by a woman or child may be reproved by an elder: "They are asking about markets, not about beer drinks"; or, "That is not our market."

"Our market," in such a context, refers to the fact that each neighborhood—indeed, each compound—sees itself as the center of five markets, each associated with a day. These markets are the ones most frequently attended by the people of that area, for Tiv prefer to attend the market that is socially and spatially the closest. If the two conflict, social distance usually outweighs spatial distance. Thus, when Ticha market and Iyon market occurred on the same day, few MbaDuku people went to Iyon and few Shangev people went to Ticha. Each referred to the other's market as "far" (*cha*). We have, in fact, heard MbaDuku wives who lived spatially a good bit farther from Ticha market than from Iyon insist that Ticha was closer. It *was* closer in social distance. The exceptions were Shangev women married into MbaDuku, who found it convenient to claim that Iyon market was close; close kinsmen in the neighborhood of a far market may make it socially closer or just as close as one spatially nearer.

Market names provide the only names for days among the Tiv. Udam have day names forming a five-day week, for which they also have a word, and can say something like "X market is held on Y day." In Tiv such a statement is meaningless: Iyon market is both a place and a date.

It is possible to single out market neighborhoods by denoting the area covered by a certain repeating cycle of market-day names. For the people who live close to Ticha this cycle was: Ticha day, Aichwa day, Chukwan day, Angur day, Tsar (Obudu) day. For people who live close to Iyon, the same days were called: Iyon day, MbaHura day, Ajio day (or Usan day—Usan is held every tenth day; hence, every second market day was Ajio; every first and third were Usan), Adikpo day, Tsar/Asawa day (Obudu day).

There is no Tiv word for these market cycles and none for the neighborhoods to which they correspond. Tiv count the cycles by counting markets, just as they count seven-day weeks by counting the Sundays. The two cycles given above contain only one day name in common, Tsar day, though Iyon and Ticha are only six or seven miles apart.

When one moves away from either market in any direction the cycles change and the neighborhoods change as well. If, for example, one moves into southern Ute, six or seven miles to the north of either Ticha or Iyon, the same days are listed as: Pev day, Aichwa day, Ajio day, Ukoo day, Tsar day. Ute people share one day (Ajio) with Iyon people, one day with MbaDuku people (Aichwa), and one day (Tsar) with both.

When we plot the locations of these markets on a map, we can see

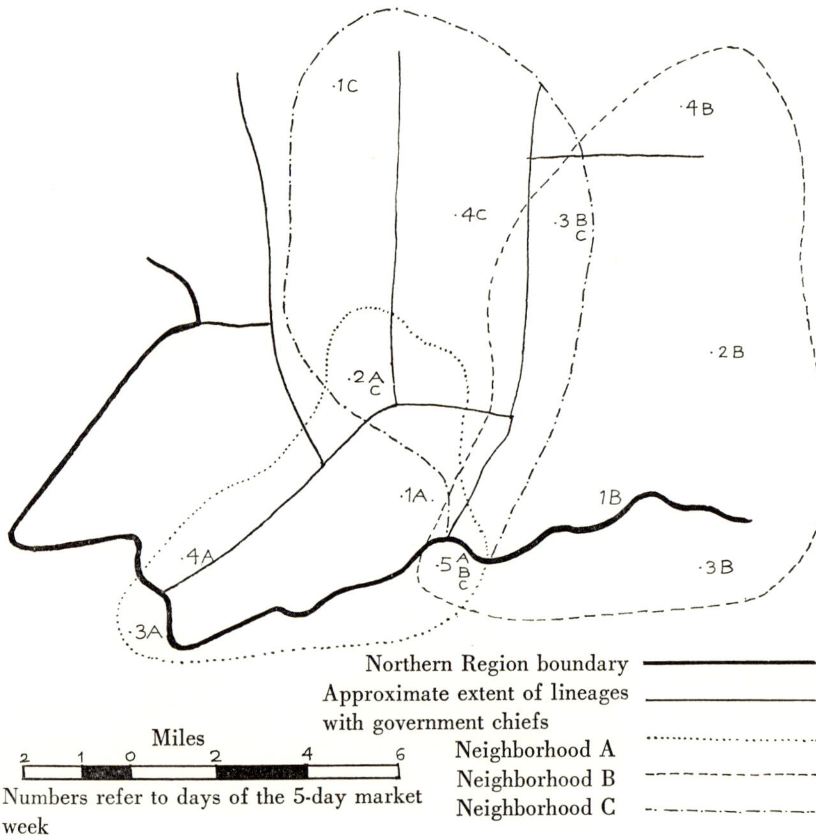

Fig. 24.—Overlapping market neighborhoods in southern Tivland.

that the neighborhoods demarcated by the three cycles overlap (Fig. 24). Each neighborhood (i.e., people who share a complete cycle) is marked by a greater degree of social interaction than is found between people of any two neighborhoods who share only one or two days. It is within the market-cycle neighborhood that one is most likely to go to settle one's disputes directly. Beyond it, there is a tendency to use intermediaries or the courts. It is within the neighborhood that one sees the same people most often. And it is within the neighborhood that most petty trading is done; it is even possible to do some professional trading within it. An Iyon man can make a profit buying guinea corn at MbaHura market and selling it at Tsar. Many MbaDuku women buy yams at Angur and resell them at a higher price at Tsar.

The event system created by the five-day market cycles thus leads to informal neighborhoods which overlap one another and eventually stretch across central and southern Tivland in chain-mail fashion. The chain-mail structure of market neighborhoods is not to be confused with the market network seen in terms of goods. The latter will be explained later. Here we shall first examine the market cycles of Ticha and Iyon, and then we shall describe the northern and eastern portions of Tivland where markets are less highly developed and do not form cycles or neighborhoods in this way.

The Ticha Cycle

Aichwa day. In 1949, the market held the day after Ticha was Aichwa, which is located in MbaGbera lineage, north of MbaDuku. Aichwa market place is about four miles from Ticha. A good many Tiv from eastern MbaDuku go to Aichwa because it does not really "begin to scream" until about noon. They can do a full morning's work on the farm, go to market, and still be back in time for evening chores. Animals butchered for meat were killed about 11:00 A.M. in Aichwa—an hour or so before the market got under way. MbaDuku people associated this late hour with the fact that Aichwa is a small market. They said categorically that small markets do not begin early. One MbaDuku woman told us that Aichwa was not small enough to be a woman's market, but on the other hand it did not begin early enough to be a man's market.

On 8 December 1949, we recorded the following at Aichwa: 3 people selling dried fish, 11 selling cotton, 8 selling cake salt, and 4 people using sewing machines. There was only one person (a Tiv woman) selling bulk salt. In all big markets of southern Tivland almost all bulk salt was sold by Ibo women. Ibo seldom venture very far into the "bush" in Tivland. Aichwa was considered to be "in the bush." Very little bulk buying occurs at Aichwa. A few Hausa and Udam did come to Aichwa to buy yams, but again the absence of Ibo (the chief exporters of yams from Ticha and Iyon) was apparent. Yams were the last items sold at Aichwa market (about 2:30 P.M.) before the place was completely left to the beer drinkers.

Aichwa market has a market *akombo* but is seldom used as a political meeting place. The Native Authority organization which governs MbaGbera and its neighbors supplies the lines of communication. Furthermore, the descendants of the man who established the market *akombo* are not interested in the market, and the market chief is a "stranger." Occasionally, however, the fact that Aichwa has an *akombo* is important; it is sometimes used as a meeting place for government chiefs from the surrounding area.

Chukwan day. Chukwan is the Tiv name for the Otukwang tribe. As was noted above in the description of Tsar market, MbaDuku have an old agreement with the Otukwang which they compare with an *ikul* treaty. Chukwan market place is about twelve miles from Ticha, although it is only four or five miles from the western end of MbaDuku. In 1949–50, western MbaDuku and MbaYongo women seldom attended; men did so only to sell Tiv cloth or to buy palm oil. Otukwang, like the rest of the Udam tribes, do not weave; today they wear either imported cotton prints or Tiv cloth. Chukwan market is a center where Tiv cloth is bought in large lots by traders for resale to the Ibo of Abakaliki and to Udam tribes such as the Boki. Tiv know almost nothing about the organization of Chukwan market save that the market master and the government chief are always different persons.

Angur day. Angur market place is in MbaYongo lineage area about four miles from Ticha. It is attended by many MbaDuku people. A smaller market than Ticha, it was, at the time of our study, a favorite at which women bought food for resale at Ticha or Tsar. Angur is held in the central cleared space of a homestead. The chief of MbaYongo, not the compound head, levied market tribute every three or four months. He sat at the entrance of the market dressed in his best, with drummers and singers surrounding his chair, while his minions took the tribute.

On 5 December 1949, we counted: 7 sellers of dried fish, 5 sellers of cotton, 2 tailors with sewing machines, 4 Hausa stalls, and 8 sellers of cake salt. There was no bulk salt to be bought: Obudu Ibo seldom "came in" this far, but Obudu Hausa came in small numbers. Angur was so "far in" that one often could not change a shilling for a full complement of pennies. The rate varied from seven to eleven, apparently depending on the affluence ascribed to the holder of the shilling.

Angur is a consecrated market which figures prominently in the history of wars and negotiations between MbaYongo and MbaDuku. During our stay it had little political importance.

Tsar day. The last day of this cycle was Tsar day. Before the revival of Tsar day people had spoken of Asawa day, in reference to the market at Obudu.

One more point must be made about the market neighborhood centering on Ticha market: there were several tiny markets within it that did not lend their names to days. One of them was even consecrated. They were Kasua Tsuwa, Kasua Shiliki in central MbaYongo, and Kasua Betse in southwestern MbaDuku.

Kasua Tsuwa was one of those markets of which Tiv say, "It is only a beer drink." It was begun by a man named Tsuwa, just outside his compound, on the main path. He cleared an area under a big locust bean tree and built five small thatched shelters on it. He announced that there would be trading on Chukwan day when few people in the immediate area went to any market. Tsuwa market gathered about noon. On 4 December 1949, at 11:50 A.M., there were eight people there. After about an hour and a quarter there were forty people, but they soon began to drift away because the women wanted to sell their beer rather than give it away.

Tsuwa's market did not catch on. Only his neighbors came. Women did not bring yams or cassava to sell, but perhaps a handful of greens or a bit of pepper as an excuse for coming at all. Since it was a neighborhood affair, they sat and talked, swapped sauce greens, drank whatever beer could be had without much payment, and then went home.

Kasua Shiliki, in MbaYongo, met on the same day as Aichwa. It was larger and more important in passing food along the goods routes than was Tsuwa's market, although it did not lose the quality of a neighborhood gathering for recreation. Some MbaDuku women, particularly those who had kinsmen in MbaYongo, went to it instead of to Aichwa. It was about four miles from Ticha, and a woman could make as much as 50 per cent on an outlay, we computed (Tiv did not), by carrying yams home to sell a few days later at another market. Ibo and Udam did not attend. However, on 18 November 1949, two Hausa had brought small loads of their usual trinkets and medicines. On that day there were 3 men selling cotton, 1 selling dried fish, 2 selling tobacco, and 7 selling cake salt. The exchange rate on the shilling that day was six pennies. It is interesting that it does not seem to occur to MbaDuku people to buy shillings at Shiliki or some other bush market with pennies. So far as we know, the discrepancy, though well known, has never led to any form of arbitrage.

The major activity at Shiliki market was women's selling and buying of sauces, greens, a spool or two of cotton yarn, yams, and small amounts of sweet potatoes, cassava, and millet. No rice was sold at Shiliki. An occasional levy was made on Shiliki market by the chief of MbaYongo or his assistants. At such a levy in December, 1949, no one questioned the chief's right, but each tried to avoid paying anything himself.[1]

Kasua Betse, in southwestern MbaDuku, was a small market that met late in the afternoon of the day of Asawa market. Since it was some seven miles from Obudu, one could go to Asawa market in the morning, buy produce, and return with it to Betse for resale in the later afternoon. Women usually returned with palm oil; men with meat, kolas, and palm wine. Betse was not large: at most, three or four hundred people attended, usually fewer. It served almost entirely as a retail center and local social event.

Betse market was consecrated. The men who claimed its *akombo* said that theirs was the oldest market *akombo* in MbaDuku. Betse had been founded by a man named Tortyar, and was a thriving trading point between Udam and Tiv before the arrival of the Europeans. Then, in a "big fight" between Nyumamgbagh and MbaItyo, the two major segments of MbaDuku, MbaItyo refused to come to Nyumamgbagh's market, established one of their own on the same day, and consecrated it. By the time the Europeans arrived, Tsar market, in MbaItyo, had eclipsed Tortyar's market in importance. Nyumamgbagh people said that Tsar himself, and the other men of MbaItyo, "stole" their market.

1. Coming to market with one of us was an excuse for not paying levy; it was often attempted but seldom successful. We have even heard women claim that they should not pay levy because they were pregnant.

The Iyon Cycle

THE CYCLE of markets which give their names to the days in the area of Iyon were very much like those found near Ticha, except that there were two major markets in the cycle—Obudu (later, Tsar) and Adikpo. Interest of Iyon people centered on their own market, not on the cycle, and the cycle seemed of less immediate importance than in MbaDuku.

MbaHura day. MbaHura market is about six miles northwest of Iyon. On 2 April 1950, there were 6 men selling dried fish, 2 men selling salt cakes, 4 selling cotton, 4 selling coconuts, 2 tailors using treadle sewing machines, 2 Hausa, and no Ibo. The main trade was in palm oil, which came from Tiv areas to the southeast. Iyon people considered MbaHura important for trade in guinea corn and baskets.

The market at MbaHura has an *akombo* which was erected long ago; it was at one time under the direction of a son of Saleor, renowned in Shangev folk history. However, after Saleor's death, well before the arrival of the Europeans, the market had shrunk in size and importance.

Usan day. Usan, the largest Uge market, was held every ten days. We visited this market once (it is on the motor road which runs east out of Obudu) but were unable in this brief time, without an adequate interpreter, to discover much about its organization.

Tiv went to Usan market early in the morning, got their marketing done, and returned early. Usan was said by local administrative officers to be a "very drunk market." This judgment was reiterated by Tiv, who usually did not choose to drink with Uge, whom they considered unpredictable when in their cups; the drunkenness of this market and its location in enemy territory were the main Tiv reasons for leaving it early in the day. Tiv women attended Usan market only in groups of three or more. Tiv bought palm products and kolas at Usan, either for use or for resale at Iyon market or even MbaHura market. They generally sold only cloth at Usan. Uge do not weave; they wear Tiv cloth almost exclusively. Usan market is one of two main points (the other is Kasimbila) where the "bush tribes" (*atoatiev*) to the southeast of the Uge get their Tiv cloth.

Business between Tiv and Uge was transacted at Iyon market, not at Usan. No Tiv acted as adviser or judge in Usan market. Except for a few who had kinsmen among Uge, Tiv went there specifically to trade and knew little about the market organization.

Ajio day. On alternate five-day cycles Usan day was called Ajio, for it had been associated with the market of the late government chief of Shangev by that name. Iyon people seldom attended Ajio market except in connection with cases in the Native Court. It was about nine miles from Iyon market place; goods went from Ajio north toward Adikpo, not south toward Obudu or Tsar. Therefore, Ajio was not really a "good" market for anything so far as Iyon people were concerned.

Adikpo day. Adikpo, in Nanev, was the seat of a very old market. It had resident Hausa (a sure sign of an important market), and it was important enough that a government road several miles long was specially built to allow lorries to reach it. Iyon people generally sold rather than bought at Adikpo; kolas and palm oil, sometimes camwood, were the main items they offered for sale. Prices on these items were very much higher at Adikpo than at any of the southern markets; prices on Tiv produce were also higher. Adikpo market was consecrated, but it was quite outside our ambient, and we know nothing of its history or of its *akombo.*

Asawa/Tsar day. Asawa and Tsar markets have already been described. For Iyon people both were primarily places to sell livestock, to buy imported goods from Ibo and Hausa, and to resell the grain which had been bought in the northeast, at MbaHura market and even further afield.

There are, of course, small markets in Shangev, as there are in MbaDuku, which are little more than social events and "beer drinks." When we first asked about markets in the Iyon cycle we were told by a child that MbaJu market existed and that it met on Adikpo day. The child was reprimanded for bringing up "matters of the roadside; we are talking about old established markets."

Seven-day Cycles

SINCE THE ARRIVAL of the European missionaries, a new system of seven-day markets has appeared. The Dutch Reformed Church missionaries, who opened their own markets, always held them on Saturdays. In eastern Tivland, where the missionary influence was felt earliest, there are today some markets held on other days of the week as well. Missionary markets at Mkar, Zaki Biam, and Tor Donga are held on Saturday. Unne's market in Tongov is held on Monday. On Tuesday is

Kado market near Wukari. We were assured that there were small beer-drink markets held on the other days of the week, but we could not get names for them and found no one who claimed to have attended them. We doubt that they exist; Tiv think of markets in cycles, associated with neighborhoods. If they find some days of the cycle unfilled, the usual explanation is that they are so far away they cannot form neighborhoods and therefore are unknown. The existence of a cycle is not doubted; the informant merely considers himself ignorant of its details.

Eastern Tiv, in Ukum, used the method of day names devised by the Dutch Reformed Church missionaries: Monday is "the first day of work" and so on through Friday. Saturday may be either "the sixth day of work" or *Saadi.* Sunday was the Hausa word, *Laadi* in Tiv pronunciation. Needless to say, none of the seven-day markets was consecrated, at least not in the Tiv manner.

Except that they are held every seven days and are usually overseen by the missionaries or by officials appointed by the mission, these markets differ little from ordinary Tiv markets; they do not differ at all in the goods offered, nor do the Tiv have any different attitudes or values in mind when they attend them. Missionaries, like most Europeans living in this part of Nigeria, find it convenient to have a market nearby. If there is not one already in existence, they start one—their motives in so doing are to have at hand a source of produce for themselves and their servants and followers.

Marketless Areas of Tivland

AMONG IHAREV PEOPLE in northwestern Tivland, markets occupy a much less important place in everyday life. They are not consecrated and play only a minor social role. Iharev compounds are in fact self-sufficient in most items: salt, palm oil, and camwood are the only important exceptions. These were obtained—when they were obtained—from house-to-house peddlers. Thus, markets never became a vital institution of allocation. Even today, all the necessary functions fulfilled in other areas by markets are fulfilled by some different institution in this one. Even if markets are begun as local social gatherings, they seldom catch on as anything more than that.

Since the railway has crossed the eastern portion of Iharev, trade has increased. Today, if one studies trade in this area, one must note that markets exist but that they are not the main institutions of trade.

The railway introduced several small groups of Ibo and other foreigners into Tiv country. When we studied the area, most of the station masters and service personnel in the way stations along the railway were Ibo. Many of the train employees on regular runs were linked as trading partners with people who lived along the railway. Our camps in Ityoshin/Iharev were too far from the railway for us to observe these trade routes minutely. Fifteen miles from the railway, however, their effect could still be felt. If a man needed, say, 3*s*., he took a chicken and walked toward the railway with it, offering it to anyone he might encounter until he found a purchaser. The new owner carried the chicken closer to the railway, where it would bring a higher price. In 1951–52, a 3*s*. hen sold for 5*s*. or 6*s*. along the railway, a full-grown cock for as much as 8*s*. These fowls, and other food as well, went into the trains bound for Ibo country.

Itinerant traders were more active in northwestern Tivland than in other parts. Most of them were Kparev youths who went south to Udam markets where camwood was cheap, and then carried it north. Once, while in Makurdi taking a rest from the Ityoshin Tiv, we met two MbaDuku friends on the Benue bridge. They had bought camwood at an Udam market 20 miles south of MbaDuku and headloaded it some 140 miles to Udi, a railroad stop north of Makurdi. They said that they could make a great deal of money this way, and that they could have made even more had they taken the camwood around to the various compounds "in the bush" instead of hurrying home to their agricultural work.

Iharev people told us that they could "learn all the things of Kparev sitting at home," because eventually all Kparev would come to sell them camwood. Iharev people bought with money; on only a few occasions we heard of Kparev traders getting chickens or goats as exchange. The traders, when they got money, tried to buy livestock to sell in central Tivland or to the railway. Although livestock was plentiful in the area, it could seldom be bought for money. We found that we often could not buy food for money in this part of Tivland; we imported salt for our cook to trade for eggs, and tobacco to trade for vegetables.

Ityoshin is not, however, completely without markets. A five-day market called Kpaka is held at Moi Igbo, on the railway south of Makurdi. Although our compounds were half a day's walk from Moi

Igbo, at least some people from our area were to be found there at every meeting. They sold chickens, cotton, and such foods as corn, pepper, millet, and guinea corn. They bought cloth, if anything. Usually they went to Makurdi, about twenty-five miles by bush path, to sell their beniseed and to buy cloth. Kpaka, a "big market" in Ityoshin, was about the size of Angur market in MbaYongo: that is, what the southern Tiv would call a very small market.

Kpaka market was not consecrated. It became active about two o'clock in the afternoon and was usually over by five. The people in our area seldom attended except at full moon, because they did not get home until several hours after dark. The Native Authority Grade-D Court sometimes met in the market place on market day.

Kpaka market was also used as a primary supply point by groups of Ibo housewives on an outing from Makurdi. There were two "suburban" trains a day. One left Makurdi in the morning, going south; the other arrived at Moi Igbo, returning to Makurdi a little after 5:30 P.M. It was a Tiv contention that these housewives, and a few traders who did the same, raised prices at Kpaka market. "Ibo always spoil a market," they said, "because they do not buy only for themselves, as Hausa do, but export food to their brothers at home."

A market similar to Kpaka is Jande, at Agana station, between Makurdi and Moi Igbo. Like Kpaka, Jande market is held in the afternoon. It is different from most Tiv markets in that trading cannot be started until the signal is given by the market chief, a Tiv.

Attendance at markets other than Kpaka was used by people in this area primarily as an excuse for visiting their kinsmen. The women from Asanyi's compound in Raav/Sengev, for example, often went to Agir market in MbaKpa/MbaKu in the north, because some of them came from there and because a daughter of Asanyi was married into that lineage. Nobody else from the area, however, attended this market.

If one tried, in Iharev, to collect a cycle of five-day names associated with markets, the people knew what one was talking about and generally tried to oblige, although without success. Some might remember a market that used to be held on such a day and add, "It was a very big market, but it died." Days in Ityoshin are indicated not by market names (except the day of Kpaka) but by the names of days in the seven-day brewing process. Sunday was the day of drinking.

Ityoshin women within a radius of seven or eight miles of Makurdi do all their marketing there. They take a load of wood to sell for money.

They say that it weighs very little considering its high price and that it can always be sold in a place where Hausa women live. People of Ningev lineage of MbaKpa, just west of Makurdi, say that there was formerly a system of five-day markets in that area, but that Makurdi market is bigger and better and runs every day, so their old markets have "died."

Away from Makurdi the main reason given for the demise of the markets is "hunger." It is the same reason given for the foundation of most of them. In Ukusu, where we had a base camp for six months in 1951, a market called WanTsum seems to have existed some forty years earlier. It was established during a famine "so that people could come with food for Ukusu to eat." It ran for several years, then "it died because of famine and there was no food." It was never resurrected. When we asked why, we were told that WanTsum market had a fetish containing *swem*. We believe that this statement is not to be taken literally; rather, they put their statements into the language and values of Kparev lineages which were familiar to us and of some prestige value to them. There was no consensus about which day of the cycle, taking Kpaka as a fixed point, WanTsum had occupied.

The important fact in Ityoshin is that trade occurs in the absence of market places. Trade and market are, obviously, two different things which may or may not coincide.

To summarize, we have found that when markets are held at five-day intervals and different markets meet on different days, a repetitive market "cycle" is formed; the fact of sharing several markets, each fulfilling jural and political, as well as economic, functions, creates an informal neighborhood of increased social interaction. Market cycles change as one moves about the country. The neighborhoods that are associated with the cycles thus also change and overlap. What we have, in effect, is a series of neighborhoods, marked by shared market cycles which overlap something in the nature of chain mail.

A market neighborhood of the sort we have described does not have a formal structure of social roles. The importance of this fact will not be overlooked when we come to consider the reasons that the market system in Tivland was stringently contained and not allowed to spread. The event sequence of activities in several neighboring markets gives rise to "neighborhoods," but the neighborhoods are not recognized social units; they comprise informal knots of increased interaction relevant to the working of political and jural institutions, but are not themselves institutionalized.

Markets Linked by Flow of Goods

MARKET PLACES may be linked by the flow of goods as well as by the institutions—created by neighborhoods and lineage groups—centering on them. Movement of goods can be effected by trading as well as by marketing. Tiv traders have been mentioned several times. Here we will investigate them further, and distinguish trading from marketing.

Kyagba, the old and experienced chief of Jechira "sept," once said that "sauce-buying markets are completely different from money-buying markets." At the time, we thought he referred to the difference in size and the distinction between consecrated and unconsecrated markets. In the course of analyzing the data, we have decided he was distinguishing marketing from trading. In the first sort of markets—"sauce markets"—the emphasis is on social relationships and on marketing. Both women and men, but predominantly women, come to sell what they have produced and to buy what they need. In markets of the second sort—those of concern primarily to men who buy in a cheap market and sell in a dear one—the emphasis is on trade. As we have seen, some women do engage in local trade within the market neighborhood; several examples have been cited. Most trade is, however, in the hands of men. Tiv thus characterize a small market in which marketing predominates and trade is minimal as a "woman's market"; a large market with a considerable volume of trade is called a "man's market."

A similar distinction was drawn by an Iyon elder in explaining a term we did not understand. He said that Ibo "journey *kpenga kpenga*." We took this term for an onomatope and asked to be told its meaning. He explained that the term came from the word for a carrying tray (*kpenga*) and that the Ibo leave their compounds with only a carrying tray; they buy in one market and sell in another and live on the profit. This, he said, was a very bad thing. We asked about the Bafum and Bamenda, whose trade routes lay through Iyon. The elder replied that they did not "journey *kpenga kpenga*" because they started from home with crops rather than money; they grew sweet potatoes and bananas, brought them and their cattle down into Tivland and Udam, exchanged them for other items, and eventually went back home to continue their farming. The Bamenda are thus marketers regardless of the distance they travel. Ibo are, on the other hand, traders.

Three sorts of marketing and trading activities can be distinguished: (1) sauce and sauce-plant exchange, the province of women; (2) marketing based on agricultural or handicraft products in which both men and women engage; and (3) trade based on money, not on direct subsistence products, predominantly the province of men. The notion of "market luck" pervades all three types of activity.

In setting out to market, Tiv marketers say they start with a notion of what they want and look about to see what they already have. If one wants a cloth and has two chickens, one takes the two chickens and starts to market. If the two chickens realize enough to pay for the cloth, and he finds the kind of cloth he wants, he says, "My *kasua* was good today." If, on the other hand, he cannot get his price for the chickens or if he cannot find a piece of cloth which meets his demands for the amount he got for the chickens, then he says, "My *kasua* was bad today."

The ability to drive a good bargain is, generally speaking, considered part of one's personal charm: "Today Wanuvia was irresistible—she bought three yams for only seven pence." "Market" is a means of having power over other people. Therefore, women, and men as well, very often economize not only to get the most for their money but also to prove that they are irresistible.

Good market luck means getting what you want when you want it and for what you want to give. The best market luck means that you get something a little more—that, in fact, both parties to the exchange get something a little more—for in the best *kasua* both people may have good market luck.

As a method of saving, a woman may buy a bag (or part of a bag) of salt. She may even have saved money over several months to buy it. After that, she is set up for market trading for several months. She makes her salt into cakes, fires them, and sells a few cakes at every subsequent market she wishes to attend. Thus, she will probably have an income of 3*d.* or 4*d.* per market with which to buy palm oil or other condiments. The salt can be seen as a means of storing wealth, or as the basis of a pin-money economy.

It is difficult to discover how much profit these women made or hoped to make. They did not think in terms of profit when they engaged in sauce trade or in what might be called the pin-money trade (Reining 1959). None could say how much her goods would sell for; the point was rather that she now had a commodity to sell in the market which would produce a fairly "steady income." Tiv women, and most of the men, were not interested in total profits when they engaged in this sort of trade; at least, they did not compute them.

Buying a bag of salt is sometimes compared by Tiv women with having a part-time specialty. Women who are potters may make as many as two dozen pots for a single firing. Then they have pots to take to market, two or three at a time, for several weeks, and from this income they buy their palm oil and sauces. It is true, of course, that each woman tries to get as good a price as possible for each item, but they emphasize the value of the small steady income rather than the total profit.

Other women make food to sell for snacks in the market. One day we watched MbaAdzan making fried cassava cakes for this purpose. She said that one small mortar of pounded and prepared cassava made about 120 fried cakes, after samples had been given to everybody who came by during the frying. She claimed that she would sell them two for a tenth-penny, eleven for a halfpenny. We asked how much she would make. She supposed she would make about a shilling. By our computation, if her total was right and she sold all her cakes at the higher price, she would get only sixpence. She did not make the computation, although she was perfectly capable of doing so had she wanted to make it. We believe that she had not even thought about the total amount until our questions brought it up. She named a figure it would be nice to have, not one she expected to make in any economic sense.

Tiv women change this notion of economizing when they enter the subsistence trade. Then their notion is nearer our own, though even then they do not compute profits. We have given several previous examples in which women buy yams or grain in a cheap market for sale in a dearer market, most often within the neighborhood. Pedantically, if a woman takes money to buy a yam to sell, it is trade; if she sells one of her own yams and replaces it at a lower price, it might be considered marketing. The fact that Tiv women are not pedantic in this particular distinction is one of the reasons they sometimes suffer when trade and market meet. In terms of their own institutions they are marketers; but traders deal in the same goods. In the absence of production for the market, this situation may lead to near disaster because they can sell their subsistence.

The point here is, however, that Tiv women seldom make the computations necessary to know what they will gain in their marketing or what profit they can expect from their petty trade. Seldom do they know whether or not they have actually replaced the amount of subsistence they have sold, though most claim (as a matter of virtue, not of economics) that they have done so.

When Tiv men engage in subsistence marketing, they seldom make computations; but when engaged in trade, they almost universally count their profits and losses, even though they do not keep any records and are unable (or unwilling) to give the investigator any figures.

Sauce marketing is, like other marketing and trade, carried on in terms of money. Money is always the standard and sometimes it is the means of exchange. In instances in which one product is bartered for another, both parties to the exchange agree on a money value of the goods bartered. Many women sell to one person for money and buy from another with money. Money that they get for their sauces is usually used in buying other sauces, although some of it may be saved for exercise books and pencils for the children or perfume for themselves.

Sauce marketing itself is seldom considered primarily a means of earning a profit. As we have seen, the sauce plants are available all over Tivland. Sauce marketing is carried on not only in connection with the local sale of such items as salt or pots, the purpose of which is to give a small, steady income, but also in connection with other activities which definitely are considered primarily as a means of earning a profit: beer making or trade in small quantities of staples. Thus, although the goods involved in sauce marketing seldom go beyond the market neighborhood, subsistence goods may be bought by traders and be sent much farther afield.

Some men carry on marketing similar to that of women but they do not compare it to sauce marketing. All tobacco is grown by men; it is cured, braided into cords which are rolled into coils, and then cut into penny lengths and sold in the market. A man expects to sell from 4*d.* to 1*s.* worth of tobacco on a market day. He usually spends this money, or some of it, in the market place on the same day, on beer or—if he has had good market luck that day—on a piece of meat. It is difficult to buy tobacco in quantities larger than a few pennies' worth at a time. We once asked to be shown a pennyworth; when the seller indicated the amount, we said we would have a shilling's worth. He began to cut his roll of tobacco at a point far short of 12 pennyworth. We complained and measured out 12 pennyworth on the roll. The seller agreed that a shilling had 12 pennies and that the measurement was correct, but he would not sell that much tobacco all at once for a shilling.

Tiv men treat tobacco as women treat salt, yams, or pots. Like women, they value regular, small income. One of the biggest rice traders in the Ticha area put the notion succinctly: he said he did not want to sell his rice all at one time because if he did so it would not "last me for a long time."

A man can go to Udam and Uge markets either with Tiv cloth and livestock or with money. He returns with kolas or camwood and sets himself up with a regular income very much as women do, although somewhat higher sums are involved. Kolas demand a fairly quick turnover because they are perishable, although they will last three or four market cycles if kept wrapped and moist.[2] Camwood, on the other hand, lasts indefinitely; a man may take several months to sell 10*s.* worth of camwood, using the proceeds to buy beer at the market and meat to hand to his wives on his return from market.

Men from southern MbaDuku often take southern produce north and return with northern produce. One Iyon man wove four pieces of cloth, sold them at Usan market for enough money to buy two kerosene tins (four imperial gallons each) of red palm oil at about 10*s.* per tin. He headloaded the palm oil some sixty miles north to an old Jukun riverside port called Bajimba at the confluence of the Katsina Ala and Benue rivers. He sold his oil for 19*s.* per can and bought dried fish. He sold some of the fish on the way home, gave some to his wife, and sold the rest of it in Iyon market on his return. From the proceeds he paid his tax, purchased some seed yams, and bought his wife a new Manchester print cloth. He had had a "good market," but he did not know the total amount of money he had handled.

Marketing activities of this sort are considered like, and often carried on in conjunction with, specializations characteristic of young men. The man described above began by weaving cloth, as do many young Tiv. Slightly younger boys make raffia bags or winnowing trays. Older men may make baskets or sell game they have hunted. Young men often make raffia-palm beds for sale at markets; there are other specialists who make chairs and plank beds, though the latter are usually ordered and never brought to the market. Marketing based on subsistence farming or on a craft is an activity of most of the community from time to time.

Comparatively few Tiv engaged in "money trade" in 1952, but the number was growing. There were several men in MbaDuku and southern Shangev who made more or less regular trips to Onitsha market, the hub of African trade in eastern Nigeria. They took orders for items not locally available. The carpenter in Iyon, for example, commissioned

2. The kola trade between Ogoja and Benue provinces is largely independent of that between the kola forests of the south and the Hausa in northern Nigeria.

a new saw while we were there, and a trader once bought a Hong Kong flashlight for us. Such a commission means saying to the trader, "I want such and such an item of such and such a quality, and I am willing to pay so much for it." The trader then accepts or declines the commission. If he accepts, he will get as much as possible of the sales price before he goes. The remainder is paid on delivery of the item. The buyer never knows how much the trader paid for the goods.

Some of these money traders also make regular purchases of produce in Tivland and either ship it or take it to larger markets for resale. These traders have what they call "market friends" (*boki kasua*) scattered about the country. A man can send to his market friend any goods to sell; he can be fairly sure that the money realized on them will ultimately be collected. The man who has several such friends scattered judiciously around the country can become a big operator. These networks of traders bound into formalized relationships are not so highly developed among Tiv as they are among Hausa, Ibo, or Yoruba traders. Nevertheless they exist. The best market friend is one who is your kinsman or someone already connected with you in some other way.

Money traders not only take orders and send produce to the large market centers like Onitsha, they also bring back large lots of trade goods from the centers—cloth, pins, and the sort of things sold by young men in local markets. A few will then sell them wholesale; more often a younger "brother" takes over the stock and sells it in the local markets. Traders themselves seldom sell in the market places; they buy there. They make a distinction between "trading" (*yamen a yam*) and "spreading things out in the market" (*teen kasua*), between trade and market.

Most money traders among Tiv live within a subsistence economy, keeping farms in their lineage areas. However, they almost never sell primarily their own produce, and they seldom do their own farm work, finding it more convenient to have it done for them. Thus, the same conditions that created money trade have sown the seeds for what may some day become a full-scale labor market. However, even here labor is not fully subject to market criteria. Most often the traders' "brothers" perform the heavy farm work, not for any stated amount of pay, but in a roughly reciprocal arrangement whereby they get cash or trade goods to sell for income. Even the biggest Tiv traders retain their farms and keep some of their wives tending them. The sign of a really successful trader is that he has managed to buy a lorry. We know of four Tiv traders who owned lorries in 1952 (two of them lived in Gboko, one in Makurdi, and one in Taraku).

The Tiv that one finds outside Tivland are seldom traders; they are much more often laborers or workmen, sometimes highly skilled. We found Tiv electricians in Lagos, skilled mine laborers in Jos, and knew of many agricultural workers in the Eastern Region and the Cameroons.

Therefore, although Tiv engage in a money trade, they had, at the time we studied them, shown an amazing tenacity for their subsistence economy, using trade only as a supplement. Even the biggest traders insisted that "farming is the important work of the Tiv." They noted that Hausa make a good life trading—"at least they like it"—and told an apparently indigenous version of the Jacob and Esau story to account for the fact that Hausa are traders and Tiv are farmers, though Hausa may make gardens and Tiv are ardent marketers.

Trading as well as marketing is thus an activity of everyday life. Many Tiv do some trading. Subsistence agriculture is also an activity of everyday life, and almost everyone engages in it. No one will willingly give up either. For this reason, trade has in only a very few cases been made the sole basis of livelihood; Tiv values place great importance on the subordination of trade. A trader is merely enterprising until he ceases to be a subsistence farmer in however attenuated a way. Then he ceases to be Tiv. To be Tiv is not merely to have one's place in the agnatic genealogy of the Tiv tribe; it is also to take up one's place on the ground, and to subscribe to and live by Tiv culture. Trade is a valued part of that culture, but subsistence agriculture is of greater value. Money, Tiv say, can never be trusted as one's farm can be trusted.

Market Patterns

Almost all of the items that men market go beyond the market neighborhood. The obvious exception is tobacco, which may, but usually does not, do so. The goods may travel beyond the market neighborhood in one of two ways we have mentioned: they may change hands, continue to change hands, and so travel long distances; or traders themselves may attend markets far beyond the market neighborhoods.

Because market neighborhoods overlap, it would be possible for an item to move across Tivland and out of it by marketing and local trade. No one ever need go beyond his market-cycle neighborhood. In fact, before the days of the *Pax Britannica* this was precisely the way in which goods did move across the country. However, with peace and improved transportation, Tiv men often move outside the market neighborhood in their subsistence marketing, and money traders move very great distances outside it.

When the movements of goods occasioned by these various forms of marketing and trade are brought into focus, a new set of market patterns emerges. When the staples traded by women change hands, they move from a market where the price is low to one where it is higher. In the Ticha cycle, Skiliki, Angur, and Aichwa are markets which, while primarily sauce markets, are also sources of produce. The produce is then resold at a market like Ticha. There is, thus, a situation in which several small markets surround a larger market and feed produce into it. We shall call these two types of markets "feeder markets" and "bulking points."

Men seldom do more than sell tobacco at feeder markets. However, at bulking markets some of them enter the produce trade. Here both they and the women buy produce for transport to a larger market such as Tsar or Obudu. Goods thus arrive at Tsar market in one of two ways: they can be brought there directly by the producer or they can be brought there by traders.

A market like Tsar or Obudu is fed not merely by one bulking market but is surrounded by many. Before the Tsar dispute, Obudu market was the central point for a number of bulking markets: Ticha, Iyon, Usan, and Chukwan, to name those which we have examined; there were others to the south. Such a market, surrounded by bulking markets, will be called a "major market." The major market is a shipping center; that is, the major shipments out are by truck. The shippers are either the owners or drivers or else passengers who accompany their goods. In a few cases in which the driver or his "clerk" is a market friend of the trader, goods are sent in care of the driver to still another market friend. There are about a dozen major markets in all of Tivland.

The trucks then proceed from the major markets to the centers of Nigerian trade: Bansara, Abakaliki, Onitsha, Makurdi, even Jos. All of them, with the exception of the relatively minor trading center at Makurdi, are outside Tivland. All are points at which European trading companies have established large installations. Hence, the trading centers are the points at which Tiv economy intersects with world trade.

Two features emerge from this description: the first is that goods form routes or trails across the countryside, and the other is that a system of markets based on the movement of goods, and to some extent on the movement of money traders, is brought into existence. This system of markets is similar to the one that W. E. Moore (1951:227) found in Mexico and called a "solar system" of markets. That is to say, several feeder markets (*D* in Fig. 25) surround a bulking market (C); several bulking markets surround a major market center (B), several of which surround the trading center (A).

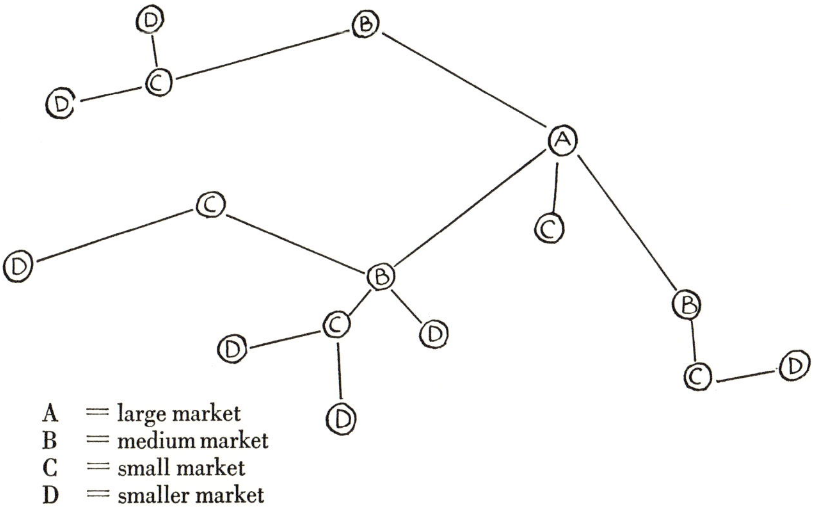

Fig. 25.—The solar system of markets.

Goods move, then, either toward the center or away from it; to oversimplify, produce moves inward toward the center, while imports move outward. Obviously, produce may at some stage also flow outward: in the case in hand, some of the produce which reaches the major markets flows back down along the line.

Some of the larger major markets are almost indistinguishable from towns: Katsina Ala, Obudu before the Tsar affair, and Gboko are examples in or near Tivland. The difference between a "town" (*guli*) and a market is that in a town, goods are sold every day, while in a market, they are sold only every fifth day. However, in most major markets a good deal of trade takes place on the day before market day, as we noted in the case of Tsar. A town was, in fact, springing up around Tsar market before the administration moved the Hausa settlement that was its core.

To summarize, there are four types of markets in Tivland: The feeder markets are marked by sauce marketing, overlapping into subsistence marketing and local trade in very small amounts. The bulking centers are marked by subsistence trade in which the small amounts bought at feeder markets, or else marketed directly at the bulking center by the producer, are combined into fairly large loads by the traders who take them to the third sort of market, the major market. Major markets are distinguished by the presence of a number of traders. Large traders buy the bulked lots from the local traders and ship them by lorry either to the trade centers of Nigeria or else to major markets in other areas, where the produce is distributed out along similar market lines.

Movement of Goods

THE RESULT of the combination of marketing within overlapping market neighborhoods and of trading within the solar system of markets is that any particular goods can be traced on a route through eastern Nigeria, including Tivland. A few of these routes can be examined here.

Although the amount of yams grown per person would seem to be about the same in all parts of Tivland save those seriously affected by land shortage, the amount of grain grown is very much greater in the northern sections. Southern Tivland, where most of the major markets are to be found, exports a greater proportion of its yams. The resulting food shortage is made up (to some extent at least) by cassava or by grain from the north. Grain thus flows south through the markets. By headload, grain comes from MbaHura market to Iyon or from the bulking markets around Adikpo and thence to Tsar or Asawa. Women often combine a trip to their kinsmen to the north with grain-buying expeditions, especially to buy millet for making beer. MbaDuku men told us that it was possible to go from MbaDuku north about twenty miles to Ihugh market, buy millet, spend a shilling "on the road," and still make money by selling in markets of the Ticha cycle.

Trade in livestock, like trade in grain and tubers, establishes a flow from north to south. In southern Tivland the price of chickens rises as you go south, and hence there is a tendency for people to take their chickens south. An Ute woman once came to see her "sister" who was married into MbaDuku. She brought two chickens with her. She asked exhaustively about the markets at which chickens fetched the highest price and was told that they were the same price at Usan and Obudu; eventually she sold them at Usan because oil was cheaper there. Tiv estimated, at our request, that a chicken fetched 3*d.* more at Obudu market than it did at Ticha market. Goats, sheep, and pigs also sold for far more at Obudu or Tsar than at any other market in Ticha or Iyon cycles (when they were available anywhere else at all). Prices rise both as one goes south and at the major markets.

Of far more importance are the cattle routes which cross Tivland. Fulani cattle are driven south every year to the major markets of southern Nigeria. Several of their routes cross Tivland, although because of tsetse, Tiv seldom engage in cattle trade. The main cattle tracks across Tivland run: (1) Yola, Katsina Ala, VandeIkya, Ogoja, Enugu, and Onitsha (this route runs well north of the area of southern Tivland on which this report is concentrated); (2) Bamenda and Mamfe, Obudu, Ogoja, and south, crossing the part of southern Tivland dealt with here; (3) Bamenda, Kasimbila, Adikpo, VandeIkya, Ogoja, and south; or (4) Katsina Ala, MbaRa, Gaav, Ogoja, and south. The cattle from north of the Benue are either taken across the bridge or are swum across the river at one of several fords, the one most commonly used being just downstream from Ibi.

Occasionally Tiv butchers buy Fulani cattle as they cross Tivland. One Tiv butcher who lived in Obudu bought enough to have one on hand to be butchered every market day, although he seldom had more than three or four at a time. He sold the meat at Obudu market, and any remainder at subsequent markets of the cycle in the next day or two. Tiv dwarf cattle, unlike the Fulani humped cattle, are primarily prestige symbols. They are seldom butchered.

Dried fish are carried south from the Katsina Ala and Benue Rivers and north from the Cross River. The two routes meet in southern Tivland. Occasionally, Tiv from MbaDuku acquire fish in Bansara or some other southern point instead of from the north, but that is the exception.

The divide between the Cross and the Benue rivers is also the approximate divide of the direction of flow of palm produce. The palm oil from

Obudu District, for example, goes in both directions. It is possible to buy palm oil there at a price ranging from 10*s*.–15*s*. for a four-gallon drum. Taken north as far as the Benue, it can be sold for 19*s*.–22*s*. for the drum. Further north it is said to fetch even more. At the same time, the four-gallon tin of red palm oil is said to be worth as much as 30*s*. in Onitsha. This last price seemed high to us, although it is one which Tiv quoted. Tiv themselves seldom traded palm oil south, but they said that the Ibo lorries that came to Obudu and Tsar often did so.

Most of the oil that goes north is headloaded, because it is necessary to have a fairly large amount of capital in order to buy enough to make truck travel worth while. Tiv do not, when figuring profit, compute any cost for transportation if they carry their own loads, but only take actual expenditure into consideration. They say that taking a lorry "spoils too much money." When we once asked if carrying eight or twelve gallons of palm oil fifty miles did not "spoil their bodies," they said that a man might get tired but would get over it. They did not make the connection of any price for expended effort.

The trade routes for cloth are some of the most erratic in West Africa. Cloth goes in every direction. However, each sort of cloth follows definite routes. A trader may take Tiv cloth to Bansara or Abakaliki, where it is bought by Ibo, and he may return with Manchester prints or with Yoruba or other African cloth. Tiv often sell their own cloth south and buy Hausa-woven cloth from the north.

Fashion enters into the cloth trade, and as a result, that trade shows many more vicissitudes than trade in subsistence produce. During 1949–50, Tiv fashion favored resist-dyed Yoruba cloth. A cloth of the sort which we bought (at Europeans' prices) in Ibadan market for about £1, sold in Tivland for 35*s*. There was a trader in Obudu market who took orders for these cloths and sent directly to Yoruba country for them.

Tiv sell to Ibo what they definitely consider to be their own inferior cloth. We once asked a weaver what he intended to do with a cloth on his loom. He replied, "If it turns out to be a really good one, I shall give it to my mother-in-law. If it turns out badly, I shall sell it to the Ibo. If it is so-so, I shall keep it for myself."

Tiv say that Uge buyers are more conscious of quality in Tiv cloth than are Ibo buyers. As a result, weavers in Iyon sell good cloth at Usan market where its quality is recognized, rather than at Tsar where (they say) the Ibo would give no more for it than they would for a poor one.

Summary

In this chapter we have followed the movement of material goods from one market to another and have found that they move either through the markets of overlapping market neighborhoods or in an inward and outward flow from tiny sauce markets to the trade centers of Nigeria, which are linked, through the trading companies, with the world economy. Sometimes the movement of goods reflects both patterns. Goods, in process of changing hands, give rise to a solar system of markets.

There is no formal system of roles that parallels the solar system of markets. Incipiently the linkage of traders with market friends or their relationships with commercial trading companies may form such an organization for linking markets, but that situation had not occurred at the time of our study. Although Tiv recognize that there are good places to buy various commodities and to sell others, they do not perceive the existence of the solar system. This system does not coincide with any other institution. Although Tiv market places are overtly institutionalized, the market system is not.

16

The Principles of Tiv Economy

ECONOMIC ANTHROPOLOGY deals with the practices and institutions by means of which people provision themselves. Its goal is to elicit the theoretical principles that best explain instances of such provisioning.

Economics as a science got off the ground in the separate but converging work of Menger and Marshall when the idea of price superseded the idea of value as the central issue. The market was the central institution of allocation in the societies that these men knew, and economic theory of market societies made great strides. In changing the focus from value to price, another shift occurred, but was not discerned at the time because economists considered only the market principle of social integration, with factors of production entering the same market as products. That shift was from a central concern with production to one with allocation.

Among the Tiv the market principle is not central. For all the vigor of the market place, the market principle of social integration is peripheral. In order to theorize about Tiv economy, it is necessary to re-insert an idea of "value" that supersedes "price," and yet not go back to focus on production. That is to say, we must deal with the allocation of things given value either by the price mechanism of supply and demand or by some other means. In order to do this, we have created an ugly compound term, the "man-thing unit." Men value things for themselves but even more for the part the things play in social relationships. From one point of view, a man and any thing he values highly can be seen as a dyad—*not,* be it noted, as a relationship (or at least not as a social relationship).

The legal mode of viewing the man-thing unit explains institutionalized "property" (itself a Western idea that needs a kind of cross-cultural scrutiny it has never received); economics investigates the man-thing unit by explaining the social allocation of the things (obviously including things like "services"). Both law and economics deal with the same unit.

We shall first examine some of the aspects of the man-thing dyads that would usually be called "ownership of the means of production." In order to avoid misconstruing the Tiv ideas we will discuss four "controls": control of land, control of labor, control of products, and, in the modern world, control of money. That is, we shall view four kinds of man-thing dyad, differentiated by the "thing." After that, we shall look at the allocational aspect of the man-thing unit and see how men pass things from one to another. That is to say, we will examine the social relations that exist when there is a single "thing" to be found in two or more man-thing dyads.

Control of Factors and of Products

We saw in Chapter 10 that most agricultural products are in the hands of women, that a man has a right to small quantities of his wife's food crops for legitimate entertaining and ritual expense, and that, aside from a husband's small claims, only a senior wife can control a woman's food produce.

Neither Tiv women nor their husbands hand over any portion of their crops as tribute, rent, or tax to anyone else (except, of course, the direct money taxation imposed by the colonial government). When a Tiv woman gives her husband yams, the amount is never stipulated and she may well argue about it. In no case does she give enough to enrich her husband or compound head. If she herself sells food, she may—but is unlikely to—give her husband a portion of the money. She spends the money for food, clothing for herself or her children, or soap and perfume. Men do not, indeed cannot, exploit the labor or the produce of their women. Women from the nearby Udam tribes say they like to marry Tiv men because they treat their wives well.

Since each woman or group of linked women controls the produce of her own fields, women control the major portion of subsistence food. Since they pay no tribute, and no person has a right to any but infinitesimal proportions of their food, wealth within the subsistence system cannot accumulate to any individual. Subsistence production and consumption do not lead to any sort of ranking or hierarchy.

We saw in Chapter 5 that a compound head controls, to a small extent, the work of the men and women living in his compound, but that no other person can do so. Control of people, and *ipso facto* their labor, is acquired in terms of marriage, filiation, and common residence. Calling age-sets is a means of controlling some labor for short periods; it is open to all members of an age-set. In 1952, there were no other established means for acquiring control of labor, although a few well-to-do traders were beginning to hire assistants other than kinsmen, and labor by contract was starting to appear.

Political office or authority does not proffer control of labor. There is no tribute labor among Tiv. In no conditions do Tiv perform labor for the benefit of a chief, a ritual specialist, or any other person except their fathers and compound heads. Communal labor is sometimes performed; our own huts, the huts and market place of an immigrating Hausa community, the "clan" schools, all were built by communal labor, for which some payment was made. Tiv express such work in terms of the people of such and such a lineage segment. Our huts were built by "the people of MbaDuku," as were those for the Hausa settlement. Such labor can be organized along either lineage or age-set lines, sometimes both. Building a new building at the MbaDuku "clan" school was done by age-sets, called out in this case by elders, after agreement among all, including the men who did the work. In some cases we know, dissenting elders have withheld the labor of the men of their compounds and have prevailed on other compound heads to withhold the labor of their people. Elders, even as a group, cannot enrich themselves by controlling labor of a younger age-set. An age-set can refuse to work and sometimes does so; a member who does not contribute his fair share of labor to a project on which it is working can, however, be fined by the set. Money from such fines is spent by the age-set for food or beer at markets and is consumed by all members present.

No person can gain in wealth by controlling and exploiting the labor of others. Tiv correlate a man's wealth and influence with the size of his family and the number of his dependents. More dependents mean more labor; more labor means more farms. Tiv never, even in the areas in which land is shortest, make overt statements of this sort in terms of obligations: they never say that one more dependent means one more mouth to feed. In response to questions they admit that this is so, but they stress the fact that the compound has gained an additional person's labor, thereby making additional farms both possible and necessary. In Iyon, where land is very short, they still speak in this idiom. Economic differentiation based on control of people is not operative in terms of so-called "surplus" production but only in terms of prestige arising from a large compound.

Men of political importance are usually men of substantial subsistence means who have large compounds. The division of labor which is possible in large compounds is recognized to be time- and labor-saving. Only one woman of a set of linked wives cooks; in units containing several women, some do farm work while others do domestic work. Tiv say that a single wife must work very hard because there is no one to share the work with her; both men and women cite this as an advantage of large compounds, of polygyny, and of wife-linking.

We saw in Chapters 6 and 7 that both men and women control whatever land they need by virtue of birth, marriage, and residence—in short, within terms of the kinship system and its spatial dimension. Only in Iyon, where no one has enough land, does the man who has the most land under cultivation have as much as nine times that of the man who has the least. Every compound head controls some land: ideally enough for his needs, with fallow to allow revitalization of the soil. There is no landless man among Tiv, for jural right to land (more correctly, to exploitable space) can be acquired by birth alone. There is no landless married woman. There is no person or individual, and no group of persons, with rights to a piece of cultivated land superseding the rights of the compound head and his people who are farming it.

Since the early days of the colonial administration, money has been regarded as a necessity. All adult males must pay taxes in money. Almost all Tiv men grow and sell beniseed or another cash crop to pay that tax. Though Tiv say that they do not have enough money, almost all of them handle some money.

As a general rule, money belongs to the person who has earned it or sold a commodity. When young boys sell coconuts, the money they make is theirs; when girls sell pepper or sauces or rice which their mothers have given them, the money belongs to the girls. When young men sell cloth, the money is their own; and when women sell food, the money belongs to them. When men sell beniseed, it is their money (unless it is sold on a commission arrangement for a kinsman).

There are, however, rules of giving and receiving which cut across this principle. No man would take the small sums of money which his young children have received at market. No man demands money which his wife has earned selling food, calabashes, or cotton yarn. We have heard of a few cases in which money was given to him, but the husband has no right to expect it. However, an older boy—about puberty and after—who makes money selling cloth which he has woven, or who makes a profit buying and selling grain, is expected to give part of the money to his father or compound head. If a father or compound head asks his adult son for money, it must be given if it is available. Fathers and compound heads who receive this money make themselves responsible for the tax of these young men and supply them with money for fines, if needed. Perhaps most important of all, the father or compound head provides a man with his first wife. His other wives he may acquire for himself, but Tiv feel strongly that a man's first wife should be given him by his father.

Control of money follows the same patterns as control of subsistence; the money belongs to the person who earned it. But as part of the reciprocal rights and duties of family and compound life, the father or compound head sometimes has an overriding right to some of it. Tiv say that during the hungry months a compound head should control the money; they equate this with the fact that he should control the food in the compound when it is very short.

To say all this in political rather than economic terms, Tiv is an egalitarian society. The legal and economic rights of individuals depend on their status: such matters as sex, kinship connections, and age. Only minimal social and political differentiation is made among persons in terms of property—that is to say, the man-thing units are astonishingly homogeneous.

If a government chief must entertain officially, he gets food from members of his own compound or else sends his guests to other compounds. When all the government chiefs of Kunav met in MbaDuku after the Kunav-Udam affray (see Chap. 13), only two actually lived in the compound of the chief of MbaDuku. The others lived with their nearby kinsmen or else with elders of MbaDuku who enjoyed the prestige of having an important guest in their compounds. In this way they did not constitute a heavy drain on the resources of the chief's compound. Some of these guests, but by no means all, brought uncooked food, though never enough even for their own parties.

Part of being an influential man or a man of political power is having a large immediate family following; political power does not, however, give a man the right to any portion of the labor or food crops of those whom he would not control by family and residential sanctions.

This short analysis of the control of produce, labor, land, and money has been sufficient to reinforce our point: that vast economic distinctions do not exist within Tiv society. Wealth is measured by numbers of dependents, by industry, and by prestige. In Tiv economy, the control of resources or of capital cannot be made the basis of differentiation and rank. Clientage, tribute, and hoarding of goods are absent. Both social and technical factors assure their absence.

Allocation

THE MAN-THING unit is stable, though its components are not: both role-player and thing may change from time to time. From the legal point of view, contract and kinship are the principles used to fix both the "man" in the man-thing unit and the social relationships among successive or concomitant "men." From the economic point of view, all forms of allocation—all the ways that items can change hands—provide a mode for viewing the changes that are constantly occurring in the stable structure of man-thing units.

Allocation and exchange must be kept separate; economists do not confuse them, but non-economists sometimes do. Allocation is the dispersal of the aggregate factors and products among members of society, regardless of the specific channels or mechanisms employed. Exchange is an act of reciprocal giving and receiving. The idea of allocation takes the goods—the things—as its starting point, whereas the idea of exchange takes the act of reciprocal giving as its starting point.

Three modes of allocation have been considered: exchange (either in a market place or outside it), gift-giving reciprocity, and inheritance. In the following few pages we shall explain the Tiv words for viewing allocation.

The word *musan* means exchange. The word can be used for trade and marketing, but Tiv seldom do so except in explaining the differences between barter and money trade. It is also possible to use it of exchange marriage. *Musan* implies a replacement or reciprocation; it is the most general term for all acts of exchange in which goods of any sort change hands.

Every item that is exchanged has an exchange value or "equivalent" (*ishe*), but no gift or inheritance has such an exchange value. In a market situation *ishe* means exchange equivalent; one might translate it as price, with the warning that it may not be extended to a more general idea of a universal standard of value. Tiv seldom ask or quote equivalents in their own trading; rather, they effect bargains, usually without recourse to this word. However, an expensive item is a thing of high equivalent (*kwagh u kehe ishe*), and to haggle is to dispute the value-equivalent (*kperen ishe*). The general term both for subsistence trading and for exchange marriage means more or less "buying value" (*yamen ishe*).

In every market transaction there is a man who sells (*te*) and a man who buys (*yam*). These words do not exactly parallel their English equivalents. *Te* means to spread out on the ground to public view, as in a market place. By extension it means to sell; there is no other way to say "sell" and no other verb to designate that half of an exchange in which one releases or gets rid of an article. *Yam*, on the other hand, means "trade" in the widest sense, but refers primarily to that half of an exchange in which one takes or gains an article. It can, therefore, often be translated by the English word "buy." Activities of traders are called *yamen a yam;* exchange marriage sometimes is called woman trading (*kwase yamen*), but more often by a polite euphemism, equivalent value trading (*ishe yamen*).

The etiquette of trading is informal, but fairly circumscribed. An informant told us, "You [the buyer] first select the article you want from those spread out before the person with whom you are dealing. You pick it up and place it on the ground between the two of you. The seller then asks for money [stating an amount]. In reply, you give him, in exact change, the amount of money you think best. He hands it back to you. You may add something and return it. Then he must pick up and hand you the item; you must not reach down and pick it up." To do so might be construed in any eventual litigation as the seller's nonconcurrence in the arrangement; it could be called buying or trading "by force" (*yamen sha apera*). Actually, of course, buying and selling is noisier and more informal than the description: Tiv description always emphasizes the quick, quiet, and seemly procedure. Haggling is sometimes said to be in bad taste but is nevertheless almost universal.

Tiv are extremely conscious of quality and have decided opinions on what goods they want. It is, for example, difficult to sell matches of British or Belgian manufacture in a Tiv market unless Swedish ones have been out of stock for some time. Patterns and quality of cloth are watched carefully. The United Africa Company, in its publications about the Coast Trade, repeatedly stressed that a great deal of market research must go into supplying the right goods to the right areas in West Africa.

The Tiv purchaser is outspoken in his criticism of goods offered for sale. If kolas are too dry, he will say so. If a piece of homespun cloth is badly woven, or the yarn of inferior quality, he will insist on the point and not buy the product. Tiv are not always able to distinguish quality in imported products; nor do they always want what Europeans consider good quality.

It is obvious, thus, that in the sphere of market exchange Tiv economize in the same way as Westerners. They try to maximize their gains in the best tradition of economic man. We have noted that the word *ishe* means equivalent or value and sometimes has to be translated as price. However, the "price" of an item that is traded in the market place is also its *kasua* or market. Thus, *kasua* must also sometimes be translated as price.

In sum, the market in which subsistence goods and imported items are sold is a competitive, uncontrolled market. Prices are uniform over a limited time and area, and are arrived at only by the working of the law of supply and demand.

Spheres of Exchange

THE MARKET MECHANISM does not operate in any sphere of exchange save in subsistence, however. We have found it necessary to refer to the sphere of exchange marriage. Many of the ideas and words that Tiv use in market transactions are also used in exchange marriage transactions, but the idea that exchange marriage is a *kasua* is repugnant, even shocking. Morally, the two spheres are completely separate.

In calling these different areas of exchange spheres, we imply that each includes commodities that are not regarded as equivalent to those commodities in other spheres and are hence in ordinary situations not exchangeable. Each sphere is a different universe of objects. A different set of moral values and different behavior are to be found in each sphere. There are several spheres of Tiv life in which some sort of exchange takes place, each of them marked by a definite category of goods, a set of moral principles, and characteristic behavior patterns. They are united, indigenously, not by money or by any institution, but by a system of ranked moral values.

Three spheres of exchange could be recognized in the early 1950's. With the introduction of money and European goods, an incipient fourth sphere was created, but the general-purpose money contained within it is contradictory to the system, and is destroying it.

The first sphere is that marked by the commodities Tiv call *yiagh* (see Chap. 4). Allocation within this sphere is carried out either through gift-giving (Chap. 11) or by marketing and ultimately by trade (Chaps. 12–14). The commodities are all locally produced foodstuffs (imported European food is not *yiagh*), including chickens, goats and sheep, household utensils and other craft products (mortars, grindstones, calabashes, baskets, pots, beds, and chairs), some tools (particularly those used in agriculture), and raw materials for producing any items in the category. To trade pepper for locust-bean sauce or yams for guinea corn is a common *kasua*. For a woman to sell yams to buy a pot, or to make a pot and sell it to buy yams, is considered normal marketing. Within this sphere those goods that are exchanged go through a free market, and the bargain is the dominant value.

The second sphere of exchange is a matter of prestige (*shagba*)[1] and is in no way associated with the markets. Whenever one of its component items enters a market situation, special explanation is required. The goods within this sphere include slaves, cattle, horses, ritual offices purchased from the Jukun, that type of large white cloth known as *tugudu*, medicines and magic, and brass rods. In 1953, one could still use the present tense, for ideally the category still existed in spite of the fact that slavery had been abolished, that colonial government offices had replaced Jukun offices (and could not be bought), that much European medicine

1. This translation creates some problems. There is some prestige, in the Western sense, attached to such imported articles as enamel basins. However, no imported consumer goods of this sort enter the *shagba* category. Hence, when we use the word "prestige" in the next few pages, it must be considered a translation of *shagba*.

had been accepted, and that metal rods had been rare since 1940. In historical contexts, the price of slaves is quoted in cows and brass rods and the price of cattle in brass rods and *tugudu* cloth. The price of magical rites, as it has been described in the literature, was in terms of *tugudu* cloth or brass rods (though payment might be made in chickens); payment for Jukun titles was made in cows and slaves, *tugudu* cloth and brass rods.

The rods are dowels about a quarter of an inch in diameter and some three feet long; they can be made into jewelry and were also used as a source of metal for castings.

We have here not price—which is a delicate adjustment of supply to demand—but rather what Polanyi has called equivalences.[2] Equivalences are overtly different from prices in that they are traditionally established, and are said not to change; they are covertly the same in that "packaging to price," as American manufacturers call it, is practiced.

Akiga, in a hitherto untranslated part of his book, says, "You could buy one iron bar (*sokpo*) for a *tugudu* cloth. In those days five *tugudu* cloths were equivalent to a bull. A cow was worth ten *tugudu*. One brass rod (*bashi*) was worth about the same as one *tugudu* cloth; thus five brass rods were worth a bull." Some of our informants disagreed about the equivalences of these various commodities. None, however, disagreed with Akiga's grouping of them. It is commonly acknowledged that the value of the brass rod declined considerably just before the arrival of Europeans in Tivland. We interpret this to mean both that brass rods became more plentiful and that they entered the market (see Chap. 17), whereas indigenously they were only exceptionally part of the market. Akiga also notes (1939:339) a dispute between two men: one had exchanged a slave to the other for two cattle; one of the cattle had not been paid for and was the subject of litigation. He also recounts (p. 380) the case of a man who was kidnapped, put into stocks, and held for the ransom of a cow. When his lineage did not pay the cow, he was sold as the slave he had become. The prestige sphere is seen to be intact in all these examples.

2. "Equivalences are set out in simple unit relations; in principle, trade is one-to-one. Higgling and haggling is not part of the proceedings; equivalences are set once and for all. But since to meet changing circumstances adjustments cannot be avoided, higgling-haggling is practiced only on *other items than price*, such as measures, quality, or means of payment" (Polanyi 1957: 262).

Fees for ritual specialists are computed in terms of items in this category. Kenneth Dewar, in an unpublished manuscript found in the Gboko files, noted that the Tiv paid the Jukun in slaves for certain titles and insignia of office. Fees for payment of *akombo* officiants are usually in terms of *tugudu* cloth. A *tugudu* is twenty narrow strips of white cloth, each about four inches wide, sewn together to form a sort of toga. Although it is the unit of evaluation for reimbursement to ritual specialists, it is, as we shall see below, seldom actually used for payment. Because of its twenty strips it is called a "twenty," and a unit of lesser value is called a "ten." Thus "the twenty," "the ten," and "the five" (the latter rare), having reference to the whole and the divisible parts of the *tugudu* cloth, are a means of evaluating certain goods and services which never enter the market. For example, the value of a ritual performance or office is quoted in twenties, but it has no market equivalent. Other items in the category may also be used as the evaluating unit of ritual office, however. Akiga notes (1939:190) that anyone who wants the *ihambe* (an important *akombo*) established for him must pay the *ihambe* officiant a ten (of *tugudu*). He adds that if the man himself becomes a master of the *ihambe*, the cost is a cow. Akiga gives the key for understanding this system. In the case of the paternal *ihambe*, the ten comes to 2*d*., but in the case of the maternal *ihambe*, the ten amounts to 1*s*. This statement, and others like it which abound in Akiga's book, should be read as ample proof that ten and twenty are not expressible in terms of money value or in terms of any items in the subsistence category. Only in a market situation need items be equal in terms of a price mechanism to be equivalent.

The fees of professional thieves, used in the old days to recover stolen goods or to seize items in payment of debt, were also computed in terms of twenties.

Fees for medicines must always be paid before the medicine is effective. When the doctor is a stranger, he will extract fees in chickens and goats; when he is a close kinsman, a token twenty may be no more than a straw from the thatch. The twenty is independent of the market price of the object used for payment.

Although the twenty is a unit of measure of non-market value in terms of analogy to a *tugudu* cloth, twenties are, in fact, usually paid in terms of chickens or goats. In some cases a very small chicken will do for a twenty; in others, it takes a goat to make one. The value in animals of any particular twenty is determined on each separate occasion. Fees of ritual officiants, quoted in twenties, are never questioned, but the particular items used to pay them may be haggled over at some length. The twenty is the unit of value in this second prestige sphere; payment may be made *ad hoc* in goods not otherwise classified within the sphere.

This second sphere is associated with prestige (*shagba*) in the same way that the first is associated with subsistence. Except for metal rods (and on occasion *tugudu* cloths) none of these items ever entered a market situation. To "buy" a slave at market was not done. Tiv will not buy cows or horses at a market. We noted above that an Ibo part-time trader tried for several weeks to sell two horses in a Tiv market before he discovered that no Tiv would buy a horse there, because horses are not *yiagh*, but rather are *shagba* or prestige. Tiv claim today that none of them would ever have sold a slave for food; there were other, and more honorable, ways of getting food, especially that of "sending hunger" along the extended web of kinship.

The third sphere is the supreme and unique sphere of exchangeable values. It contains a single item: rights in human beings other than slaves, especially dependent women and children. Even twenty-five years after official abolition of exchange marriage, it was the category of exchange in which Tiv were emotionally most entangled. Its values are expressed in terms of kinship and marriage.

Tiv marriage is an extremely complex subject. Though economists might find basic market principles at work, Tiv adamantly separate marriage and *kasua*, and so must we. Tiv say that before the coming of Europeans all real marriages were exchange marriages. In its simplest form, an exchange marriage involves two men exchanging sisters. Actually, this simple form seldom or never occurred. In order for every man to have a ward (*ingol*) to exchange for a wife, small localized agnatic lineages formed ward-sharing groups ("those who eat one ingol"—*mbaye ingol i mom*). There was an initial allocation of the women of the group as wards among the men of the group, so that each man became the guardian (*tien*) of one or more wards. The guardian then arranged the marriage of his ward, exchanging her with outsiders for another woman (her partner or *ikyar*), who might become the wife of the guardian or one of his close agnatic kinsmen, or—in some situations—become a ward in the ward-sharing group and be exchanged for yet another woman who could become a wife.

Tiv are, however, sensible people, and they know that successful marriages demand that women be consulted. Elopements occurred, and sometimes a woman in exchange was not forthcoming. Therefore, a debt existed from the ward-sharing group of the husband to that of the guardian. These debts sometimes lagged two or even three generations behind actual exchanges. The simplest way of paying them off was for the eldest daughter of the marriage to return as ward to the ward-sharing group of her mother, thus canceling the debt. Often, however, practical considerations were barriers to the smooth functioning of this system.

Because of its many impracticalities the system had to be buttressed in several ways if it was to work: one way was to provide what might be called (though there is no Tiv word) "earnest" during the time of the lag; another was to recognize other types of marriage as binding to limited extents. These two ways are confused, because right up to the abolition of exchange marriage in 1927, all non-exchange marriages were treated as if they were lags. When lags in exchange occurred they were usually filled with earnests of brass rods or, occasionally, of cattle. The brass rods or cattle in such situations were *never* exchange equivalents (*ishe*) for the woman. The only *ishe* for one woman is another woman.

Although Tiv decline to grant it antiquity, another type of marriage was found at the time Europeans arrived: "accumulating a woman/wife" (*kem kwase*). Since the terminology of this union was adapted to describe bridewealth marriage, declared by administrative fiat of 1927 to be the only acceptable form, Tiv are understandably confused on some points. As we understand it, *kem* marriage consisted in acquisition of sexual, domestic, and economic rights in a woman, but not the rights to filiate her children to the lineage of the husband. In exchange marriages, both rights *in genetricem* (rights to filiate a woman's children) and rights *in uxorem* (sexual, domestic, and economic rights in a woman) were automatically acquired by husbands and their lineages. In *kem* marriage, only rights *in uxorem* were acquired. To affiliate the *kem* wife's children, additional payments were made to her guardian. These payments were for the children, not for rights *in genetricem,* which could be acquired only by exchange of equivalent rights in another woman. *Kem* payments were made in brass rods. Rights in women, however, had no equivalent (*ishe*) in brass rods or in any other item—save, of course, identical rights in another woman. *Kem* marriage, in which payment in brass rods for rights *in uxorem* was the norm, resulted in minor social disabilities for the children, because they did not have the full ritual and legal rights that were accorded children of completed exchange marriages. Their disabilities were, however, minor compared to those of children born to women with no recognized husbands, who therefore lacked a full complement of kinsmen. *Kem* marriage was similar to, but showed important differences from, bridewealth marriage as practiced in South and East Africa. There, rights in women and rights in cattle formed a single economic sphere and could be exchanged directly for one another. Among Tiv, however, conveyance of rights in women necessarily involved direct exchange of another woman. By Tiv custom "bridewealth" was not an exchange of equivalents but payment in a medium that was specifically not equivalent. (Akiga 1939:120–217 presents a different but accordant view.)

The three chief economic spheres of Tiv life, each marked by a category of exchangeable items, are considered of more or less equal practical importance, but are arranged in a hierarchy on the basis of moral values. Stealing subsistence is the most heinous theft. Akiga notes (1939:377), and our informants agree, that stealing cows or brass rods is less serious. "Stealing a woman," or, as we would say, eloping with someone else's wife or daughter, was cause for fighting and litigation, but it was apparently not until the era of European-dominated courts that it became a "criminal" offense.

Moral values also follow the hierarchy: the category of prestige is superior to (but no more important than) the category of subsistence; great prestige assumes adequate or ample subsistence. The category of dependent persons is superior to (but no more important than) the categories of prestige and subsistence. A large number of dependent persons, demonstrating success in attracting and keeping dependents, assumes adequate or ample prestige and subsistence goods. But, conversely, many dependents give one prestige and enable one to produce ample, even generous, amounts of subsistence wealth.

Ethical values reinforce the hierarchy. The ethics of kinship are more compelling than the ethics of mere prestige and always take precedence; ideally, one must always sacrifice prestige or hope of gain to aid a kinsman. The ethics of prestige are more compelling than the mores of markets and exchange of subsistence wealth; a man foregoes mere gain in subsistence wealth for the sake of prestige or to fulfill kinship obligations.

As Tiv see it, every man, so long as he works, has a right to sufficient *yiagh*—and *yiagh* is the result of hard work and non-interference of witches or other unseen forces. However, acquisition of prestige and women takes more than hard work—it takes a "strong heart," by which Tiv mean both courage and attractiveness.

"Conveyance" and "Conversion"

THAT TIV CONCEPTUALIZE exchange articles as belonging to different categories and rank the categories on a moral basis gives rise to two different kinds of exchanges: exchange of items contained within a single category, and exchange of items belonging to different categories. For Tiv, these two different types of exchange are marked by separate and distinct moral attitudes.

To mark the distinction between the two types of exchanges which Tiv maintain by different behavior and different values, we shall use separate words. We shall call those exchanges of items within a single category "conveyances" and those exchanges of items from one category to another "conversions." Conveyances are morally neutral; conversions have a strong moral quality in their rationalization.

Exchanges within a category excite no moral judgments beyond comments regarding market luck or the qualities of a bride. Exchanges between categories are different. The man who exchanges lower category goods for higher category goods does not brag about his market luck but about his "strong heart" and his success in life. The man who exchanges high category goods for lower rationalizes his action in terms of high-valued motivation (most often the subsistence or ritual needs of his kinsmen).

Conveyance is a simple matter—the two institutions most intimately connected with it are markets and marriage. Tiv markets are, as we have seen, free markets with price (*ishe*) determined by supply and demand; Tiv marriages are exchange marriages, with all transferable rights in one woman exchanged for the precise equivalent rights (*ishe*) in another woman. Both are called "trading equivalents" (*ishe yamen*). Conveyance in the prestige sphere seems, to the latter-day investigator, to have been less highly institutionalized. It centered on slave-dealing, curing, and the acquisition of status.

Conversion is more complex. Conversion depends on the fact that some items of every sphere could, on certain occasions, be used in exchanges in which the return was not equivalent (*ishe*). Given the moral ranking of the spheres, such an exchange leaves one party in a good position and the other in a bad one. Tiv say that it is a good thing to trade food for brass rods, but that it is a bad thing to trade brass rods for food; that it is good to trade your cows or brass rods for a wife, but very bad to trade your marriage ward for cows or brass rods.

Thus, it is possible for an individual to invest wealth if he "converts up" into a morally superior category: to convert subsistence wealth into prestige wealth and both into women is the "economic" aim of individual Tiv. Such conversion is the ultimate type of maximization. "Converting down" is to be avoided except in cases of serious want.

To convert his brass rods up to a wife, a man could *kem* a wife and *kem* her children as they were born. Her daughters, then, could be used as wards in his exchange marriages. It is the desire of every Tiv to "acquire a woman" (*ngoho kwase*), either as wife or ward, in some way other than sharing in the exchanges within the ward-sharing group. A wife whom one acquires in any other way is not the concern of one's ward-sharing group, because the woman or other property exchanged for her did not belong to the group. The daughters of such a wife are not divided among the members of a man's marriage-ward group, but only among his sons. Such a wife is not only indicative of a man's ability and success financially and personally, but rights in her are the only thing he owns not ethically subject to the demands of his agnates.

Wives may sometimes be acquired by means of more elaborate conversions. We discovered one case (in the course of a witchcraft moot) in which a man two generations before had traded a slave for a cow, which he in turn gave to a kinsman for a marriage-ward to exchange for a wife; the distribution, as marriage-wards, of the daughters of this marriage was called into question. Sometimes Tiv acquire foreign wives for cattle, from tribes whose custom it is to use cattle as bridewealth. The daughters of such women are considered Tiv and can be exchanged in regular Tiv fashion—they do not go into the "pool" of wards in the marriage-ward group unless a man's agnates force him by threats of witchcraft to share his property (*dyako*) with them.

Conversion up, from the prestige sphere to the kinship sphere, was thus fairly common; it consisted in all the forms of marriage save exchange marriage. Usually payment was in brass rods, occasionally in cows.

Upward conversion from the subsistence sphere to the prestige sphere was also usually in terms of metal rods. On occasion, they entered the market place. Tiv say that it was possible in the old days to buy brass rods for food, but most of them add that it was so only if the owner of the brass rods was short of food or required an unusually large amount to give a feast, making too heavy a drain on his wives' food supplies. In case of famine, of course, people converted down. Implied threats from kinsmen that a man did not have "a strong enough heart" to withstand the envy directed against him sometimes made him voluntarily forego prestige goods or wives in order not to be conspicuous. The motivation to convert down is never material gain, but either a sign of grave material want or what Tennyson somewhere called the "craven fears of being great."

Captain R. M. Downes told us that in the early days of colonial government in Tivland, and on into the 1920's, one saw Tiv going to market with brass rods, which were used in some market exchanges. Brass rods could not possibly have been a general currency, however. They were not divisible; although some Tiv said in the 1940's that they could be cut into parts, none claimed that it was often done. Therefore, one could not receive change from a brass rod. Moreover, a single rod was worth much more than the usual market purchases for any given day of most Tiv subsistence buyers. Although it might have been possible to buy chickens with brass rods, one would have to have bought a very large quantity of yams to equal one rod, and to buy an item like pepper with rods would be laughable. So far as we are able to determine, there was no monetary unit used in subsistence trading until the introduction of British coinage. Brass rods overlapped from the prestige to the subsistence sphere on some occasions, but only on special occasions and for large purchases.

Conversion direct from the subsistence category to the kinship category was and is denigrated and is extremely rare. Although all Tiv with whom we discussed the matter denied emphatically that Tiv would ever sell a kinsman, wife, or ward to get food (there were other reasons for selling them), Akiga—in an untranslated section of his book—mentions a famine which was so severe that, as a last resort, men sold their daughters to foreigners in exchange for food so that they could keep their sons alive. Akiga used this point as an argument to prove conclusively that the famine was all but lethal to the tribe.

Not only is conversion possible, but upward conversion is encouraged; it is, in fact, the measure of a man's worth. Tiv are scornful of a man who is merely rich in subsistence goods or, in 1953, in money. If, having adequate subsistence, he does not seek prestige in accordance with the old values or does not strive for more wives and hence more children, the fault must be personal inadequacy or unattractiveness. They also admit that they try to keep a man from making conversions; jealous kinsmen of a rich man are said to bewitch him and his people by *akombo* in order to make him expend his wealth on sacrifices, thus maintaining economic equality. However, once a conversion has been made, demands of kinsmen are not effective, or at least they take a new form.

Therefore, the man who can convert his wealth into higher categories is successful—he has a strong heart. He is both feared and respected. If he is strong enough to resist excessive demands of his kinsmen, but still fulfills his kinship obligations generously, he is feared as a man of special, potentially evil, talents (*tsav*). Most autobiographies collected from Tiv contain variants of this story: "When I was a very small child my kinsmen gave me a baby chicken. I tended it carefully and when it grew up it laid eggs and hatched out more chickens; I exchanged these chickens for a young nanny goat, who bore kids which I put out with various kinsmen until I could exchange them for a cow. The cow bore calves, and eventually I was able to sell the calves and procure a wife." Every successful man considers such a story one of the important sequences of his life; it proves he has been successful.

Metal rods hold a pivotal position in this entire process, and it is not surprising that early administrators considered them money. Originally imported from Europe, they were used as "currency" in some parts of southern Nigeria in the slave trade (but see Polanyi 1966). In Tiv ideology, slaves and brass rods are associated. For conveyances within the prestige sphere brass rods were used as a standard of equivalence, and they were a medium of exchange; they were also a mode for storage of wealth and were used as payment. But brass rods were a general-purpose currency *only within the prestige sphere.*

Brass rods sometimes entered into conversions, when they fulfilled only one of the functions of money—that of payment—and cannot be considered a general-purpose currency. We have examined in detail the valuational reasons why equivalency could not exist between brass rods and rights in women, between brass rods and food. Tiv never valued either wives or food in equivalences of brass rods. Thus the function of a standard was withheld from the brass rods that entered conversions. The perishable nature of Tiv subsistence goods, the fact that storage facilities beyond a single season were never developed, and the prevalence of trypanosomiasis combined to make storage of subsistence goods impracticable. The only way to "store" one's wealth was to convert it to prestige goods; the only way to store "wealth" in the kinship sphere was through women and children. Finally, there were no items or commodities (including brass rods) used as means of exchange for conveyance within either the subsistence sphere or the kinship sphere. Therefore, outside the prestige sphere, brass rods were in exceptional circumstances used as a means of payment in conversions. They did not serve any other money uses in conversions.

Payment

TIV NOTIONS of payment elucidate the matter more fully. The English verb "to pay" covers two Tiv verbs: *kimbi* and *nambe*. If we oversimplify for a moment, we can say that *kimbi* means to pay for goods, while *nambe* means to pay for services.[3] Within the subsistence sphere, I *kimbi* a bundle of guinea corn for a hoe, or I *kimbi* spinach greens for mushrooms. Within the prestige sphere, I *kimbi* brass rods for a cow. But, still within the prestige sphere, I *nambe* the *akombo* practitioner who performs rites for my benefit though I *kimbi* him (or anyone else) for the medicine he gives me. Within the marriage sphere of exchange the word *kimbi* is never used. I *nambe* my wife's guardian by giving him rights in my ward. If I do not have a ward, I *nambe* him in brass rods until I have a daughter to give him as his ward, thus ensuring that I *nambe* him properly.

Ideas of payment are best illustrated by anticipating some of the data from the next chapter and examining Tiv evaluation of the all-purpose currency introduced by the British. Currency is called *inyaregh*, but the word is older. Both Tiv dictionaries give "money" as the basic meaning of this word; both extend it to "property" (though it is our opinion that such extension obscures more than it clarifies). Both note that the notion is subdivided into "seed money" (*sam nyar*), "wet money" (*inyaregh ki ndorough*), and "basis of money" (*ityo nyar*). Seed money is said to be "property consisting of money or brass rods" by Abraham, while Malherbe (1934) says that it is "all possessions like the brass rod, camwood, cloth, money, etc., contrasted with *inyaregh ki ndorough* (wet money) which is possessions like livestock." Abraham also says that "wet money" refers to "property consisting of living things." Both also add another term, "basis of money" (*ityo nyar*), for "personal property, capital."[4]

Our information differs from both dictionaries. Our informants contrasted "wet *inyaregh*" with "dry *inyaregh*." We believe that no Tiv

3. So far as we are aware, Firth (1939) was the first to point out that this distinction is a very common one in languages used by people who practice a subsistence economy.

4. Malherbe adds the interesting note, "Originally a man's cotton cloth, and afterwards money, got from the sale of the cotton, etc., then a cow which he bought from money he earned himself (the cow being the means wherewith he could secure a wife)."

would ever say "A goat is wet *inyaregh*." He might, however, say, "I owed him a goat and paid him a dry goat," meaning he paid currency instead of the goat; he might also pay a currency debt with "wet *inyaregh*." As we saw was the case of the tens and the twenties, there is no fixed rate of exchange between a "dry goat" and a "wet goat," but the amount of currency or the size of goat must be fixed *ad hoc* on each occasion. Further, it seems to us that "seed *inyaregh*" (*sam* means not only seed, but also any small, miscellaneous items that come in collections) is approximately the equivalent of petty cash, while "basis of money" means wealth which enables one to make a payment. The two could not possibly be mutually exclusive or contrasted. Rupert East, in translating *Akiga's Story*, had in several places to translate *inyaregh* as price—that which must be given in payment.

Taking all these factors into consideration, it becomes evident that *inyaregh* means "means of payment." We have, thus, one money use (equivalence) expressed by the Tiv word *ishe*, while another (means of payment) is expressed by the word *inyaregh*. Either of these terms, as well as *kasua* or market, must sometimes be translated "price." The English word has all three implications.

Summary

THIS CHAPTER has demonstrated that market exchange is one form of exchange among others; that exchange is one sort of allocation of goods among others; and that distribution of goods is the economic mode of perceiving the relationship between men and things, while "property" concepts are the jural mode of perceiving such a relationship. In looking further into the symbolic means of discussing the relationship between men and things, we have discovered that Tiv value concepts are more diffuse than those of Westerners because Tiv lacked general-purpose money. The fact that equivalence may be suggested with one set of symbols and payment with another leads to evaluation in multiple modes. The valuational aspect of the man-thing unit is diffused in many institutions and subject to many moralities. So long as these values are held, and their institutionalization is adequate to maintain them, any single institution such as the market is held in check.

17

Growth of the Market Sector

IT IS OBVIOUS from the foregoing chapters that the market in Tivland is—or at least was—a contained institution. Each market place was highly institutionalized, but the system of market places was not. The free market operated within the narrow sphere of subsistence goods, but householding overshadowed the market principle even there, and the factors of production never entered the market. In short, the market principle did not reach many corners of Tiv life that it dominates in our own.

Three tasks remain: first, to examine the way in which the Tiv market system was held in check. We shall, in the process, be led to the second point: a more detailed exposition of the differences between Tiv and Western markets. Finally, we shall examine the impact of the highly institutionalized Western market system on the economy of the Tiv.

Containing the Market

PERHAPS THE MOST CHARACTERISTIC FEATURE of the Tiv market is that it is extraordinarily contained and shows little tendency to invade the other institutions of the society. When the West unleashed the market in the nineteenth century, it was soon necessary to erect new barriers against it (Polanyi 1944). The market, based as it is on priority of self-interest, is a most pervasive institution; only definite boundary mechanisms can keep the market principle from swamping the whole culture.

The market was contained among Tiv because they do not—or, in the past, did not—confuse marketing with trading. Marketing means selling one's products and then buying one's requirements. Trading, on the other hand, means transporting goods from a distance with the overt intention of making a profit—buying in the cheaper market and selling in the dearer. Indigenously, Tiv are marketers, not traders. Even as late as 1950 there were comparatively few large-scale traders although all Tiv were marketers. This confinement of marketing operations to one's produce and one's immediate needs meant that one of the most important characteristics of the Western type of market system was lacking. In the West, the distinction between trade and market has been recognized (Marx 1867:Part II, Chap. 4), but the two are not institutionally separable.

In the second place, trade could and did take place independently of market places. Even if the volume of trade had been large, it need not have influenced market places greatly. We have seen that today Tiv traders go to the northwestern areas of Tivland, where there are no market places, and peddle their wares from one compound to another. In the nineteenth century most of the peddlers were Hausa. They traded luxury goods for goats and a few other items, but they did not trade extensively in the staple commodities that Tiv marketed. Therefore, trade and markets were kept distinct from the standpoint of goods as well as institutions. Thus, Hausa traders—because they were few and because they did not export much of the produce that Tiv marketed—did not occasion significant changes in production or upset the standard of living. We have already noted the problems which accompanied the introduction of Ibo traders, specifically because Ibo traded in the produce which Tiv marketed among themselves.

Third, Tiv did not have a general-purpose money. Marketing operations were, indigenously, carried on largely by barter. Only "big deals" were sometimes carried out in terms of brass rods, which were of such great value that their use was necessarily restricted. Even in the absence of ethical considerations, brass rods could not have been used for the "ha'penny ha'penny exchange" characteristic of most of the transactions of Tiv markets.

A fourth reason that the market was contained was that, indigenously, even that trade which took place in the market was limited to the neighborhood. Before the British administration, when Tiv traded (i.e., bought in a cheaper market for sale in a dearer), they did so almost entirely within the neighborhood market cycle. If goods moved beyond these neighborhoods, it was a result of local trade and the overlapping of neighborhoods.

In the fifth place, the political aspects of market places were of immeasurable importance in maintaining the singularity of each individual market place and inhibiting the institutionalization of a more general market system. Markets were—and to some extent still are—political plums; thus any attempt to organize several markets into a unit is a political act. Political activity in Tivland can be seen as a pull between the lineage principle and non-lineage principles such as the market. When one principle became tyrannical, the other could be used to halt it. Therefore, the fission and fusion of groups inherent in a lineage system could be called into play whenever any individual, or any lineage, began to organize several markets, or whenever the market of one lineage began to dominate those of other lineages. A market system was truncated in Tivland by the political functions of market places. As we have seen, the political functions became less important after the establishment of foreign administration, and another check on the market was removed.

Finally, the force of kinship and lineage institutions, and the security that living in them provided, held in check any desire to engage in entrepreneurial activity in the market, for to enjoy the proceeds of trade one had to share it among one's kinsmen or else be ostracized by them.

In the presence of all these features, Tiv economy was marked not by the spread of the market, but by the development of conversion among ranked spheres of goods.

Western Markets

In Europe and America, matters are different. The West has factor markets, and one market influences the next through an organization of commercial firms accordant with goods, events, and values. One of the reasons that Western economy had any impact on African economies at all is that, with Western-type institutionalization, markets "grow" and the economy "develops." One of the processes of growth is the ingestion of institutions, either domestically or in extrinsic societies. Internally, our institutions can often erect legal or other barriers for containing the market and maintaining their own integrity. But the extrinsic societies against which the Western market expands often lack the institutional means for resisting the spread of the market once it has begun.

Although Tiv markets are essentially free markets, market participants lack a full-blooded market orientation. We noted a lack of arbitrage in a situation in which profit could be made by taking advantage of differential exchange rates among different coins. This lack would never have occurred in a thoroughly "marketized" society. We noted that Tiv, when they can get the amount they need for a specific purpose, will sell chickens at lower prices than they know they can get elsewhere. Westerners may do this too, but with them it would be a matter of overtly computing costs. Tiv deprecate living on profit and are scornful of a man who is rich in money but lacks the basic Tiv goals for social maximization. Importantly, they do not compute profit; only when Tiv have become thoroughly imbued with Western notions do they do so—this is especially indicative because in the West profit is the essence of the market orientation. They compute cost of transporting produce only if they expend lorry fare. Finally, the most valued "goods" are not marketable subsistence goods at all, but rights in women.

All of these characteristics, which have been mentioned in the course of our description and analysis of Tiv markets, stand at odds with the situation one finds in Western markets. It is these characteristics which have been most influenced by the impact of Western economy.

The Impact of the West

THE SPREAD of the market was, of course, only one aspect of the impact of the West on Africa. Colonial administrations and Christian missions have been as important. One of the most startling administrative innovations was a general peace, which was, of course, one political result of the European desire for an expanding market. As the Europeans intended, this policy had a far-reaching effect on trade and markets. Tiv stress the fact that with peace came both the development of traders and a vast increase in the amount of marketing. Whereas formerly to trade beyond the bounds of one's kinsmen and friends was to court death and slavery, in colonial times it became a safe occupation.

Road building was also begun, making it even easier to move about the country. Peace and communications led also to a considerable increase in the number of market places, organizationally no different from the old ones. Although it is doubtful that new routes for goods appeared, the quantities of goods increased and those that formerly had been handed along a chain of middlemen in overlapping market neighborhoods were now moved by long-distance traders.

At the same time, the institutions of colonial government took over some of the most important functions of market places: political activities that in the past had been handled solely by market organization were now often handled through the "Native Administration." [1] This change resulted in two situations that led to the spread of the market. When office-holding overshadowed markets as a political plum, the spread of a market was no longer "automatically" countered by the segmentation of Tiv lineage groups. Second, with the diminution of the political functions, the functions of trade became a more pronounced aspect of a market place. This increase in the relative importance of trade in the market institutions would thus have occurred even had the number of traders and the amount of goods not increased. Although, as we have seen, Tiv markets are still far from being merely institutions for carrying on trade, they are more nearly that than they were before the arrival of the British.

Not only was the internal marketing system perturbed by the introduction of alien institutions, the economic institutions of the Tiv were put into touch with world economy. Northern Nigeria, like much of the rest of the colonial world, was taken over by trading companies with governing powers. The close linkage of government and trade was evident when taxation was introduced into Tivland. Tax was originally paid in produce which was transported and sold through Hausa traders who were government contractors. A few years later, coinage was introduced; taxes were demanded in that medium. It became necessary for Tiv to "go into trade" or to make their own contracts with foreign traders in order to get cash.

1. Iyon market was something of an anachronism. It was located on an important boundary across which the Native Administration bureaucracy did not function smoothly. Therefore, it had retained its political functions and was a living example of what Tiv said all markets had once been.

The trading companies cooperated with the new government in introducing a "cash crop" to be bought by the traders in return for cash to pay taxes and, incidentally, to buy imported goods. The crop which proved best adapted for this purpose in Tivland was beniseed, which Tiv already grew.

The trading companies established buying centers throughout Tivland. Usually these points were—or became—the major markets. The buying centers were put into the hands of African traders, usually Kalabari and Ibo, the only Nigerians then able to keep records and speak to the traders in English. These African traders gained control of beniseed marketing in the 1920's, by offering Tiv advance and extremely low payment. This system was upset when, after a few years, Tiv did not have enough money at harvest to pay the tax, since the money had long since been spent. The government thereupon outlawed middlemen and supervised all sales. It was made a heavily finable offense to sell or buy beniseed anywhere else. The Tiv Native Treasury at the time of our field work contained a beniseed marketing board similar to the Cocoa Marketing Board in Yoruba country (Galletti, Baldwin & Dina 1956). We know little about it because the Tiv whom we knew cared little about it—a sign both that it was alien and that it worked no hardships.

There is still another way in which the trading companies link Tiv economy to the economy of the outside world: they stake African traders to imported goods as part of a program specifically designed to develop what they called "native entrepreneurs." The companies said in 1950 that they found retail trade unprofitable, even in the centers of heaviest European population, and that they maintained retail stores only as a gratuitous service to European residents. The cloth, enamelware, and ironmongery which they imported for the African market were sold through a network of dependent—not to say client—African traders.

Introduction of Western administrative and trading institutions thus had the effect of reducing the non-economic functions of the market place at the same time that the economic elements of the market principle were reinforced and enlarged.

Probably no single factor has been so important as the introduction of general-purpose money. Neither cash crops and taxes nor extended trading affected the basic congruence between Tiv ideas and institutions to the same extent that money did. With the introduction of money, indigenous ideas of social maximization—conversion of all forms of wealth into women and children—no longer led to expected results.

Several factors must be examined. Two of the economic spheres—slavery and exchange marriage—had been legally destroyed. Moreover, general-purpose money provides a common denominator among all the spheres, thus making all items expressible in terms of a single standard and, ultimately, exchangeable. Coinage has been misunderstood by Tiv. They call it *inyaregh* or "means of payment." At the same time, however, it is a standard of value and a generalized means of exchange—something present in the indigenous Tiv economy only within the prestige sphere.

A moment's consideration makes it obvious that the category of prestige goods, centering about cattle, slaves, ritual office, and brass rods, has ceased to exist in material terms, even though the sphere is maintained ideally. Slave-dealing was prohibited from the first effective European control (between 1910 and 1914). Brass rods were "withdrawn" by the administration and "replaced" with coinage. *Tugudu* cloths are still to be seen, but they have lost their importance because they can now be bought with money. Money has largely replaced animals as payment for ritual services, and the old system of the twenties and the tens is largely a memory. Determining the size of an animal used as a twenty has been replaced by haggling over pennies.

European and African traders have introduced many new commodities to Tivland, of both European and African manufacture, and have increased manyfold the quantity of familiar commodities. These goods, particularly European goods, were introduced concurrently with coinage and are considered part of the money complex. They do not enter into any formerly existing economic sphere, yet they form their own sphere only imperfectly. Thus, there are now many commodities that do not fit precisely into traditionally structured exchange situations.

It is in the nature of a general-purpose money that it standardizes the exchangeability value of all items on a common scale. It is precisely this function which brass rods, a commodity used as limited-purpose money in the old system, did not perform outside the prestige category. They were used only as a means of payment (but specifically not as a standard) in conversion.

Early administrative officers interpreted brass rods as "money," by which they meant general-purpose money. It seemed an obvious move to establish an exchange rate between brass rods and a new coinage, to "withdraw" the rods, and hence to "replace" one currency with another.

The actual effect was, of course, to introduce a general-purpose currency in place of a limited-purpose money. Today all conversions and most conveyances are made in terms of coinage. Yet Tiv constantly express their distrust of money. They compare the monetary system to the subsistence system at great length, always to the disfavor of the former. Money does not (they say) reproduce itself or bear seed. You spend (*vihi,* literally "spoil") money, and it is gone; a man cannot spend a field, and though he sacrifices a goat, it has already borne kids. Money, Tiv contend, is the root of much of their present-day trouble. Elders say that money should be exchanged only for those European goods which were introduced more or less concurrently with it. Obviously it is a view that cannot successfully be put into practice.

Concurrence of the introduction of money, introduction of tax, introduction of cash crops, and imposition of general peace meant that trade expanded rapidly. Men's trade, like women's, had been based on subsistence goods, but unlike women's, on goods that could be procured and traded over fairly long distances. Even though the produce they handled did not change, much of their marketing became money trade: men traveled long distances starting with money and ending with money.

Tiv do not condemn money trade in all contexts. They condone it so long as the trader maintains a farm. As a means to get money they consider it as legitimate and sensible as cash-cropping or migrant labor. Yet, with general-purpose currency, a new money use has been added: means of exchange. The old limited-purpose brass rods currency seems seldom to have been used specifically to facilitate exchange, but only as equivalent-evaluation and payment. This and the fact that a single means of exchange entered into all of the economic spheres broke down the distinctions among the spheres. Put in another way, a single cultural item (coinage) had been introduced into several institutions; it had the same function and purpose within each institution, and therefore the major distinguishing features among these institutions were blurred.

When semiprofessional traders, using money, began trading in the foodstuffs which were formerly the province of women doing marketing, they became, in fact, the most active buyers and sellers of grain at Tiv markets, buying in small quantities, selling in bulk to exporters. Women engaged in subsistence marketing while men engaged in money trade; the fact that a single cultural item—currency—was used broke down this distinction as well as the separate spheres.

Both trade and marketing are approved, yet Tiv see truckload after truckload of foodstuffs driven away from their major markets every fifth day. They say that food is less plentiful today than it was in the past, although more land is being farmed. Tiv elders deplore this situation and know what is happening, but they do not know where to place the blame. In attempts to do something about it, they sometimes announce that no women are to sell any food at all. But when their wives disobey them, the men do not feel they were wrong to have done so. In their condemnation of the situation, which is depriving them of food faster than they are able to increase production, Tiv elders curse money, the Europeans who brought it, and the Ibo traders it brings.

Of even greater concern to Tiv is the influence money has had on marriage institutions. This highest sphere of economic activity was dealt a crippling blow by administrative abolition of exchange marriage, and substitution of bridewealth marriage, payable in money, as the only legal form. This act turned all marriage gifts and exchanges into conversion payments, at the same time making direct conveyance of rights in women illegal. The administration had been under great pressure from the missions (Bohannan & Bohannan 1957:384–445) and acted in response to what appeared superficially to be popular demand. It is our opinion that, at the time we were working among them, both Tiv and the administration believed this action to have been precipitate and ill-advised.

Today every woman's guardian, in accepting money as bridewealth, feels that he is converting down. Although attempts are made to spend money which is received in bridewealth to acquire brides for one's self and one's sons, it is in the nature of money, Tiv insist, that it is difficult to do. A good man spends his bridewealth receipts for brides; but a good man is hard to find. Tiv deplore the fact that they are required to "sell" (*te*) their daughters and "buy" (*yam*, but more euphemistically, *kem*, "to accumulate") wives. There is no dignity in it since the possibility of making a bridewealth marriage into an exchange marriage has been removed.

General-purpose currency has reduced all the various spheres of exchange to a single sphere; it has homogenized, so to speak, the separate transactional spheres. The resulting institutional arrangement has taken many of its characteristics from the market. Rights in women were exchanged; but they were not a *kasua* until they were exchanged in the same medium as that used in the market place. Ritual specialists were rewarded for their services, but they did not sell rituals until coinage made it possible. Even gift-giving is threatened with coinage reciprocity. The market has become pervasive; it is no longer merely an institution—it threatens to become a way of life.

Yet the basic Tiv value of social maximization—converting one's wealth into women and children—has stood fast against institutional changes. In this discrepancy between values and institutions, Tiv have come upon what is to them a paradox, however familiar and comprehensible it may seem to Westerners. It is easy to sell subsistence goods for money to buy prestige articles and women, thereby aggrandizing oneself at a rapid rate. The food so sold is exported, decreasing the amount of subsistence for consumption, and bridewealth gets higher. Rights in women have entered the market; and since the supply is finite and the demand infinite, the price of women has become inflated.

It is apparent that Western institutions have unified the institutional aspects of Tiv markets and other institutions into a pervasive system. Through the growth of transportation facilities, markets have been brought closer together; therefore price responses cover a larger geographical area than in the past. The price mechanism also applies to a wider range of commodities than it did before the introduction of general-purpose money.

Gathering price data is difficult in an economy like that of the Tiv because price has to be in terms of weight, volume, pieces, or something of the sort. As we have seen, such measures did not exist indigenously. If one were to collect prices over a very wide range of markets, one would have no opportunity to do anything else. We make these statements not in apology but to explain the lack of price data which an economist needs to examine the growth of the market sector. Economic theory would predict that the closer the integration of the market places, the narrower the price differentials for a single commodity. We can say with conviction—but without figures—that such integration of market places has increased, that price differentials have become narrower over a wider area, and that the prediction is undoubtedly correct.

To generalize, every institution is held in check by other institutions or activities within the society in which it is found. When that system of checks and boundaries is disturbed, the institution is likely either to be swamped or to spread to cultural spheres from which it was formerly excluded. In the impact of Western culture on Tivland indigenous checks on the market institution were nullified, the market institution itself reinforced. Hence, the area of relevance for the market institution has expanded. The expansion has been brought about by simple means: formerly the event sequences, the material culture, and the value systems which formed part of the market institution led out of it each into different institutions. They are now all leading into one another. In addition, the European trading firms, the African trading firms which have followed them, and the staking of traders have combined with the spontaneous formation of new social relationships and groups to form the skeleton of a role system uniting all the markets of the land—something which was not present in the indigenous system.

In the past Tiv practiced a multicentric economy. The system was one in which several institutions were integrated in terms of a moral hierarchy: by a value system not directly accompanied by a social or material cultural system. Subsistence, prestige, and kinship institutions were linked by a system of moral values and by a goal for action which, in carefully controlled ways, disengaged material items from one of these institutions and placed them into another. With the introduction of the fourth sphere—currency and imported goods—Tiv sought to interrelate it with the other three spheres on the same principle of moral ranking. However, the method of disengagement of material items—that is, of conversion—was affected by money. When the shatteringly simple idea of general-purpose money was introduced into the multicentric economy, a compelling force was exerted toward the development of a unicentric economy.

Therefore, the appearance of foreign rule in Tivland had two early effects on the Tiv economy: the introduction of a further pluralizing factor and the simultaneous introduction of an extremely powerful simplifying factor. The ensuing struggle is one between the ideology of the unicentric economy, implicit in general-purpose money, and the multicentric economy, dominated by multiple standards of evaluation and payment and marked by conversion among them, as it is known to the Tiv.

In a unicentric economy dominated by general-purpose money, if any part of the economy displays market characteristics, everything which enters an exchange situation will do so, even taxation and fines. Therefore, the singularizing process in a formerly multicentric economy is likely to be seen as a spread of the market.

The situation we have found among Tiv is not unfamiliar to students of primitive economies. Recognition of the fact of ranking and non-equivalent spheres goes back at least as far as Thurnwald (1932:105) and figures in more recent accounts (Herskovits 1952:257; Boeke 1953: Chap. 6). The ranked hierarchy of exchangeable goods and services is an example of what Firth (1939:5–7) has called "multiple standards" in the absence of a "true money or a price mechanism" of general applicability. He noted that these multiple standards are most commonly found when attempts are made to measure services and goods against one another. The Tiv material supports this thesis but demands some modification: Tiv differentiate payment for goods (*kimbi*) from reimbursement for services (*nambe*), but this is not their fundamental basis for division into multiple standards; rather, that basis is formed by the institutionalization of subsistence, prestige, and kinship.

Multicentric economy should not, however, be confused with what Boeke (1953) and his followers (Furnivall 1944) have termed a dual economy, which is a direct result of the impact of Western institutions on indigenous institutions and the continued functioning of each, with a rather uneasy overlap between them in a common market.

Neither should a multicentric economy be considered, as Firth seems to do, merely the absence of price mechanism. It seems to us reasonable to believe that a unicentric economy may or may not be characterized by a price mechanism based on supply and demand. Certainly, the multicentric economy is not the logical opposite of the supply-and-demand, price-mechanism economy, for as we have seen among the Tiv, it may actually contain it (or, in Western economy, be contained by it).

There would seem to be a second principle at work here. Whereas in some unicentric economies the organizing principle is the supply-and-demand price mechanism, in some multicentric economies the organizing principle is what we have called "internal conversion": internal because the multicentricity makes possible conversion from one sphere to another within a single economy. In some cultural circumstances, internal conversion is the working economic principle just as, in other circumstances, the price mechanism is the working economic principle. That is to say, in any market exchange sphere only specified goods may be exchanged, only specified means of payment may be used. Further, in the market exchange sphere, prices are determined in a special way—by the supply-and-demand mechanism. But in the prestige and kinship spheres, equivalence ratios are established, not via the market mechanism, but rather by ritual or prestige rules. Internal conversion seen in this way is an institutionalized means of bringing together, by moral ranking in the absence of a general standard, items which are otherwise not in the same sphere of thought or action, thus achieving distribution.

Postscript

IN WORKING THROUGH our material on religion and witchcraft among the Tiv, Paul Bohannan has come to realize that the "flesh debts" of Tiv religion are carefully tied in to the system of ranked spheres in Tiv economic life; some flesh debts are "paid" through conversions to lower-ranking goods. This material will be included in our forthcoming monograph on Tiv religion.

While teaching economic anthropology after this book went to press, Laura Bohannan came to think that consecrated markets among the Tiv should be compared with the ports of trade considered by Polanyi and his associates. This idea will be investigated and published as an article.

Bibliography

Abraham, R. C.
1940 *The Tiv People* (2d ed.). London: Crown Agents for the Colonies.

Akiga (B. Akiga Sai)
1939 *Akiga's Story*. Trans. Rupert East. London: Oxford University Press for International African Institute.

Betts, T. F.
1941 "The Tiv Plantations, 1939–41," *Farm and Forest,* Vol. II.

Boeke, J. H.
1953 *Economics and Economic Policy in Dual Societies.* New York: Institute of Pacific Relations.

Bohannan, Laura
1952 "A Genealogical Charter," *Africa,* XXII, 4:301–15.

Bohannan, Laura and Paul
1953 *The Tiv of Central Nigeria.* London: International African Institute.

Bohannan, Paul
1954 *Tiv Farm and Settlement.* London: Her Majesty's Stationery Office.
1955 "Some Principles of Exchange and Investment among the Tiv," *American Anthropologist,* LVII:60–69.
1957 *Justice and Judgment among the Tiv.* London: Oxford University Press.
1963 *Social Anthropology.* New York: Holt, Rinehart & Winston.

Bohannan, Paul and Laura
1958 *Three Source Notebooks in Tiv Ethnography.* New Haven: Human Relations Area Files.

Bohannan, Paul, and George Dalton
1962 *Markets in Africa.* Evanston: Northwestern University Press.

Briggs, G. W. G.
1941 "Soil Deterioration in the Southern Districts of Tiv Division, Benue Province," *Farm and Forest,* Vol. II.

COLE, C. W.
1949 *Report on Land Tenure, Zaria Province.* Kaduna: Government Printer.

COLSON, ELIZABETH
1954 "Ancestral Spirits and Social Structure among the Plateau Tonga," *International Archives of Ethnography,* XLVII, 21–68.

DALZIEL, J. M.
1948 *The Useful Plants of West Tropical Africa.* London: Crown Agents for the Colonies.

DEWAR, K.
1935 "Intelligence Report on the Ukan District of the Tiv Division of the Benue Province." (Unpublished Ms.)

DOWNES, R. M.
1933 *The Tiv Tribe.* Kaduna: The Government Printer.

EAST, RUPERT (trans.)
1939 *Akiga's Story.* London: Oxford University Press for International African Institute. (Items of the text are footnoted "Akiga 1939"; those from the notes, "East 1939.")

EVANS-PRITCHARD, E. E.
1949 *The Sanusi of Cyrenaica.* Oxford: Clarendon Press.

FIRTH, RAYMOND
1939 *Primitive Polynesian Economy.* London: Routledge and Kegan Paul.

FORTES, M., and E. E. EVANS-PRITCHARD (eds.)
1940 *African Political Systems.* London: Oxford University Press for International African Institute.

FURNIVALL, J. S.
1944 *Netherlands India, a Study of Plural Economy.* New York: Macmillan Co.

GALLETTI, R., K. D. S. BALDWIN, AND I. O. DINA
1956 *Nigeria Cocoa Farmers.* London: Oxford University Press.

HALL, E. T.
1963 "A System for the Notation of Proxemic Behavior," *American Anthropologist,* LXV, 1003–26.
1966 *The Hidden Dimension.* New York: Doubleday.

HERSKOVITS, M. J.
1952 *Economic Anthropology.* New York: Alfred A. Knopf, Inc.

JONES, G. I.
1949 "Ibo Land Tenure," *Africa,* Vol. XIX.

LÉVI-STRAUSS, C.
1963 *Totemism.* Boston: Beacon Press.

MCINTOSH, D.
1940 "The Tiv Plantations, Their History, up to 1939," *Farm and Forest,* Vol. I.

MALHERBE, W. A.
1934 *Tiv-English Dictionary with Grammar, Notes, and Index.* Lagos (n.p.).

MALINOWSKI, B.
1935 *Coral Gardens and Their Magic.* 2 vols. London: George Allen and Unwin, Ltd.
1945 *The Dynamics of Culture Change.* New Haven: Yale University Press.

MARX, K.
1867 *Capital.* New York: Modern Library.

MOORE, W. E.
1951 *Industrialization and Labor.* Ithaca: Cornell University Press.

POLANYI, K.
1944 *The Great Transformation.* New York: Rinehart & Co.
1966 *Dahomey.* Seattle: University of Washington Press.

POLANYI, K., C. ARENSBERG, AND H. PEARSON (eds.)
1957 *Trade and Market in the Early Empires.* Glencoe: Free Press.

REINING, C. C.
1959 "The Role of Money in the Zande Economy," *American Anthropologist,* LXI, 1:39–43.

ROWLING, C. W.
1949 (a) *Report on Land Tenure, Kano Province;* (b) *Report on Land Tenure, Plateau Province.* Kaduna: Government Printer.

SLUITER, G.
1960 "Kikuyu Concepts of Land and Land Kin." Master's thesis, University of Chicago.

THURNWALD, R.
1932 *Economics in Primitive Communities.* London: Oxford University Press.

VERMEER, D. E.
1966 "Geophagy Among the Tiv of Nigeria," *Annals of the Association of American Geographers,* LVI, 2:197–204.

Index

Index